METHODS IN MOLECULAR BIOLOGY

Series Editor
John M. Walker
School of Life and Medical Sciences
University of Hertfordshire
Hatfield, Hertfordshire, UK

For further volumes:
http://www.springer.com/series/7651

For over 35 years, biological scientists have come to rely on the research protocols and methodologies in the critically acclaimed *Methods in Molecular Biology* series. The series was the first to introduce the step-by-step protocols approach that has become the standard in all biomedical protocol publishing. Each protocol is provided in readily-reproducible step-by-step fashion, opening with an introductory overview, a list of the materials and reagents needed to complete the experiment, and followed by a detailed procedure that is supported with a helpful notes section offering tips and tricks of the trade as well as troubleshooting advice. These hallmark features were introduced by series editor Dr. John Walker and constitute the key ingredient in each and every volume of the *Methods in Molecular Biology* series. Tested and trusted, comprehensive and reliable, all protocols from the series are indexed in PubMed.

Mass Spectrometry Data Analysis in Proteomics

Third Edition

Edited by

Rune Matthiesen

Computational and Experimental Biology Group, CEDOC, Chronic Diseases Research Centre, NOVA Medical School, Faculdade de Ciências Médicas, Universidade NOVA de Lisboa, Lisboa, Portugal

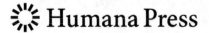 Humana Press

Editor
Rune Matthiesen
Computational and Experimental
Biology Group, CEDOC, Chronic
Diseases Research Centre, NOVA
Medical School, Faculdade de Ciências
Médicas
Universidade NOVA de Lisboa
Lisboa, Portugal

ISSN 1064-3745   ISSN 1940-6029   (electronic)
Methods in Molecular Biology
ISBN 978-1-4939-9746-6   ISBN 978-1-4939-9744-2   (eBook)
https://doi.org/10.1007/978-1-4939-9744-2

This Humana imprint is published by the registered company Springer Science+Business Media, LLC, part of Springer
Nature.
The registered company address is: 233 Spring Street, New York, NY 10013, U.S.A.

Preface

The third edition of *Mass Spectrometry Data Analysis in Proteomics* starts as the previous editions with a basic introduction to mass spectrometry-based proteomics. The following chapters are written by experts in specific subdomains of proteomics. The aim is to provide detailed information on each topic to allow newcomers to follow the content but at the same time present novel ideas and views that can influence future developments in mass spectrometry-based proteomics. The authors have been offered to provide complete coverage of their domains with no page restriction. The first edition of the book was published in 2007, and we have witnessed a strong improvement in sensitivity and reproducibility of mass spectrometry-based proteomics. Furthermore, quality of computational methods has also improved. Consequently, all chapters have been updated to provide the newest and most relevant information on each topic. In contrast to the previous editions, this edition aims to also provide instructions on how to perform the most relevant computational methods such as database-dependent searches, although the focus is still on computational concepts.

There are now several examples demonstrating that clinical proteomics is reproducible, and we and many others believe that mass spectrometry will play a major role in clinical research in the coming years. Criticism of mass spectrometry often revolves around lack of throughput and it will never compete with genomics because of the lack of a method to amplify the analyte. However, clinical proteomics provides complementary information to genomics which we refer to as proteogenomics, and several methods exist to amplify the signal of protein-based targets. The analogy that the genome is like a cookbook for the organisms is justifiable. Furthermore, the sample preparation steps are often more straightforward for mass spectrometry, and new workflows allow analysis of more than 300 blood samples per day using a single instrument. However, to cook a delicious cake more is needed: a good chef, functional working tools, and fresh ingredients. To answer questions about the quality of the cake, it is often more informative to taste it rather than reading the recipe. In fact, many diseases have a major environmental component that potentially is better reflected in the proteome than the genome.

Lisboa, Portugal   *Rune Matthiesen*

Contents

Contributors

RUEDI AEBERSOLD • *Department of Biology, Institute of Molecular Systems Biology, ETH Zürich, Zürich, Switzerland; Faculty of Science, University of Zürich, Zürich, Switzerland*

HANS CHRISTIAN BECK • *Department of Clinical Biochemistry and Pharmacology, Odense University Hospital, Odense C, Denmark*

JAKOB BUNKENBORG • *Alphalyse, Odense M, Denmark*

ANA SOFIA CARVALHO • *Computational and Experimental Biology Group, CEDOC, Chronic Diseases Research Centre, NOVA Medical School, Faculdade de Ciências Médicas, Universidade NOVA de Lisboa, Lisboa, Portugal*

JELENA ČUKLINA • *Department of Biology, Institute of Molecular Systems Biology, ETH Zürich, Zürich, Switzerland; Ph.D. Program in Systems Biology, University of Zurich and ETH Zurich, Zürich, Switzerland*

BARNALI DEB • *Institute of Bioinformatics, International Technology Park, Bangalore, India; Manipal Academy of Higher Education (MAHE), Manipal, Karnataka, India*

MICHAEL J. EGGERTSON • *Waters Corporation, Milford, MA, USA*

NAGORE ELU • *Department of Biochemistry and Molecular Biology, Faculty of Science and Technology, University of the Basque Country (UPV/EHU), Leioa, Spain*

JOHN R. ENGEN • *Department of Chemistry and Chemical Biology, Northeastern University, Boston, MA, USA*

KEITH FADGEN • *Waters Corporation, Milford, MA, USA*

IRENE A. GEORGE • *Institute of Bioinformatics, International Technology Park, Bangalore, India*

HARSHA GOWDA • *Department of Genetics and Computational Biology, QIMR Berghofer Medical Research Institute, Brisbane, QLD, Australia; Institute of Bioinformatics, Bangalore, India; Manipal Academy of Higher Education, Manipal, India*

ALEX HENNEMAN • *Center for Proteomics and Metabolomics, Leiden University Medical Center, Leiden, The Netherlands*

ROBERTO HENRIQUES • *NOVA Information Management School (NOVA IMS), Universidade NOVA de Lisboa, Lisboa, Portugal*

PRASHANT KUMAR • *Institute of Bioinformatics, International Technology Park, Bangalore, India; Manipal Academy of Higher Education (MAHE), Manipal, Karnataka, India*

BENOIT LECTEZ • *Department of Biochemistry and Molecular Biology, Faculty of Science and Technology, University of the Basque Country (UPV/EHU), Leioa, Spain*

MARIA PAULA MACEDO • *Centro de Estudos de Doenças Crónicas (CEDOC), NOVA Medical School-Faculdade de Ciências Médicas, Universidade NOVA de Lisboa, Lisboa, Portugal; Department of Medical Sciences, Institute of Biomedicine, University of Aveiro, Aveiro, Portugal; APDP-Diabetes Portugal Education and Research Center (APDP-ERC), Lisboa, Portugal*

RUNE MATTHIESEN • *Computational and Experimental Biology Group, CEDOC, Chronic Diseases Research Centre, NOVA Medical School, Faculdade de Ciências Médicas, Universidade NOVA de Lisboa, Lisboa, Portugal*

UGO MAYOR • *Department of Biochemistry and Molecular Biology, Faculty of Science and Technology, University of the Basque Country (UPV/EHU), Leioa, Spain; Ikerbasque, Basque Foundation for Science, Bilbao, Bizkaia, Spain*

YASSENE MOHAMMED • *Center for Proteomics and Metabolomics, Leiden University Medical Center, Leiden, The Netherlands; University of Victoria–Genome British Columbia Proteomics Centre, Victoria, BC, Canada*

SONALI V. MOHAN • *Department of Genetics and Computational Biology, QIMR Berghofer Medical Research Institute, Brisbane, QLD, Australia; Institute of Bioinformatics, Bangalore, India; Manipal Academy of Higher Education, Manipal, India*

D. S. NAYAKANTI • *Institute of Bioinformatics, Bangalore, India; Manipal Academy of Higher Education, Manipal, India*

NEREA OSINALDE • *Department of Biochemistry and Molecular Biology, Faculty of Pharmacy, University of the Basque Country (UPV/EHU), Vitoria-Gasteiz, Spain*

MAGNUS PALMBLAD • *Center for Proteomics and Metabolomics, Leiden University Medical Center, Leiden, The Netherlands*

NICOLAI BJØDSTRUP PALSTRØM • *Department of Clinical Biochemistry and Pharmacology, Odense University Hospital, Odense C, Denmark*

PATRICK G. A. PEDRIOLI • *Department of Biology, Institute of Molecular Systems Biology, ETH Zürich, Zürich, Switzerland; ETH Zürich, PHRT-MS, Zürich, Switzerland*

ANA PINA • *Centro de Estudos de Doenças Crónicas (CEDOC), NOVA Medical School-Faculdade de Ciências Médicas, Universidade NOVA de Lisboa, Lisboa, Portugal; ProRegeM PhD Programme, NOVA Medical School/Faculdade de Ciências Médicas, Universidade NOVA de Lisboa, Lisboa, Portugal; Department of Medical Sciences, Institute of Biomedicine, University of Aveiro, Aveiro, Portugal*

GORKA PRIETO • *Department of Communications Engineering, Faculty of Engineering of Bilbao, University of the Basque Country (UPV/EHU), Bilbao, Spain*

JUANMA RAMIREZ • *Department of Biochemistry and Molecular Biology, Faculty of Science and Technology, University of the Basque Country (UPV/EHU), Leioa, Spain*

ALAN L. ROCKWOOD • *Department of Pathology, University of Utah, Salt Lake City, UT, USA*

GAJANAN SATHE • *Institute of Bioinformatics, Bangalore, India; Manipal Academy of Higher Education, Manipal, India*

JYOTI SHARMA • *Institute of Bioinformatics, International Technology Park, Bangalore, India; Manipal Academy of Higher Education (MAHE), Manipal, Karnataka, India*

JESÚS VÁZQUEZ • *Laboratory of Cardiovascular Proteomics, Centro Nacional de Investigaciones Cardiovasculares (CNIC) and CIBER de Enfermedades Cardiovasculares (CIBERCV), Madrid, Spain*

JUAN ANTONIO VIZCAÍNO • *European Molecular Biology Laboratory, European Bioinformatics Institute (EMBL-EBI), Cambridge, UK*

THOMAS E. WALES • *Department of Chemistry and Chemical Biology, Northeastern University, Boston, MA, USA*

MATHIAS WALZER • *European Molecular Biology Laboratory, European Bioinformatics Institute (EMBL-EBI), Cambridge, UK*

Chapter 1

Introduction to Mass Spectrometry-Based Proteomics

Rune Matthiesen and Jakob Bunkenborg

Abstract

Mass spectrometry, a technology to determine the mass of ionized molecules and biomolecules, is increasingly applied for the global identification and quantification of proteins. Proteomics applies mass spectrometry in many applications, and each application requires consideration of analytical choices, instrumental limitations and data processing steps. These depend on the aim of the study and means of conducting it. Choosing the right combination of sample preparation, MS instrumentation, and data processing allows exploration of different aspects of the proteome. This chapter gives an outline for some of these commonly used setups and some of the key concepts, many of which later chapters discuss in greater depth. Understanding and handling mass spectrometry data is a multifaceted task that requires many user decisions to obtain the most comprehensive information from an MS experiment. Later chapters in this book deal in-depth with various aspects of the process and how different tools addresses the many analytical challenges. This chapter revises the basic concept in mass spectrometry (MS)-based proteomics.

Key words Data formats, Proteomics, Mass spectrometry, Sample preparation

1 Introduction

1.1 The Mass Spectrometer Instrument

A mass spectrometer is a device for measuring the mass-to-charge ratio of ionized molecules. The output from all mass analyzers is intensity versus mass-to-charge ratio (m/z, see **Note 1**). This simple device provides a wealth of qualitative and quantitative information ranging from elemental composition to detailed structural information. All mass spectrometers consist of three main parts; an ion source, a mass analyzer, and a detector (see Fig. 1). For MS/MS (aka MS to the second) an additional instrument component for peptide fragmentation is needed, which is discussed further in Subheading 3. The analyte ions are produced in the ion source. The ion source generates ions by transferring molecules from the condensed (liquid or solid) phase to gas phase and ionizing them in the process (either positive or negative charge state see **Note 2**). The most commonly used ionization methods in proteomics are electrospray ionization (ESI) [1] and matrix-assisted laser desorption and ionization (MALDI) [2]—two soft ionization techniques

Rune Matthiesen (ed.), *Mass Spectrometry Data Analysis in Proteomics*, Methods in Molecular Biology, vol. 2051,
https://doi.org/10.1007/978-1-4939-9744-2_1, © Springer Science+Business Media, LLC, part of Springer Nature 2020

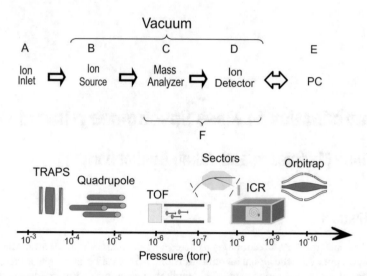

Fig. 1 Simple mass spectrometer overview. (**a**) Ion inlet: In shotgun proteomics peptides are separated by liquid chromatography either off line or on line. (**b**) The ion source converts molecules from solid or liquid phase into ionized species in the gas phase. (**c**) The mass analyzer separates the ions according to their mass-to-charge ratio. (**d**) The detector records a signal that is electronically amplified and stored as m/z versus intensity. (**e**) A computer typically controls liquid flow, vacuum, electrostatic fields, records data, and processes data during a sample run. (**f**) The mass analysis of ions takes place at low pressures typically 10^{-5}–10^{-10} mbar depending on the mass analyzer used

that can deliver fragile biological macromolecules intact into the gas phase. Electric and magnetic fields in the mass analyzer can control the charged biomolecules in the gas phase. The two laws of physics that govern dynamic movement of charged particles in electrical and magnetic fields are Newton's second law $[F = m * a]$ and Lorentz force law $[F = Q(E + v \times B)]$ where F is the force applied to the particle, m is mass, a is acceleration, Q is the charge state of the particle, E is the electric field, and $v \times B$ is the cross-product between the velocity of the particle and the magnetic field. Equating these two equations leads to an equation that basically governs and enables us to move, trap, and determine the mass of the charged particles in a mass analyzer. The ions that are produced in the ion source are transferred to the mass analyzer where they are separated according to their mass to charge ratio (m/z). There are a number of different mass analyzers operating by different principles and having different properties that are described later. Mass analyzers are often used in combination. For example, ion fragmenting by tandem mass spectrometers are widely used because it elucidate the ion composition through the obtained fragmentation pattern. The measured m/z pattern from a fragmented peptide can reveal the amino acid sequence of the peptide and pin point modifications and their location in the sequence.

Table 1
Overview of different mass spectrometry instrument components and methods

Ion source	Mass analyzer	Fragmentation	Detector
ESI [5]	Quadrupole [6]	CID/CAD [7]	Electron multiplier [8]
MALDI [9]	Time of flight [6]	HCD [10]	Inductive [11]
SELDI [12]	Ion-traps	ETD [13]	HED [8]
PD [14]	3D quadrupole [6]	ECD [15]	MCP [16]
ESSI [17]	Linear quadrupole [18]	PSD [19]	Faraday cup [8]
FAB [14]	Orbitrap [4]	IRMPD [20]	Scintillation counter [8]
LDI [21]	FT-ICR [11]	BIRD [22]	
	Magnetic sector [23]	SID [24]	

The different components combine in various ways to create different types of instruments. The references are suggested as entry points for further reading. The most common instrument components for MS-based proteomics are highlighted in bold. Abbreviations are spelled out in Appendix 1

The ions are finally recorded by a detector. The mass analyzer and detector are always within the high vacuum region [3] and typically also the ion source. Ion sources combined with different mass analyzers gives mass spectrometers such as MALDI-TOF (time-of-flight), ESI-IT (ion trap), and ESI-Orbitrap, as described later [4]. A more complete overview of different mass spectrometry components is presented in Table 1.

Different mass analyzers operate by manipulating the ions using different principles but common for all is that the results are transformed to intensities as a function of m/z (mass over charge) values. For example, Orbitrap and FT-ICR measure an AC image current induced by ions trapped in an electric or magnetic field that can transformed to m/z values by Fourier transformation. Time-of-flight instruments accelerate ions and measure the flight time between acceleration and hitting a detector. The measured flight times enables calculation of m/z values. The output from the instrument is ion intensity at different m/z values. The result is visualized by an m/z versus intensity plot also referred to as mass spectrum (*see* Fig. 2).

Hyphenating liquid chromatography with MS (LC-MS) allows acquisition of a whole series of MS spectra as the molecules elute from the chromatography column. Each MS spectrum in this array is also referred to as a survey scans because the tandem mass spectrometer can further interrogate the selected ions from this spectrum. In a typical MS setup for protein identifications the survey scan is analyzed on the fly by the instrument software to select ions that are isolated, fragmented and analyzed by a mass analyzer to generate an MS/MS spectrum (*see* Fig. 3). This type of data collection is also referred to as data-dependent acquisition

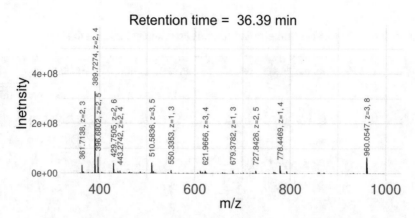

Fig. 2 The MS spectrum or survey scan displays the abundance of ions (intensity) as a function of mass-to-charge ratio (m/z). This spectrum displays ions in the m/z range from 350–1000 Th and the y-axis displays the intensity of the signal from ions. The isotope pattern allows inference of the ion charge state (z), and the charge state together with the m/z values enable calculation of the mass of the ions. The most abundant peaks with a well-defined isotope distribution are annotated with m/z, charge state and number of additional peaks in the isotope distribution

(DDA). In the MS/MS spectrum of peptides, sequence information is obtained by correlating residue masses to mass differences between peaks (*see* **Notes 3** and **4**). Typically, the most prominent features of the spectrum are extracted and used to query a protein sequence database using different software tools. These tools work by comparing the observed fragments to fragments obtained by in silico enzyme digestion and fragmentation. Software tools that compares the experimental spectra to in silico generated spectra based on a sequence database are also known as database dependent search engines. In contrast, de novo sequencing algorithms aims at building amino acid sequences from scratch only matching the information in MS/MS spectra.

Figure 3a summarizes the observed ions and gives a convenient overview of the sequence coverage. The annotated raw spectrum in Fig. 3b contains more information but is more complex to read especially if many multiple charged fragment ions are observed.

The proteomics field applies many different instrument setups and fragmentation techniques. A glossary of some of the most commonly used is given in Table 1 and a selection of these is discussed in further detail later in this chapter.

1.2 MS-Based Proteomics

Proteomics is the global analysis of all aspects of proteins and MS has in recent years become one of the most informative methods for studying proteins. Mass spectrometry offers complementary information to the detailed structural information obtained by NMR and condensed phase methods such as X-ray crystallography. Among the advantages are that mass spectrometry is much easier to automate, more sensitive, easy to hyphenate with different

A

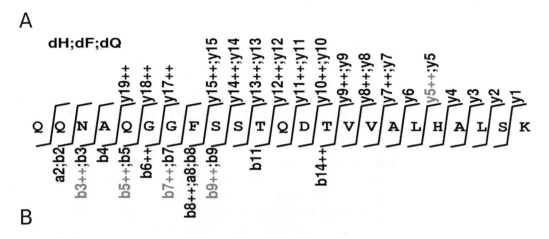

B

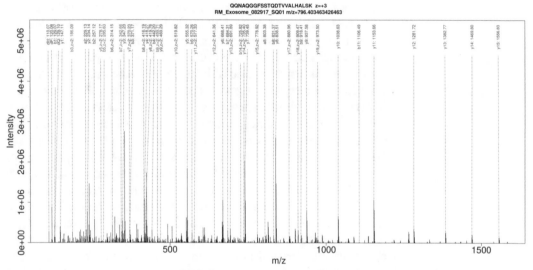

Fig. 3 Extracting information by tandem mass spectrometry. Using a Q-Exactive to analyze peptides, a plus tree charged ion with observed m/z 796.4034 was isolated and fragmented by high-energy collisional dissociation (HCD). (**a**) Summary of the observed ions in the MS/MS spectrum (*a*, *b*, and *y* ions). "d" indicates diagnostic ions. Fragment listed with "++" are doubly charged and otherwise single charged. The annotation of observed MS/MS peaks was performed with a mass accuracy of 0.005 Th. The ions labelled in red matched within mass accuracy, but the second isotopic peak was not observed for possible charge states +1–3. (**b**) The resulting MS/MS spectrum displays the intensities and m/z values of the resulting fragment ions. As described later, the mass differences between peaks correlates with amino acid residue masses and the peptide sequence is in this case deduced to QQNAQGGFSSTQDTVVALHALSK

separation techniques and useful information are routinely obtained from complex mixtures. Alternative methods such as protein arrays are more sensitive but also expensive and suffer from technical problems such as poor reproducibility. A discussion on some of the main applications of MS-based proteomics is provided in the following paragraphs.

One of the main aims of MS-based proteomics is to identify and quantify proteins and their post translational modifications in either a purified, enriched or complex protein mixture. MS-based proteomics consists of three approaches bottom-up [25], middle-down [26], and top-down approaches [27] (*see* Fig. 4). In bottom-up proteomics proteins are digested into peptides that serve as input to the MS equipment. Middle-down proteomics proteins follow the same schematic outline as illustrated in Fig. 4 for bottom-up but use a protein chemical or enzymatic cleavage that generates peptides longer than tryptic peptides (e.g., the enzyme Lys-C has been used). On the other hand, in top-down proteomics an ion trapping mass spectrometer isolates a full-length protein that subsequently fragments inside the mass spectrometer and the masses of the fragments are recorded. Top-down proteomics was traditionally restricted to FTICR instruments but within the last decade increasingly replaced by orbitraps [28, 29]. Top-down proteomics is still limited to purified proteins or simple protein mixtures and therefore require some form of protein prefractionation. Shotgun [30] proteomics is a special case of bottom-up proteomics, where typically lysine or trypsin digest a complex mixture of proteins, followed by multidimensional high performance liquid chromatography online coupled to the mass spectrometer. The difference between bottom-up and shotgun proteomics is that bottom-up strategies does not necessarily have LC separation of peptides prior to MS, whereas in shotgun strategy LC and typically multidimensional LC is always used to separate a complex mixture of peptides originating from many different proteins. The main disadvantage of bottom-up proteomics is that the shorter sequence tags frequently occur in several proteins. Furthermore, if the peptide contains a modification then it is impossible to quantify the co-occurrence with another modification occurring on another peptide from the same protein.

Site identification of post translational modifications is another major use of MS-based proteomics. Amino acid residue modifications, such as phosphorylation, acetylation, and methylation, regulate various cellular processes. MS can pinpoint the site of these modifications on a large scale. Other modifications such as glycosylation, ubiquitinylation, and sumoylation are more complicated due to their size and complexity. Although the site mapping of some of these modifications is possible on a large scale they are much harder to fully characterize. Relative and absolute quantitation of peptide and proteins together with their associated post translational modifications is regularly performed by MS (*see* Chapter 7 for more details on quantitative proteomics).

MS-based strategies refered to as N- and C-terminomics determine the N- and C-terminal of the proteins, respectively. These strategies therefore allow elucidation of primary protein structure and to map proteolytic cleavage sites. By chemically modifying the

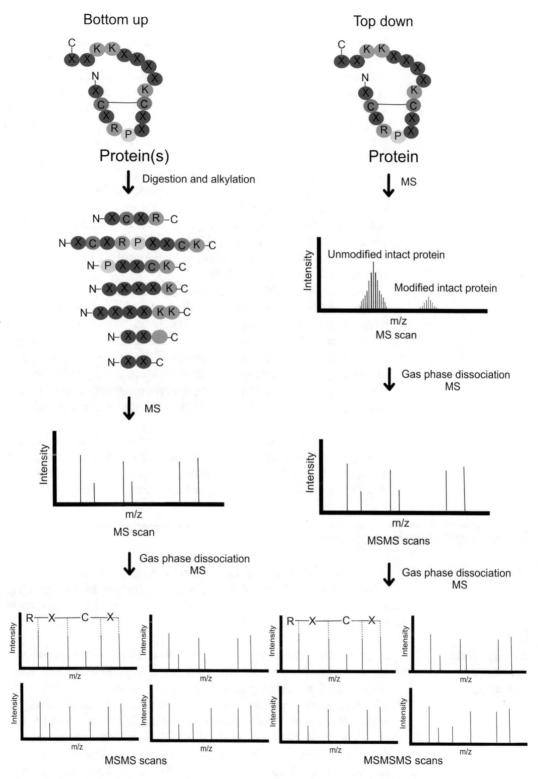

Fig. 4 Outline of the different analytical strategies named bottom-up, top-down, and shotgun proteomics. Intact proteins are analyzed in top-down experiments, whereas the proteins are processed to peptides prior to MS analysis in bottom-up and shot-gun experiments. In top-down proteomics simplified protein mixtures are analyzed, for example, obtained by affinity purification

C- and N-terminus at the protein level these termini become distinct from the C- and N-termini generated at later stages in the analysis. Several methods based on negative enrichment have been proposed for enrichment of peptides that define the N- and C-terminus of proteins. These methods are referred to as N-TAILS [31] and C-TAILS [32], respectively. N-TAILS has also been combined with iTRAQ quantitation the so-called iTRAQ-TAILS [33]. Combined fractional diagonal chromatography (COFRADIC) provides an alternative approach to N-terminomics [34]. The lysine amines can also be blocked by guanidination followed by biotin tagging of terminal amines and positive enrichment on a streptavidin column [35]. N- and C-terminomics find use cases in the study of proteolytic signaling and for identification of protease labile sites in enzymes, which might allow the design of enzymes with improved stability.

Artificially modifying accessible surface amino acid residues followed by MS allows inference about protein conformation. For example, hydrogen deuterium exchange can be used to examine if protein has folded correctly, to measure conformational changes after buffer exchange or modification of the protein, identification of binding sites of antibodies, other proteins and ligands, identification of protein sites stimulating protein aggregation and to validate that different protein production strategies end up with the same protein fold.

Mass spectrometry has found many related applications beyond identification and quantitation of proteins and has been pitted against macromolecular analytical challenges as microorganism classification [36] and studying protein complexes [37].

2   Introduction to Methodologies

The above short introduction intended to give a quick glimpse of MS-based proteomics and form a starting point for further reading. In the subsequent section more, details of some of the methods mentioned above is provided.

2.1 The Basic Concepts of Proteomics

The term proteomics covers the analysis of all proteins expressed in an organism. One of the first tools used in proteomics was two-dimensional gel electrophoresis (2D-GE) introduced in 1969 [38] where proteins are separated by their isoelectric point (pI, the pH where the protein is uncharged) via isoelectric focusing and by size via their migration on an SDS-PAGE gel. Mass spectrometry (MS) techniques have been combined with two-dimensional polyacrylamide gel electrophoresis (2D-PAGE) for direct and systematic identification of polypeptides. It has been shown that traditional 2D-PAGE can resolve up to 1000 protein spots in a single gel [39]. This is an impressive number but compared with the number of expressed genes in various organisms, which range typically from 5000–40,000, it is clearly not sufficent. Especially if one also considers that in eukaryotes each gene can have several splice forms and each protein can have an array of posttranslational modifications [40]. Efforts have been made to optimize the standard 2D-PAGE technique by making larger 2D-PAGE [41]. The technique uses multiple narrow range isoelectric focusing gels to improve separation in the first dimension. In the second dimension, multiple long SDS-PAGE gels of different polyacrylamide concentrations are used. The large 2D-PAGE was claimed to resolve more than 11,000 protein spots. In general, the low-abundance and hydrophobic proteins are difficult to identify by 2D-PAGE based methods [42]. It is normally not possible to retain protein with extreme pI or molecular mass. In addition, 2D-PAGE has a low sample throughput [43].

Protein spots resolved by 2D-PAGE are typically cut out and the proteins enzymatically digested in-gel [44, 45] or digested during blotting onto membranes containing immobilized trypsin [46]. This is a typical bottom-up experiment where each digest yields a peptide mixture that is analyzed by MS (peptide mass fingerprinting, PMF), by MS/MS or by a combination of the two.

Multidimensional protein identification technology (MudPIT) is a method where the entire complex protein sample is digested without prior separation on the protein level. The resulting highly complex mixture of peptides is fractionated by chromatography on the peptide level, originally using strong cation exchange chromatography, and analyzed by LC-MS. This type of approach is also referred to as Shotgun Proteomics [30] and proven efficient for identification of proteins. However, shotgun technology has some problems with the confidence of the identified peptides, correlating the identified peptides to the proteins they originated from and how to accurately quantify the proteins. Ways to quantify without stable isotope labels is to integrate the UV absorbance from LC, integrate intensity counts in LC-MS scans recorded by the mass spectrometer of a peptide in the chromatographic step, integrate fragment ion counts in MS/MS spectra or count the number of times a peptide is identified (aka spectral counting). The method requires high reproducibility when two samples are compared. Reproducibility across several steps of chromatography is especially difficult to achieve if nano-liquid chromatography (nLC) columns are used. Especially frustrating are failures in the analysis due to partial blocking of columns they may occur to different extents during consecutive runs. A more precise method of quantification of peptides in LC-MS/MS experiments is to use stable isotope labeling. The quantitative methods can be divided into relative and absolute quantification methods (*see* Chapter 7). The principle is that two or more samples are labeled with different stable isotopes. The differential labeling can occur during the biosynthesis of the proteins in cultured cells (SILAC, *see* Chapter 7), by reacting residues with labels containing different stable isotopes (Chemical labeling, *see* Chapter 7), or by enzyme catalyzed incorporation of ^{18}O in the peptide from ^{18}O water during proteolysis [47].

The shotgun method gives an enormous amount of data, which requires automatic processing. Automatic computer-based interpretation of data calls for high quality statistical testing to evaluate the quality of the interpretation. In the traditional 2D-PAGE several peptides from the same protein normally confirm the identification, whereas the shotgun method often claims identification of proteins from two peptide sequence tags. This is problematic since the same tryptic peptides can occur in rather diverse protein sequences. Therefore, the shotgun experiments require that the significance of protein assignment is more precisely evaluated than for the 2D-PAGE method.

2.2 Sample Preparation for MS

The quality outcome of MS-based proteomics is heavily dependent on sample complexity and purity. The main reason for this is ion suppression (*see* **Note 5**) where many different species compete for the charges during the ionization process. This can derive from peptides that overlap in liquid chromatography (LC) retention time and m/z value dimensions. The quality of the data is also heavily affected by contaminations from different sources. It is for example not uncommon when studying proteins from cell culture models to detect proteins from cell culture medium (typically bovine proteins from the serum used to supplement growth media). If proteins from the cell culture media are not included in the searched database then homologous proteins from the target organism get identified instead of the contaminating proteins from the cell culture media [48] (*see* **Note 6**). Even if the contaminating proteins are included in the searched database, problems such as ion suppression and overlap with target peptides of interest may occur. During sample preparation human and sheep keratin from clothing can contaminate the sample and cause the same problems as protein contaminants from the cell culture medium. It is therefore essential to work in a laminar flow hood, use gloves and rinse the gloves on regular basis. Measures to minimize chemical contaminants should also be taken. Polyethylene glycol [49] from plastic tubes, volatile chemicals [50] in the oil used for mass spectrometer pumps and detergents used in buffers to solubilize proteins are possible sources of chemical contaminants. It is therefore recommended to consider the use of detergents for cleaning glass ware such as gel electrophoresis plates and as a rule of thumb prepare all solutions freshly. FASP (Filter Aided Sample Preparation) protocol can clean a detergent contaminated sample [51].

A proteomic project starts by generating a protein extract from tissues or from a cell culture. It is an advantage to chemically control the reactive cysteines at the earliest possible stage to prevent mixtures of different cysteine modifications. The reactive cysteines can for example become oxidized or react with nonpolymerized acrylamide during electrophoresis [3]. Typically the protection is done by reduction with DTT and alkylation with iodoacetamide, but a discussion of the advantages of different cysteine modifications is presented in [52].

In a MALDI-TOF MS setup for peptide mass fingerprinting to identify a protein, the protein must reach almost 100% purification before the proteolytic cleavage and subsequent MS analysis. The preferred method for partial protein purification in combination with MALDI-TOF MS is 2D-PAGE. LC-MS based approaches can both be used for simple protein mixtures purified by chromatography or 2D-PAGE or for more complex protein mixtures. An increase in protein sample complexity leads to more complex peptide mixtures requiring more efficient chromatography steps prior to MS analysis. In LC-MS the chromatography is directly coupled to the MS instrument. In the typical 1D-LC-MS a reverse phase C18 column is used for separating semi complex samples because the volatile buffers are well suited with MS analysis. In 1D-LC-MS the gradient and separation time are often varied according to the complexity of the sample.For directly analyzing whole cell extract 2D-LC-MS approaches are typically used. The first column is then frequently a strong cation exchanger (SCX) directly coupled to a reverse phase column that finally is coupled to the MS instrument. An overview of typical strategies is provided in Table 2. Gilar and coworkers tested different 2D-LC-MS setups and found that SCX-RP, HILIC-RP, and RP-RP 2D systems provide suitable separation [53]. High pH RP coupled to low pH RP was found to be superior of the tested combinations. For simplicity the first dimension is frequently made off line. Batth and colleagues recently demonstrated that High pH RP coupled to low pH RP is also superior

Table 2
Typical separation methods used in combination with MS setups

Pre-MS separation	1. step	2. step
2D-page	MALDI-MS	LC-MS
1D-PAGE	1D-LC-MS	
Chromatography	1D-LC-MS	
None (raw extract)	2D-LC-MS, e.g., combing high and low pH RP	

Notice that a common strategy when 2D-PAGE separation is used is to first attempt to identify the in gel digested protein by using half of the sample on MALDI-MS (both MS and MS/MS) and if this fails then use the remaining sample for LC-MS runs for separating complex phosphor-peptides mixture compared with TiO$_2$ and SCX [54].

The proteolytic cleavage is most often done with trypsin in a 1:50 trypsin to protein ratio. Trypsin is a serine protease that specifically cleaves at the carboxylic side of lysine and arginine residues if these are not followed by proline (*see* **Note 7**). It is worthwhile to note that "in gel digestion" requires more trypsin than in solution digestion. The abundance and distribution of lysine and arginine residues in proteins are such that trypsin digestion yields peptides of molecular mass that are well suited for analysis by MS. The specificity of trypsin is essential and the quality and price varies a lot [55]. Wild-type trypsin is subject to autolysis, generating pseudo-trypsin, which exhibits a broadened specificity including chymotrypsin-like activity [56]. Additionally, trypsin is often contaminated with chymotrypsin. For these reasons, affinity purified trypsin, reductively methylated to decrease autolysis, and treated with TPCK (N-tosyl-l-phenyl chloromethyl ketone) a chymotrypsin inhibitor, is preferable for MS-based trypsin digestion. Such trypsin preparations are commercially available under the trade name Sequencing Grade Trypsin (Sigma) or Trypsin Gold (Promega). Direct protein digestion in solution constitute an alternative approach. Trypsin cleaves rather specifically after the basic residues arginine and lysine, if not followed by proline, and is by far the most commonly used enzyme for proteomic studies. The cleavage motif for trypsin is conveniently summarized as "[RK].<P>", where "." (full stop) indicates the cleavage position. Letters in "[]" square brackets are alternative amino acids needed for recognition. Amino acids in "<>" angle brackets or followed by "^" are amino acids that prevent recognition and cleavage. Cyanogen bromide (CNBr), which cleaves after methionine ("[M]."), and endoproteinases such as chymotrypsin (cleaves after large hydrophobic amino acids "[WYF].<P>"), Lys-C (cleaves after lysine "[K]."), Lys-N ".[K]", Glu-C "[ED].<P>", and Asp-N (cleaves before aspartate ".[D]") are the most common used. Other enzymes with broad specificity such as Proteinase K ([AEFILTVWY].) are useful for generating peptide ladders with overlapping peptide fragments. Overlapping peptides are useful for de novo sequencing of proteins. Proteinase K is also frequently used for hydrogen deuterium experiments to obtain close to complete sequence coverage (*see* for example Chapter 18). The documentation for the program PeptideCutter contains a very nice list of proteases and their specificities [57]. Most of these cleavage patterns are implemented in silico in the R package "cleaver" [58]. A efficent cleavage method must be compatible with the subsequent mass spectrometric analyses, have close to 100% cleavage efficiency, and have a high specificity. The limitation of possible cleavage methods has led to the development of engineered proteases with new specificity [59]. Mass spectrometric analysis of peptide fragments can lead to the identification and partial sequencing of a protein, and identification of modifications. Such strategies based on proteolysis are referred to as bottom-up sequencing and are the most common techniques used in mass spectrometry. An alternative approach, referred to as top-down sequencing, is starting to emerge, where intact proteins are fragmented directly in the mass spectrometer [60]. After digestion, it is advantageous to concentrate and remove buffer contaminants by reversed-phase (RP) microcolumns to increase signal-to-noise ratios and sensitivity before mass spectrometric analysis. However, small and hydrophilic peptides are typically lost by using the RP microcolumns. This problem can be solved by combining the RP microcolumns with graphite powder [61]. Additionally, the graphite powder is used to remove some types of undefined biopolymers. An increasing number of methods have been developed to enrich for peptides having specific modifications. The enrichment is in many cases essential since it lowers the ion suppression from nonmodified peptides and thereby increases the signal of peptides having the specific modification. For example, hydrophilic interaction liquid chromatography (HILIC) is used for the enrichment of glycosylated peptides [62]. IMAC and titanium dioxide have proved effective for enrichment of phosphopeptides [63, 64].

2.3 The Issue of Complete Digestion

Peptide and protein identification strategies are often robust toward partial incomplete digestion. In fact, in many cases the identification of proteins is improved by having 1–2 missed cleavage sites per peptide since longer peptides have a more unique mapping to proteins. Consider also the peptide "ALKESTTR" that contains one missed cleavage site after tryptic digestion. Here we expect the possibility of having a missed cleavage site since the internal lysine is neighboring glutamic acid (E). Whether the missed cleavage is observed or not depends on the extent of tryptic digestion. "ALK" is typically outside the mass range and even though "ESTTR" in most cases is in the observed mass range, it is too short to contribute significantly to confident protein identification. However, the peptide "ALKESTTR" is detectable and contributes significantly to protein identification.

For some relative and absolute quantitation techniques the completion of digestion needs to be well controlled. A quantitative technique will be affected by uncontrolled digestion if the stable isotope label is inserted into peptides after the protein digestion. Some of the relative quantitation techniques that are affected by uncontrolled completion of digestion are iCAT, HysTag, iTRAQ, and label-free quantitation, whereas SILAC is not affected (*see* Chapter 7). For absolute quantitation AQUA (absolute quantification) [65] and QCONCAT [66] are affected, whereas PSAQ™ (Protein Standard Absolute Quantification) [67] is not. Brownridge and Beynon [68] provide a more detailed review and experimental evidence on the topic of the digestion.

**2.4   Sample
Separation Methods**

As already mentioned above, a simple peptide sample for MS analysis gives less risk of peak overlap and more sensitivity due to less ion suppression. Both protein and peptide separation techniques at different stages can be applied to obtain simple peptide mixtures. First of all, raw extract enriched in proteins of specific cellular compartments such as cytoplasm, nuclear, mitochondria, plasma membrane proteins and extracellular proteins can be prepared. A simple technique to simplify these subcellular proteomes is to use differential centrifugation [69–71] and/or differential ammonium sulfate precipitation [72, 73]. Immuno-precipitation of specific targets can isolate the target or the target together with its binding factors depending on the washing conditions. Likewise, transfection of a DNA constructs coding a tagged target protein allows enrichment of the target protein. Glycans and amino acids with azide tags can be used to metabolically label subpopulations of proteins and chemically, via click chemistry, attach biotin. Next, Streptavidin columns efficently enrich the biotin conjugated molecules. Alternatively, posttranslational modifications can be tagged with biotin after cell lysis. Special protein binding domains such as for example ubiquitin binding domains can be used to isolate ubiquitin modified subproteome [74]. Lectins can be used to isolate subpopulations of glycoproteins [75]. Proteins can also be separated by free flow electrophoresis [76] and liquid [77] or gel based isoelectric focusing. Column chromatography based on size exclusion, ion exchange and reverse phase chromatography based on hydrophobic interactions. 1D- and 2D-PAGE can further separate proteins and also cleanup the samples from detergents and salt that can interfere with the quality of the MS data.

The peptide sample is in most cases cleaned up (desalted and cleaned from gel debris) by reversed phase (e.g., ZipTip or Stage-Tip) columns before LC-MS even if the peptide sample is obtained by in gel digestion. The column material used is typically C18 (typically used for tryptic digests) for small peptides and C4 for proteins and longer peptides. It is essential that the peptides on C18 beads are stepwise eluted if only one elution in high concentration of acetonitrile is used then all peptides will not be eluted. A number of HPLC/UPLC columns or beads in addition to SCX, C18, and C4 columns can be used for peptide separation such as HILIC (glycopeptides), graphite (glycopeptides), titanium dioxide (phosphopeptide), and IMAC (phosphopeptide).

**2.5   Ionization
Methods**

The formation of gas-phase ions is required before measurement of the molecular sample masses by the mass analyzer. The generation of intact gas-phase ions is in general more difficult for higher molecular mass molecules. Early ionization methods for peptides and proteins, like fast-atom bombardment (FAB) and ^{252}Cf plasma desorption (PD) were successful and are credited for initiating the attention of the biochemists toward mass spectrometry [14]. The breakthrough in mass spectrometry of proteins and peptides came with the introduction of electrospray ionization mass spectrometry (ESI-MS) [5] and Matrix-Assisted Laser Desorption/ionization (MALDI MS) [9]. The advantage of these soft ionization techniques is that intact gas-phase ions are efficiently created from large biomolecules, with minimum fragmentation. MALDI and ESI ionization are further discussed in the next sections. Subheadings 2.5–2.9 discuss a common instrument setup using MALDI for ionization and a TOF mass analyzer, which is a common instrument setup. Subheadings 2.10–2.12 discuss instrument setups using ESI ionization with either an ion trap or a Q-TOF mass analyzer. In current MS-based proteomics mainly five mass analyzer types are used: quadrupole mass filters (Q), time of flight (TOF) mass analyzers, ion traps (IT), FT-ICR, and Orbitrap analyzers. These five mass analyzers can work as standalone but more frequently they are combined in hybrid configurations of which some combinations are more popular. Popular configurations for MS-based proteomics are Q-TOF, Q-ion mobility-TOF, triple quadrupole, ion trap-Orbitrap and Q-Orbitrap. Note, ion mobility is not considered as a mass analyzer because it separates ions based on mass, charge and size and is measured as drift time. In ion mobility the ions travel through a pressurized cell under influence of a weak electric field and the movement is slowed down by collisions with N_2 gas. The different mass analyzers are next discussed together with ionization methods.

2.6 MALDI Ionization

MALDI is an improvement of the laser desorption ionization (LDI) technique. In LDI soluble analyte is air dried on a metal surface and the ionization is achieved by irradiating with a UV laser. The disadvantage of LDI is that, in general, it has low sensitivity, the ionization method causes ion fragmentation, and the signal is very dependent of the UV absorbing characteristics of the analyte [21]. This is solved with MALDI by decoupling the energy needed for desorption and ionization of the analyte. In MALDI the analyte is mixed with a compound, termed the matrix, which absorbs the energy from the laser. The sample is co-crystallized with an excess amount of the matrix. A variety of matrices are used and the matrices are commonly small aromatic acids. The aromatic group absorbs at the wavelength of the laser light, while the acid supports the ionization of the analyte. Irradiation with a short-pulsed laser, often a 337 nm N_2 laser, causes mainly ionization of the matrix followed by energy and proton transfer to the analyte [78]. The MALDI efficiently ionizes peptides and proteins intact. Contaminants frequently encountered in protein and peptide samples, such as salts, urea, glycerol, and Tween-20 normally suppress ionization. However, MALDI based ionization is less affected by these contaminants. It is believed that these compounds are excluded from the matrix and peptide/protein crystal

[3]. Nevertheless, MALDI ionization displays sensitive to low concentrations of SDS [79].

The matrix preparation has a major effect on the quality of MALDI-MS spectra. Several different matrix-sample preparations have been developed. The earliest and most frequently used technique is the dried droplet method [9, 80]. A small volume of sample is added to a droplet of saturated solution of matrix in a volatile solvent on a target and the drop is dried to form crystals. An improvement of this method was made by applying a pure matrix surface by fast evaporation before applying the matrix-analyte mixture [81]. Fast evaporation can be obtained by dissolving the matrix in volatile solvents like acetone. The method gives a more homogenous layer of small crystals. Inhomogeneous crystals formed under slow evaporation will give lower resolution, low correlation between intensity and analyte concentration, and lower reproducibility. A number of other preparation methods have been studied by Kussmann et al. [82] using a variety of matrices: HCCA (α-cyano-4-hydroxy cinnamic acid), SA (sinapinic acid), DHB (2,5-dihydroxybenzoic acid), and THAP (2,4,6-trihydroxyacetophenone). General guidelines for matrix use are summarized in Table 3.

Some of the matrix preparation methods allow cleanup from buffer contaminants by a short wash with ice-cold 0.1% trifluoroacetic acid (TFA). No universal matrix preparation procedure performs well for all peptides and proteins. Only general guidelines have been found such as the requirement of the sample-matrix mixture to be adjusted to pH < 2 for optimal signal-to-noise

Table 3
Commonly used MALDI matrices and some general guidelines for matrix usages

Matrix	Application	Solvent	Wavelength
α-Cyano-4-hydroxycinnamic acid (CHCA)	In general, efficent but mainly for peptides below 2500 Da	Acetonitrile, water, ethanol, acetone	337, 355
2,5-Dihydroxybenzoic acid (DHB)	Hydrophobic peptides or modified peptides	Acetonitrile, water, methanol, acetone, chloroform	266, 337, 355
3-Hydroxypicolinic acid (3-HPA)	Glycosylated or phosphorylated peptides	Ethanol	337, 355
4-Hydroxy-3-methoxycinnamic acid (ferulic acid)	Proteins	Acetonitrile, propanol, water	266, 337, 355
3,5-Dimethoxy-4-hydroxycinnamic acid (SA)	Peptides, proteins, lipids	Acetonitrile, water, acetone, chloroform	266, 337, 355

ratio. Therefore, the above studies point to the conclusion that several combinations should be tested for each sample and that further optimization in some cases will be necessary.

In general, when MALDI ionization is applied to tryptic peptides, protein sequence coverage of typically 30–40% is obtained for well prepared samples. The competition for the protons in the matrix plume partly explains the limited sequence coverage. The more basic arginine containing peptides have higher affinity for the protons than the lysine containing peptide. Arginine containing peptides are more frequently observed in MALDI MS spectra because of the higher proton affinity and the higher stability. In addition, some peptides co-crystallize poorly with the matrix.

2.7 The TOF Mass Analyzer

In TOF-MS a population of ions, for example derived by MALDI, is accelerated by an electrical potential as shown in Fig. 5 and each ion acquires a kinetic energy depending on the ions mass and charge state. After acceleration the ions pass through a field free region where each ion is traveling with a speed characteristic of their m/z value.

At the end of the field free region a detector measures the time of flight. The recorded TOF spectrum is a sum of the following times: $\mathrm{TOF} = t_a + t_D + t_d$, where t_a is the flight time in the acceleration region, t_D is the flight time in the field free region, and t_d is the detection time. The drift time t_D approximates the flight time because the acceleration region is much smaller than the field free region:

$$t_D = D\sqrt{\frac{m}{2zeV_{ac}}} \tag{1}$$

where D is the drift distance, z is the number of charges of the ion, e the elementary charge, V_{ac} is the acceleration voltage, and m the mass of the ion. Therefore, t_D for an ion is proportional to $(m/z)^{1/2}$ [83].

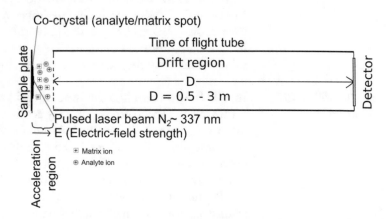

Fig. 5 Schematic representation of a linear TOF mass spectrometer

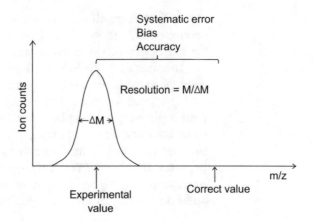

Fig. 6 Mass accuracy and resolution concepts

The mass resolution for TOF analyzers and other mass analyzers is reported as $m/\Delta m$ (Δm full width of peak at half maximal height FWHM, *see* Fig. 6). Mass accuracy is typically reported as the absolute value of ($m_{observed} - m_{theoretical})/m_{theoretical} \times 1,000,000$ (ppm—parts per million). However, as mass spectrometers have similar absolute mass accuracies in the range 0–1800 m/z, absolute mass accuracy or root mean square represents a more meaningful way of stating mass accuracies [84]. The mass difference between observed and theoretical mass is a result of systematic error, bias, and instrument accuracy.

The ion production time, initial velocity distribution, and the ion extraction time all contribute to determining resolution. The initial kinetic energy differences is compensated by adding an ion reflector (also called an ion mirror) to the TOF instrument. The reflector delivers an electric field that returns the ions in an opposite direction at an angle to the incoming ions at the end of the drift region. The more energetic ions will penetrate more deeply into the reflecting field and, therefore, have a longer flight path than ions with the same mass but less kinetic energy. The resolution of MALDI TOF MS is considerably improved by using delayed ion extraction (DE) technique [85]. In this system ions formed by MALDI are produced in a weak electric field, and after a predetermined time delay they are extracted by a high voltage pulse.

2.8 The MALDI-TOF MS Spectrum

In a TOF spectrum it is the time of flight of ions that is recorded. However, the spectrum is normally reported as m/z versus relative intensity. Figure 7 shows the MALDI spectrum of tryptic peptides of bovine serum albumin analyzed on a Bruker Reflex III TOF mass spectrometer using HCCA as matrix. The peak annotation was done automatically using a cutoff for signal-to-noise ratio of 4.

Peptide ions generated by MALDI generally occur in a low charge state, and normally only the singly charged state of a peptide

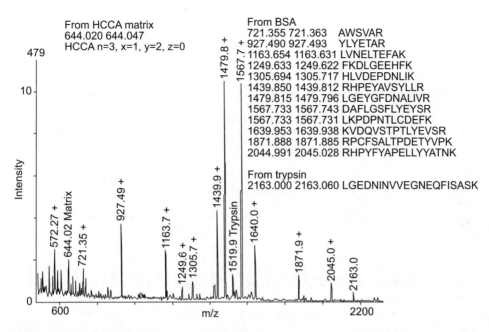

Fig. 7 Typical MALDI-TOF MS spectrum. (+) indicates mass peaks that matched the masses of theoretical tryptic peptides from BSA. The *first column* in the *upper right corner* shows the experimental masses, the *second column* the theoretical tryptic peptide masses, and the *third column* shows the tryptic peptide sequences from BSA

is observed [86]. For proteins it is common to observe also the two and three charged states.

Frequently, a MS spectrum obtained from a purified protein contains unmatched peaks. Typical reasons are tryptic missed cleavage site, modifications of tryptic peptides, nontryptic cleavages sites, tryptic peptides from other contaminating proteins, ions from matrix clusters, and fragmentation of ions during MALDI-TOF MS. The observed non-tryptic cleavages are often due to proteolysis from proteases in sample of interest or pseudotrypsin, which is generated by autolysis of trypsin. Pseudotrypsin has a broader specificity than trypsin (*see* Subheading 1.2).

The modifications of the tryptic peptides have different origins, in vivo posttranslational modification, uncontrolled chemical modifications during sample preparation, and intentional modifications such as cysteine and cystine modifications.

The most frequently observed contaminating tryptic peptides come from trypsin and human or sheep keratins from hair, skin and cloth. Often proteins purified by chromatography or 2D electrophoresis are not always 100% purified from other proteins in the sample.

For samples with high concentration of peptides and low salt, ions from matrix clusters are normally not observed in the mass region of interesting tryptic peptides (*m/z* above 500). For dilute

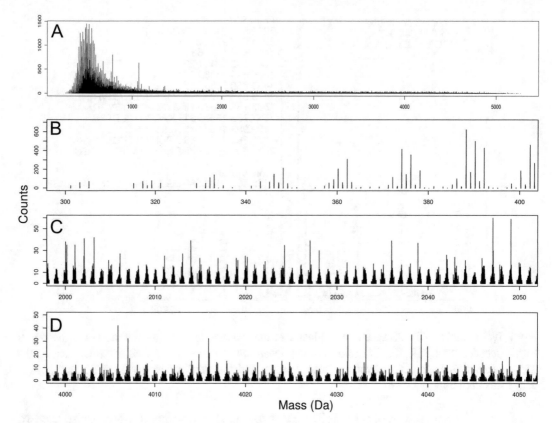

Fig. 8 Distribution of the theoretical masses between 400 and 6000 Da of tryptic peptides for all human proteins at different zoom in levels. The mass scale was divided in unique mass bins, and the numbers of tryptic peptides within each bin were counted

samples, and samples containing high concentration of salt, mass peaks from matrix clusters might be observed in the mass region up to 2000 Da for HCCA [87]. In general, peaks of matrix clusters can be observed up to m/z 1000. If the matrix cluster masses are not removed prior to the peptide mass fingerprinting (PMF) search, erroneous protein matches might end up in the search result. One solution to this problem could be to remove all mass peaks below m/z 800, as shorter tryptic peptides are in general less unique as shown in Fig. 8.

However, the low mass region still contains valuable information and some shorter peptides actually have more unique mass than longer peptides (*see* Fig. 8). However, identification of the mass peaks from the matrix clusters is preferred. Most of the matrix peaks observed in the positive mode are given by the following equations [87]:

$$M_{\text{Cluster}} = nM - xH + yK + zNa - m(e^-) \qquad (2)$$

where

Table 4

Masses (in Da) of the elements used in the calculation of ion adducts

	Mass (Da)
H	1.007825032
Na	22.98976967
K	38.9637069
HCCA	189.0425931

The elemental masses are from http://www.ionsource.com, and the mass for HCCA was calculated using the elemental masses

$$y + z = x + 1 \tag{3}$$

and

$$y + z \leq n + 1 \tag{4}$$

n, y, z, and x are integer values and $m(e^-)$ is the mass of an electron. $M_{cluster}$ is the observed mass of the ionized matrix cluster, while M, H, K, and Na stand for the masses of HCCA, hydrogen, potassium, and sodium, respectively. For the DHB matrix a similar equation where also the loss of water is taken into account, have been established [87]. Table 4 gives the masses used to calculate mass of the different matrix clusters.

The mass peak 644.02 m/z in the MALDI-TOF spectrum in Fig. 7 was assigned as a matrix peak ($n = 3$, $x = 0$, $y = 2$, and $z = 0$) using Eqs. 2–4. Thereafter only one annotated mass peak of Fig. 3 is unassigned. Zooming in on the low mass region some additional low intensity peaks were annotated (Fig. 9).

Low intensity peaks are often noise peaks and therefore better left unannotated for the first database search to avoid erroneous protein identification. Further annotation of the low intensity mass peaks identified one additional mass peak matching BSA, and four matching matrix clusters. The result of the final data analysis is summarized in Table 5. The identification of trypsin and matrix cluster peaks gives the opportunity to use these mass peaks for internal calibration (*see* Appendix 2).

Matrix clusters with cation contaminants might be suppressed either by purifying samples on reversed phase (RP) microcolumns, by using cetrimonium bromide in the matrix solution [88], or by recrystallizing the matrix in ethanol [87]. However, RP purification might lose some peptides and cetrimonium bromide lowers the intensity of sample ions although very efficient in suppressing matrix clusters.

Fragmentation of sample ions may occur due to collisions with matrix molecules during ionization. Optimal combination of

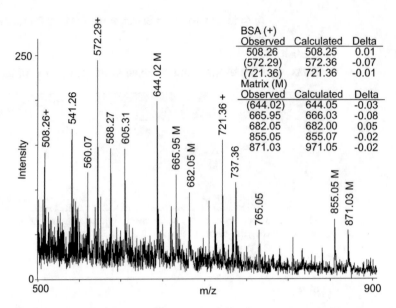

Fig. 9 Zooming in on the low mass region of the MALDI-TOF spectrum shown in Fig. 7 Peptide mass peaks from BSA are indicated by (+), and (M) indicates mass peaks that correspond to matrix clusters. The observed mass peaks in parentheses were already annotated in the first annotation round. The annotation in the second round was done manually

Table 5

Interpretation of low intensity peaks in the low mass region of a MALDI spectrum of a BSA tryptic digest

Observed mass (Da)	Calculated mass (Da)	Delta mass (Da)	Match	n	x	y	z
499.204			?				
508.257	508.252	0.005	BSA				
541.262			?				
560.071			?				
(572.291)	572.363	−0.072	BSA				
588.265			?				
605.314			?				
(644.022)	644.047	−0.025	Matrix	3	1	2	0
665.954	666.029	−0.075	Matrix	3	2	2	1
682.054	682.003	0.051	Matrix	3	2	3	0
(721.355)	721.363	−0.008	BSA				
737.361			?				
765.047			?				
855.047	855.072	−0.025	Matrix	4	2	2	1
871.026	871.046	−0.02	Matrix	4	2	3	0

The observed mass peaks in parentheses were already annotated in the first annotation round matrix, laser fluency, and extraction delays minimize MALDI frag-mentation [79].

2.9 Annotation of MALDI-TOF MS

To annotate a MS spectrum, it is essential to understand the nature of the isotopic distribution. Biological molecules are mainly composed of the elements carbon (C), hydrogen (H), nitrogen (N), oxygen (O), and sulfur (S). Some biological molecules including proteins and DNA can also bind metal ions. Natural isotopes of the central elements occur at almost constant relative abundance (Table 6). A more extensive list of biological relevant isotopes is provided at http://www.ionsource.com/Card/Mass/mass.htm.

From the values given in Table 6 one can calculate relative isotopic abundance of different biological molecules. The mono-isotopic mass is calculated as the sum of the most abundant isotope for each element. The relative abundance of the monoisotopic mass of a molecule with the composition $C_xH_yN_zO_vS_w$ is calculated using the following expression [90]:

$$P_M = P_C^x * P_H^y * P_N^z * P_O^v * P_S^w \qquad (5)$$

where, P_M is the relative abundance of the monoisotopic peak for the molecule. P_C, P_H, P_N, P_O, and P_S are the abundance of the monoisotopic masses of the C, H, N, O, and S elements, and x, y, z, v, and w are positive integer values. The expression is simply the probability that all the elements in the molecule have the mono-isotopic mass. A similar expression can be made for the monoiso-topic mass plus one:

Table 6
Masses and abundance of biologically relevant isotopes [87, 89]

Isotope	A	%	Isotope	A + 1	%
^{12}C	12	98.93(8)	^{13}C	13.0033548378(1)	1.07(8)
^{1}H	1.0078250321(4)	99.9885(7)	^{2}H	2.0141017780(4)	0.0115(7)
^{14}N	14.0030740052(9)	99.632(7)	^{15}N	15.0001088984(9)	0.368(7)
^{16}O	15.9949146221(15)	99.757(2)	^{17}O	16.99913150(2)	0.038(1)
^{32}S	31.97207069(12)	94.93(3)	^{33}S	32.97145850(1)	0.76(2)
Isotope	A + 2	%	Isotope	A + 4	%
^{14}C	14.003241988(4)	–	–	–	–
^{3}H	3.0160492675(11)	–	–	–	–
^{18}O	17.9991604(9)	0.205(1)	–	–	–
^{34}S	33.96786683(11)	4.29(3)	^{36}S	35.96708088(3)	0.02(1)

Uncertain digits are shown in parenthesis

$$
\begin{aligned}
P_{M+1} = {} & \binom{x}{1} P_C^{x-1} P_{C+1} P_H^y P_N^z P_O^v P_S^w \\
& + P_C^x \binom{y}{1} P_H^{y-1} P_{H+1} P_N^z P_O^v P_S^w + \cdots \\
& + P_C^x P_H^y P_N^z P_O^v \binom{w}{1} P_S^{w-1} P_{S+1}
\end{aligned} \tag{6}
$$

where P_{C+1}, P_{H+1}, P_{N+1}, P_{O+1}, and P_{S+1} are the abundance of the monoisotopic mass + ~1 Da of the elements. Again, the expression is the probability that one atom in the molecule is the monoisotopic mass plus one. Note, this way of calculating the isotopic distribution is an approximation, which works well when comparing with observed isotopic distribution from mass spectrometers that are unable to resolve the different elements contribution to the M + 1 ion. In a similar way one can make an expression for the abundance of the M + 2 isotope peak. How to perform such calculations in practice is described in Chapter 2. Zooming in on the mass peaks 927.49 and 2045.00 *m/z* in Fig. 7 reveals the isotopic distributions in Fig. 10.

For PMF searches it is obligatory to annotate the monoisotopic mass peak for each isotopic distribution since monoisotopic peak lists are what most PMF search engines expect as input. If the isotopic peaks are unresolved, which is sometimes observed for peptides above 2500 Da then some search engines allows the usages of the average mass.

The annotation of peptide masses is normally straight forward since the ions present in the MALDI-TOF peptide MS spectrum are predominantly single charged. Although, background noise and peak overlap might complicate the peak annotation. The charge states are determined from the distance between two peaks in the isotopic mass distribution of an ion (*see* Fig. 10). Isotopic peaks roughly differ by the mass of a neutron and are hence separated by

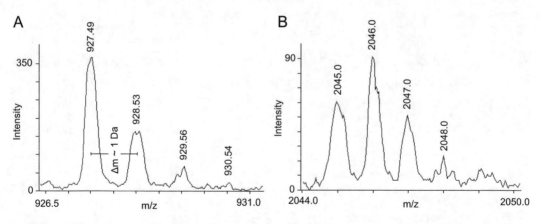

Fig. 10 Zoom in on two annotated masses from the spectrum in Fig. 7 (**a**) At low masses the monoisotopic peak is the most intense peak in the isotopic cluster. (**b**) At higher masses it is no longer the monoisotopic peak that is the most intense peak in the isotopic distribution

~1 Th for single charged ions, ~0.5 Th for double charged, etc. In general, the charge state is calculated by

$$\Delta m = \frac{m+1}{z} - \frac{m}{z} = \frac{1}{z} \Leftrightarrow z = \frac{1}{\Delta m} \qquad (7)$$

where Δm is the mass difference between the isotopic peaks.

2.10 MALDI-TOF MS

MALDI-TOF MS has been used extensively together with 2D-electrophoresis [91] and still plays a role in proteomics. One of the biggest problems with the MALDI-TOF MS for identification of proteins is the difficulty in obtaining highly significant search result in all cases. This is because not all the expected tryptic peptides show up in the experimental MS spectrum. It is additionally complicated by unknown mass peaks. The PMF by MALDI-TOF MS is limited to the analysis of purified proteins. Obtaining statistically significant PMF results for protein mixtures has proven difficult. However, there have been reports on strategies to identify protein in mixtures containing few proteins by PMF searches [92, 93]. Using more cleavage methods, in addition to trypsin, improves the significant identification of proteins identified by PMF. Ideally the search engine must allow both average and mono-isotopic mass as input for PMF searches.

Using MALDI-PSD (post-source decay) it is possible to obtain sequence information with the MALDI method by generating fragment ions [19]. However, the PSD fragmentation is rather complex. With the introduction of a collision cell in a MALDI-TOF-TOF MS it is possible to obtain low and high-energy CID fragmentation [47], which can provide more informative MALDI data and extend the MALDI method to more complex mixtures containing tryptic peptides from several proteins.

Another problem related to the MALDI method is the heterogeneity in matrix-sample crystal formation, which can pose difficulties in terms of quantifying and reproducing results. Using fast evaporation methods when applying the matrix seem to give more homogenous matrix-sample crystals. However, it is still often necessary to search for a high-yielding sample spot within a target or conduct some averaging procedure over several spectra.

2.11 Electrospray Ionization

In ESI mass spectrometry the ions are formed at atmospheric pressure by passing a solution containing peptides or proteins through a fine needle at high potential in order to generate small charged droplets [3]. The potential difference between the capillary and the orifice of the mass spectrometer, located 0.3–2 cm away from the capillary, is typically 2–5 kV [94]. The electric potential is responsible for charge accumulation on the liquid surface, leading to the production of charged droplets from the liquid cone at the capillary tip. The initial drops formed by ESI range from few micrometers to 60 μm in diameter [95]. These drops shrink by evaporation, which results in increased charge density. The increased charge density creates a Coulomb repulsion force that eventually will exceed the surface tension causing drop explosion (Coulomb explosion) into smaller drops. This process continues until the drops are small enough to desorb analyte ions into the gas phase. A efficent sample spray is depending on flow rate, liquid conductivity, and surface tension. The addition of organic solvent to aqueous analyte solution results in lower surface tension, heat capacity, and dielectric constant, all of which facilitates formation of fine droplets by Coulomb explosion.

A nebulizer gas assists the formation of droplets. The nebulizing nitrogen flows through the outside of the needle. As the liquid exits the nebulizer gas helps breaking the liquid into droplets. In addition, a N_2 drying gas is applied on the entrance of capillary to help droplet evaporation.

ESI has proven effective in producing gas-phase ions of proteins and peptides [96]. In general, the charge state distribution of a specific polypeptide increases with the number of ionizable residues and the length of the polypeptide chain. Therefore, multiple charge states are commonly observed for proteins and peptides. The distribution of these charge states will depend on the primary structure of the protein and the equilibrium between the different protein folds in solution prior to the electrospraying and events in the gas phase [97]. The net charge in solution will depend on intrinsic polypeptide properties as well as extrinsic factors. The intrinsic factors include the number, distribution, and pK_as of ionizable amino acid residues, which depend on the initial three-dimensional conformation. The extrinsic factors are solvent composition, pH, ionic strength, and temperature [97].

MS/MS spectra of polypeptides are most often acquired in positive ionization mode. The negative ionization mode applied in a scanning mode is useful to detect sulfated and phosphorylated peptides [98].

For analysis in positive mode the sample solution is often acidified, which also leads to relatively high amounts of anions in addition to the protons. The distribution of $(M + nH)^{n+}$ shifts toward larger n and decreasing the m/z values of the ions formed when the pH is lowered [99]. However, the different types of anions have been observed to have different effects on the net average charge of peptides and proteins ions in the spectra. The propensity for neutralization follows the order: $CCl_3COO^- > CF_3COO^- > CH_3COO^- \sim Cl^-$ [97]. The charge reduction is proposed to occur in two steps. The first occurs in the solution where the anion can pair with a basic group on the peptide or protein. The second occurs in the gas phase during desolvation where the protons dissociate from the peptide to form the neutral acid and the peptide in reduced charge state [97]. It has been found that acetic acid and formic acid produce improved ESI results than

TFA. Additionally, detergents should be avoided. Detergents lead to signal reduction and add complexity to the mass spectra due to the formation of noncovalent clusters [3].

Cations also affect the quality of ESI mass spectra. Cation impurities from buffers can cause a reduction of analyte signal due to spreading of the signal over multiple m/z values; it also adds complexity to the spectra. The presence of alkali metals in ESI mass spectra indicates that desalting of the sample is required.

ESI is performed at atmospheric pressure, which allows for online coupling of high-performance liquid chromatography (HPLC) and capillary electrophoresis (CE) to mass spectrometers [100, 101]. ESI can be coupled to mass analyzers such as triple quadrupoles, ion traps, quadrupole-time-of-flight (Q-TOF), and Orbitraps. A powerful and common setup is RP-LC coupled to ESI MS/MS. Nano and microbore RP-columns coupled to electrospray-MS/MS is becoming standard setup. The typical flow rates in commonly used LC systems, depending on the chromatographic setup, are segmented into categories referring to the flow rates: 100–300 nl/min, 10–100 µl/min, and 200–1000 µl/min for the nano, micro, and normal flow LC for ESI systems, respectively. However, values of 10–15 nl/min, 1–10 µl/min have also found usage [93, 102]. The low flow rate, high ionization efficiency, and high absolute sensitivity make nano-ESI-MS/MS suitable for identification or sequencing of gel-isolated proteins available in subpicomole amounts [103].

2.12 Ion-Trap Mass Analyzer

In an LC-ESI-3D-quadrupole-ion trap system a continuous flow of dissolved analytes from the LC is converted to a stream of ions by ESI and guided into an ion-trap mass analyzer by a combination of electrostatic lenses and an RF (radio frequency) octapole ion guide. First, the incoming ions are focused toward the center of the ion-trap, which is composed of three electrodes, a ring electrode and two end caps. This is accomplished by slowing down the incoming ions with helium (He) gas (typically 1–5 mbar He) and trapping them in a 3D-quadrupole field [6, 18]. The quadrupole field establishes a parabolic potential and induces an oscillatory harmonic motion of the ions at a frequency known as the secular frequency. The Mathieu equations models the ion oscillation [104].

Simultaneously trapped ions in the ion-trap are limited by the low and high m/z range. The stored ions can be ejected according to their m/z value by applying a dipolar field of a frequency proportional to the secular frequency of an ion. By detecting the ejected ions at different dipolar field frequency, a mass spectrum can be obtained.

Ion-traps can store a precursor ion of interest and eject all other ions simultaneously. Fragmentation of the stored ions produces series of ions, which allows inference of sequence information.

This is done by increasing the energy of the isolated precursor ion by excitation with the dipolar field. The amplitude for excitation is less than that used for ejection. The increased energy of the precursor ion leads to harder and more frequent collisions with the helium gas causing fragmentation of the precursor ion [18]. The fragmentation is termed collision-induced dissociation (CID). The product ions will have a different secular frequency than the precursor ion preventing further dipolar excitation and hence fragmentation. The resulting fragment ions is then ejected at different dipolar field frequencies and detected producing the MS/MS spectrum. A linear quadrupole ion-trap is similar to a 3D ion-trap but traps ions in a 2D quadrupole field instead of a 3D field [105]. The linear ion-trap is effectively combined with the Orbitrap mass analyzer [106].

In Orbitrap instruments ion packets are injected at high energies into the Orbitrap mass analyzer and the ions are electrostatically trapped in an orbit around a central spindle shaped electrode [4]. Recently the Orbitrap mass analyzer has been combined with a quadrupole, replacing the linear ion-trap and thereby obtaining fast high-energy CID peptide fragmentation because of parallel filling and detection modes [106]. The main advantage of the Orbitrap instruments are the sensitivity and resolving power. For example, the Orbitrap Elite Hybrid mass spectrometer from Thermo Scientific provides a resolving power of approximately 240,000–1,000,000 FWHM at m/z 400, which is around 10–40 times higher than current TOF mass analyzers.

2.13 Q-TOF Mass Analyzer

The Q-TOF mass analyzers are popular combinations of the low-resolution quadrupole mass analyzer with the high-resolution TOF analyses. A typical Q-TOF configuration consists of three quadrupoles, Q1, Q2, and Q3, followed by a reflectron TOF mass analyzer [71]. In some instruments some of the quadrupoles are replaced by hexapoles or octapoles operating under the same principle. Q1 is used as an ion guide and for collisional cooling of the ions entering the instrument. When acquiring MS spectra, Q2 and Q3 serve as transmission elements, while the reflector TOF separates ions according to the m/z values.

For recording MS/MS, Q1 is operated in mass filter mode to transmit only precursor ions of interest. The width of the mass window of ions allowed to pass Q1 determines the range of the isotopic cluster [104]. The ions in the selected m/z range are then accelerated to Q2, where they undergo CID usually using nitrogen or argon as collision gas. The fragment ions are collisionally cooled and focused in Q3. Finally, their m/z values are measured using the reflector TOF.

2.14 Triple Quadrupole Instruments

The triple quadrupole consists of the linear assembly of three consecutive quadrupoles and it is an interesting instrument in terms of possible experiments. Product ion scan (Fig. 11b), precursor ion scan (Fig. 11c), neutral loss scan (Fig. 11d), and selected reaction monitoring (Fig. 11e) are the four main types of experiments.

The survey scan is produced by scanning Q3 across the m/z range and can be used to detect the parent ions (all ions in an m/z range are transmitted through Q1 and Q2). In the product ion scan ions are selected in Q1 and fragmented by CID in Q2 and the products are then measured by scanning Q3 across the m/z range. This setup corresponds to the general acquisition method used in other MS setups but unfortunately the triple quadrupole is not competitive in terms of mass accuracy.

In precursor ion scan Q3 only transmits an ion of interest and Q1 is scanned to identify the precursors that give rise to the ion of interest. This is useful for identifying possible parent ions of peptides that are modified with specific modifications. For example, fragmentation of glycopeptides often results in intense low mass glycans fragments that are characteristic for only glycopeptides [107]. This strategy can also be used for phosphopeptides that gives intense ions at 79 m/z when the instrument is run in negative mode [108].

In neutral loss scanning Q1 and Q3 both transmit specific ions by scanning Q3 with an offset from Q1 corresponding to the neutral loss. Since phosphopeptides often display intense ions corresponding to ~−98 and ~−80 Da loss from the parent ion the neutral loss scan can also be used to detect phosphopeptides from a background of nonphosphopeptides.

In selected reaction monitoring (SRM) both Q1 and Q3 transmit a specific m/z interval. Q1 transmits an m/z interval that corresponds to the parent ion of a peptide of interest and Q3 transmits a narrow m/z interval that corresponds to an intense and specific fragment ion of the target peptide. Because the quadrupoles are not scanned the sensitivity is fairly high with a reasonably high degree of selectivity and SRM has gained increasing popularity and has been suggested as a sensitive method for quantification of disease biomarkers. It has also been claimed that it in some cases is cheaper and more sensitive than antibody based approaches [109] although some concerns about the sensitivity of SRM methods have been raised [110]. Further details on SRM methods can be found in Chapter 12.

3   The MS/MS Spectrum

In-depth information on the composition of analytes can be gained by tandem mass spectrometry of a parent ion by inducing fragmentation. The fragment ions in an MS/MS spectrum of a peptide depend on the amino acid sequence. For example, the immonium ions can be used as an indicator of the presence of the corresponding amino acid in the peptide sequence [111]. More importantly the mass difference between peaks can be correlated to residue masses or

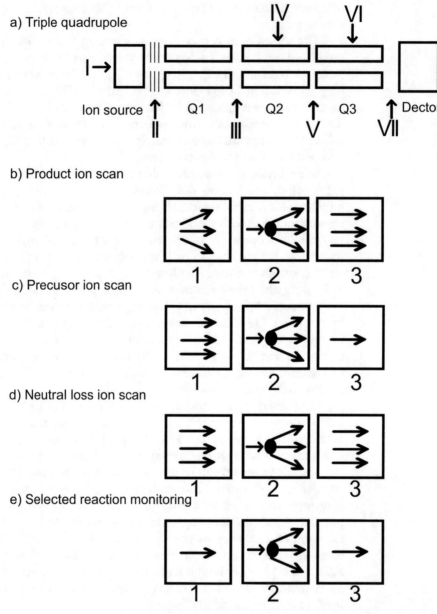

Fig. 11 Outline of triple quadrupole and the experiments performed with such instrumentation. (**a**) Triple quadrupole instrument. (I) Sample inlet, (II) all ions in gas phase, (III) selected ions based on m/z, (IV) collision gas inlet, (V) fragment ions in selected m/z interval, (VI) fragment ion selection, and (VII) detect ion. (**b**) Ions in a small m/z interval are transmitted in Q1 (indicted by the horizontal line) and transferred to Q2 for CID. The fragments are then measured by scanning m/z intervals in Q3. (**c**) Incoming ions are measured by scanning in Q1 and transferred to Q2 for CID. Q3 transmit only ions in a small m/z interval. (**d**) Q1 and Q3 are both scanning ions with a small m/z offset so that m/z interval transmitted in Q3 correspond to the neutral loss from the parent ion in Q1. Q2 functions as the collision cell. (**e**) Q1 and Q3 transmit ions in a specific but not necessarily the same m/z interval and Q2 functions as collision cell

residue masses of modified amino acid residues. Two types of tandem mass spectrometry configurations are commonly used: tandem in space and tandem in time. If the fragmentation process is separated in space it is tandem in space mass spectrometry: the first mass analyzer separates and isolates the ion of interest (the precursor ion). The isolated ion is transmitted to the collision cell where it is fragmented. The fragments are transmitted and analyzed in the second mass analyzer (tandem in space). In tandem in time precursor ion selection, fragmentation, and separation of fragments all occur in one mass analyzer at different time points.

The isolated precursor ion can be fragmented by a vast number of methods and the resulting spectrum is called an MS/MS spectrum. In most cases the mass analyzers are run in positive mode, and CID is used for fragmentation. For peptide analysis the precursor ions with the highest intensity and with a charge state of +2 or more are generally preferred since they give the most informative fragmentation spectra using CID. The precursor ion can have different charge states. Singly charged tryptic peptide ions generally do not fragment so easily as tryptic peptide ions with higher charges.

A number of computer programs such as Probid [93], Mascot [112] (Matrix science), SEQUEST (Thermo Finnigan) [113], Phenyx (Genbio) [114], VEMS [115], X!Tandem [116], Andromeda [117], and OMSSA [118] exist for automatic interpretation of MS/MS spectra. However, all these programs perform a match between theoretical spectra and the experimental data under user defined constraints. To obtain a correct interpretation of data it is necessary to understand the instrumentation used and the fragmentation pathways of peptides (and especially modified peptides) using a given fragmentation method.

If a peptide sample is ionized by ESI in the positive mode, then positive peptide ions will be generated. The protons favor attachment to all the strongly basic sites of the peptides. Examples of these sites are the N-terminal amine, lysine, arginine, and histidine residues [111]. Protons attached to the N-terminal amine may move to any of the amide linkages, whereas the protons associated with arginine, lysine, and histidine are bound more strongly. This model is termed the mobile proton model and explains many observed phenomena in MS/MS spectra of peptides [119]. The precise location of the proton after transfer to the backbone is not fully established. However, the carbonyl oxygen of a peptide bond has been suggested to be the most likely candidate [119].

According to the mobile proton model a peptide ionized by ESI is best viewed as a heterogeneous population with different locations of the charges. The population of peptides is then focused into the collision cell where it is accelerated to induce fragmentation. The kinetic energy from the collisions with the collision gas is converted to vibrational energy in the peptide ions. The peptides

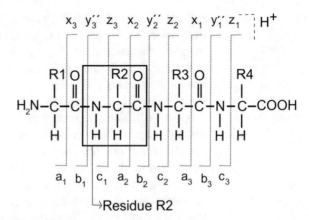

Fig. 12 The nomenclature used for peptide fragment ions. a, b, and c ions are charged fragments containing the N-terminal part of the fragment and x, y, and z the C-terminal part. The *box* indicates the composition of residue R2. Residue mass for R2 is given by summing the masses of the elements of residue R2. a, b, and c ions are numbered from the N-terminus whereas the x, y, and z ions are numbered from the C-terminus of the peptide

then release the vibrational energy by fragmentation. For most peptides fragmentation can be described as a charge directed cleavages [119, 120]. In the charge directed cleavages, fragmentation is guided by the site of the protonated peptide bonds. The charge directed cleavage is a complicated reaction where the different chemical bonds along the peptide backbone are cleaved with different probability [120]. The result is mainly *b*- and/or *y*-type sequence ions (*see* Fig. 12) [121].

If the number of lysine, arginine, and histidine residues in a peptide equals the number of positive charges of the peptide then there are no mobile protons. These stronger bound protons can be mobilized if the kinetic energy of the peptides is increased in the collision cell. However, the increased energy might lead to charge remote fragmentation where no proton is involved in the fragmentation [119].

Tryptic peptides are the most commonly analyzed species and most tryptic peptides have charge states higher than one because most peptides contain at least two basic residues—the N-terminal amine and the C-terminal lysine/arginine. The major low-energy CID pathway by proton-directed cleavages of doubly charged peptides gives mainly *b* and *y*-ion fragments. The most favored reaction leads to the formation of singly charged *b* and *y* ions. Since the mobile proton can be located at different amide bonds a whole series of *b* and *y* ions is typically generated for tryptic peptides. The mass difference between two neighboring b_n and b_{n+1}-ion mass peaks corresponds to the residue mass of the most C-terminal residue of the b_{n+1} ion (*see* Fig. 12). Similarly, the mass difference between two neighboring y_n and y_{n+1}-ion mass peaks corresponds to the residue mass of the most N-terminal residue of the y_{n+1} ion. By calculating the mass differences between peaks in an MS/MS spectrum and relating these mass differences with a list of residue masses it is possible to deduce the peptide sequence, or a shorter part of it, that is, a sequence tag. In Appendix 3 a table of residue masses is provided.

From Fig. 12 the following simple but useful relation between b and y ions is evident:

$$m_{p2+} = m_{b_i} + m_{y_{n-i}} \qquad i = 1, 2, \ldots n-1 \qquad (8)$$

where m_{p2+} is the mass of the doubly charged parent ion, n is the length of the peptide and the right-hand side of the equation is the sum of the masses of any pair of complementary b and y ions. This equation is used to verify that the complementary ion series is present in the experimental spectrum.

Calculating theoretical a-, b-, and y-ion series for a sequence tag and compare these with the experimental spectrum serves as a way to validate the match of the tag and spectrum. Consider a peptide of n amino acids $AA_1, \ldots, AA_n$ with masses $m(AA_j)$. The mass of the doubly charge peptide are calculated as

$$m_{p2+} = m(H_2O) + 2*m(H) - 2*m(e) + \sum_{j=1}^{n} m(AA_j)$$

$$= 20.025118 + \sum_{j=1}^{n} m(AA_j) \qquad (9)$$

This is the mass of a doubly charged peptide precursor ion and is sometimes observed in the MS/MS spectrum at m/z $m_{p2+}/2$. In practice it is necessary to take the mass of the electron $m(e)$ into account since the mass accuracy of common MS/MS mass spectrometers is in the range 0.001–0.02 Da and a low number of missing electrons will have an effect on the third decimal. The mass of the electron is here accounted for to assure precise mass values.

If the b_i ion is composed of the amino acids $AA_1, \ldots, AA_i$; its mass is given by

$$m_{b_i} = m(H) - m(e) + \sum_{j=1}^{i} m(AA_j) = 1.007276 + \sum_{j=1}^{i} m(AA_j)$$

$$(10)$$

The expression is the sum of all the residue masses of residues in b_i plus the mass of a proton [122].

The masses of the y-ion series are computed in a similar way [5]:

$$m_{y_{n-i}} = m(\mathrm{OH}) + 2 * m(\mathrm{H}) - m(e) + \sum_{j=i+1}^{n} m(\mathrm{AA}_j)$$

$$= 19.017841 + \sum_{j=i+1}^{n} m(\mathrm{AA}_j) \qquad (11)$$

The masses in the a-ion series (*see* Fig. 12) are calculated as

$$m_{a_i} = -m(\mathrm{CO}) + m(\mathrm{H}) - m(e) + \sum_{j=1}^{i} m(\mathrm{AA}_j)$$

$$= -26.987638 + \sum_{j=1}^{i} m(\mathrm{AA}_j) \qquad (12)$$

The frequency of the different fragmentation ions observed in ESI low-energy CID are ordered as $y > b > a$ for tryptic peptides with charge state higher than one, whereas the fragment ions c, x, and z are not observed to any significant extent [122].

The b_1 ions are seldom observed in the MS/MS spectrum due to the lack of a carbonyl group to initiate the nucleophilic attack on the carbonyl carbon between the first two amino acids; without this carbonyl the cyclic intermediate cannot form [123].

Fragmentation of a doubly charged peptide with one proton on a basic amino acid and one mobile proton will lead to formation of a b_i and y_{n-i} ion. The b_i ion would still have a mobile proton and can therefore undergo further fragmentation in a quadrupole type collision cell, and to a much lesser extent in an ion-trap (*see* Subheading 2.13). The b ions preferentially fragment to smaller b ions than to a ions [123]. Since the b_2 ion is the last b-ion fragment the corresponding mass peaks often have high intensity in MS/MS spectra generated in a quadrupole type collision cell (*see* Fig. 13).

The b_2 ions can fragment further to a_2 ions. This means that high intensity a_2-ion mass peaks are observed *see* Fig. 13. The presence of a_2- and b_2-ion mass peaks with high intensity and with a mass difference of ~28 Da is very useful in manual interpretation of MS/MS spectra. The mass of the observed b_2 ion compared with the mass of b ions of all combinations of two amino acids might reveal the N-terminal sequence of the peptide. This often gives a limited set of possibilities. The assignment of the precise ordering of the two amino acids depends on observing the y_{n-1} ion. If the y_{n-1} ion is not observed the ordering is only completed after finding a matching peptide from a protein database.

The C-terminal amino acid of a tryptic peptide is either lysine or arginine, except for the C-terminal peptide of a protein and for unspecifically cleaved peptide bonds. The y_1 is often observed in CID/HCD data with low intensity at m/z 147.11 for lysine and

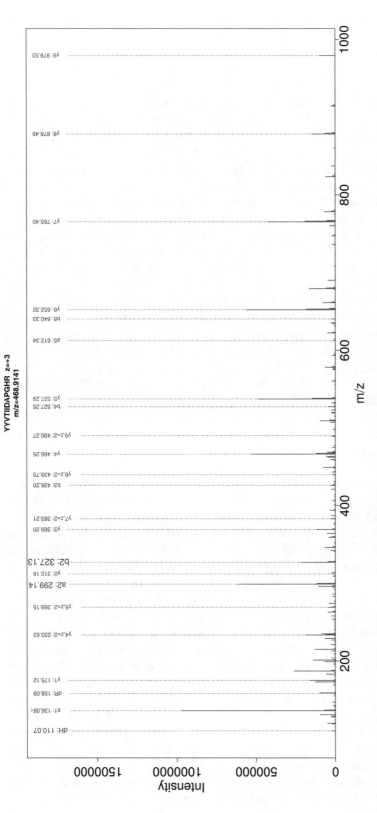

Fig. 13 An HCD MS/MS spectrum of the tryptic peptide YVVTIIDAPGHR with charge 3+. The annotation of each peak is given as ion type and mass (Da) for the peaks that correspond to fragments of the peptide. The intense a2 and b2 ions are annotated with increased fonts

175.12 for arginine. Identification of the C-terminal amino acid decreases the number of possible database matches and increases the search speed.

In the low mass region, one also finds the immonium ions, $H_2N^+ = CHR$ (*see* Appendix 4). The m/z of unmodified immonium ions ranges from m/z 30 (G) to m/z 159 (W). The immonium ions are used as an indication of the presence of certain amino acids. In practice immonium ions are observed for peptides containing L/I, V, F, Y, H, P, and W whereas S, T, R, and K are ineffective sources [124]. If L/I, V, F, Y, H, P, and W are N-terminal residues the corresponding immonium ion can be abundant [125]. However, immonium ions from internal residues are also quite frequently observed.

Other ions apart from the single bond fragmentation a, b, and y ions exist and they complicate MS/MS spectra further. The y_ib_j ions also called the internal fragment ions are one example. The y_ib_j ions are generated from b- or y-ion fragments [125]. Low-energy CID of protonated peptides also gives rise to fragmentation products, which arise from small neutral losses from b, y and y_ib_j-ion products (*see* Appendix 4). The a ions result from the loss of CO from b ions, and are seen as b ions minus 28 Da [125]. The loss of NH_3 can occur from b, y, and y_ib_j ions containing the residues N, Q, K, and R. Peptide ions containing the residues S, T, D, and E can display a neutral loss of H_2O [111]. The neutral losses also reveals information about peptide modification. An example is the neutral loss of H_3PO_4 from phosphoserine or phosphothreonine giving anhydro derivatives.

3.1 Fragmentation Methods

Fragmentations observed in mass spectrometry are divided into three main categories [126]: (1) Unstable ion fragments, formed in the ion source, with dissociation rate constants of $k_{diss} > 10^6 \text{ s}^{-1}$; (2) metastable ion fragments, formed between the ion source and the detector, with $10^6 \text{ s}^{-1} > k_{diss} > 10^5 \text{ s}^{-1}$; (3) stable ions that remain intact during their time in the mass spectrometer and have $k_{diss} < 10^5 \text{ s}^{-1}$. The list of techniques for precursor ion fragmentation includes CAD/CID (collision activated dissociation/decomposition or collision induced dissociation), HCD (high-energy collisional dissociation/higher-energy C-trap dissociation), PSD (post-source decay), ECD (electron capture dissociation), IRMPD (infrared multiphoton dissociation), BIRD (blackbody infrared dissociation), SID (surface induced dissociation), or ETD (electron transfer dissociation). IRMPD [20], BIRD [22], and SID [24] are not discussed further since they are mainly used for top-down proteomics. The most frequently used fragmentation method is CAD/CID. However, the other methods have been found useful in specific areas. In the following section, the most common fragmentation methods currently used in bottom-up proteomics is described in more detail.

CAD/CID [7]. CID is the most frequently used fragmentation method. It works by inducing collisions between the precursor ion and inert neutral gas molecules. This leads to increased internal energy followed by decomposition of the precursor ion. The activation of precursor ions is separated in time from the dissociation process. The activation time is many times faster than the dissociation which explains why CID is modeled as a unimolecular dissociation process and explained by Rice–Ramsperger–Kassel–Marcus theory (RRKM) [127] or quasi-equilibrium theory (QET) modeling.

$$[M + nH]^{n+} \xrightarrow{\substack{\text{Kinetic energy} \\ \text{increase}}} [M + nH]^{n+} * \xrightarrow{\substack{\text{Collision with} \\ \text{inertgas}}}$$

$$[M + nH]^{n+} ** \xrightarrow{\substack{\text{Unimolecular} \\ \text{fragments}}} a, b, \text{and} y \begin{cases} *\text{increased kinetic energy} \\ **\text{increased internal energy} \end{cases}$$

Both low-energy and high-energy collisions are used for peptide fragmentation. Low-energy collisions are used in quadrupole and ion-trap instruments, and they result mainly in *a*, *b*, and *y* ions; immonium ions; and ions from neutral loss of ammonia and water from the *a*, *b*, and *y* ions. High-energy CID is mainly used in sector and TOF-TOF instruments and yields similar fragment spectra of peptides as low-energy CID. The main difference is that *d*, *v*, and *w* ions corresponding to amino acid side chain cleavage are observed in addition to more intense immonium ions. CAD and CID can be used interchangeably. However, CID is used more frequently in the literature. Many variants of the CID method exist. The ion-trap-based resonant excitation can be divided into two main groups, on-resonance and sustained off-resonance irradiation (SORI) SORI-CID is the most widely used technique in FT-ICR MS [11]. On-resonance excitation works by applying an on-resonance rf pulse to increase the kinetic energy of the ions. The increased kinetic energy increases the collision energies and the number of collisions with the collision gas. However, the collision energies drop for each collision that the ion experiences. It is therefore an advantage to activate the ions several times, as is the case in multiple-excitation collisional activation [128]. SORI also use multiple excitations by applying RF pulses slightly above and below the resonant frequency that causes the ions' kinetic energy to oscillate. The advantage of SORI compared with On-resonance CID is that the distribution of the number of collisions of the ions is broader for SORI, meaning that a broader range of fragment ions are generated [11].

HCD is a CID type fragmentation obtainable on Orbitrap instrumentation by using the C-trap as collision chamber [10] in a separate octapole collision cell. HCD fragmentation with Orbitrap detection yields mainly *b* and *y* ions characteristic of quadrupole type CID. The obtained spectra a low mass cutoff at m/z 50, high resolution, and increased ion fragments resulting in higher quality MS/MS spectra [129].

ECD has turned out to have great usability for studying peptides with modified amino acid residues [15]. In ECD, trapped multiple charged ions are irradiated with a low-energy electron beam (<0.2 eV) generated by a heated filament electron gun [130]. The electron capture by the positively charged peptides leads to intensive backbone fragmentation yielding mainly c and z$^\bullet$ fragments.

$$[M + nH]^{n+} \overset{e^-}{\rightarrow} [M + nH]^{(n-1)+\cdot}$$
$$\underset{\text{Local ECD cleavage}}{\rightarrow} \quad c, \; z^{\cdot} \; \text{fragments} \; \cdot \; \text{Radical}$$

It is evident from the above reaction scheme that the protonated peptides should have a minimum charge state of 2+ to lead to detectable fragment ions. The mechanism of ECD cleavage is thought to involve the release of an energetic hydrogen atom upon electron capture. The released hydrogen atom is collisionally de-excited and then later captured at sites having high hydrogen atom affinity such as carbonyl oxygens and disulfide bonds [131]. The cleavage occurs at the site of hydrogen atom capture. A unique feature in ECD is that the ECD cleavage is faster than the intramolecular energy randomization and is therefore called a nonergodic process. The advantage of ECD compared with CID is that the fast ECD process produces peptide backbone cleavages of peptides with intact modifications such as glycans and phosphorylations that often are unstable under CID conditions [132].

ETD [13]. In ETD singly charged carrier anions (e.g., anthracene) transfer an electron to multiply protonated peptides. The transferred electron induces fragmentation of the peptide backbone similar to that of ECD. ECD requires that the precursor sample ions are immersed in a dense population of near-thermal electrons making it a technical challenge to use ECD in ion-traps because of difficulties of keeping the low mass electrons in an RF field. The advantage of ETD compared with ECD is its used in combination with mass analyzers that trap ions with radio frequency (RF) electrostatic fields such as ion-traps. It also provides more efficient electron transfer giving better fragmentation. ETD has become available for a variety of tandem MS instruments including ion-traps, Q-TOFs, and Orbitraps.

4   Detectors

It is outside the scope of this chapter to cover the hardware and software issues associated with the different detectors used in mass spectrometry but the characteristics for a given instrument imposes constraints for the data analysis. Detector saturation effects can for example skew the near Gaussian mass peaks of intense ions downward, so the experimental mass of intense ions will have negative mass error and can pose severe limitations for quantitation. Table 1 provides an overview of the different detectors in use and gives some references for further reading on detectors.

5 Data Formats Used in Proteomics

The data acquired during a typical LC-MS/MS experiment consists of a large collection of precursor ion intensities and associated tandem mass spectra from which the most essential information must be extracted. The data formats used in mass spectrometry are divided in two main groups which are as follows: (1) formats containing raw data and (2) formats containing processed data. The raw data formats are often proprietary formats from the mass spectrometry vendors. Standard raw data formats have been invented to ease data extraction for program developers such as mzDATA and mzXML.

For data base searching the most relevant information is collected for each MS/MS event: the observed m/z and charge state of the precursor ion and a selection of the most informative m/z-intensity pairs for the fragment ions. For processed MS/MS spectra several file formats exists, which are associated with various software applications such as pkl (MassLynx, Micromass), mgf (Mascot generic file), dat (Data analysis, Bruker), and dta (Xcalibur). In these files the raw MS/MS spectra have been processed to a peak list. For some of the formats the data are further processed to contain only singly charged and monoisotopic peaks. The structure of the pkl, and mgf file formats are shown below:

The mgf file format:

```
BEGIN IONS
TITLE= Cmpd 2, +MSn(371.23), 31.8 min
PEPMASS= 371.23 762929

101.07 1287
105.12 2277
...
513.22 1081

END IONS
```

In the Mascot generic file format, the fragment ions from an MS/MS spectrum are enclosed between "BEGIN IONS" and "END IONS." The title line is very useful for specifying extra information such as retention times from the chromatogram and spectrum numbers. This is especially useful if searches with Mascot are performed since the title line is associated with the peptide results in the Mascot result display. The *PEPMASS* is the precursor ion mass (*m/z*) and the following lines are fragment ion mass tab delimiter and ion intensity.

6   The pkl File Format

```
592.5793 617.0220 2
97.0468 1.378e0
112.0647 4.553e0
..
962.1050 3.278e-1
```

The pkl file format is produced by PLGS v2.05 and Masslynx v4.0 (Waters). In the pkl format the MS/MS spectra are separated by an empty line. The first line in a MS/MS spectrum is precursor ion mass, intensity of precursor ion, and charge state separated by the space character. The following lines are the fragment ion masses, and intensity separated by the space character. mzIdentML document stores identification results (discussed in Chap. 17).

7   Challenges for Current MS-Based Proteomics

MS-based proteomics is extremely powerful as witnessed by the many applications presented in later chapters. Nevertheless, it is useful to make a note on some of the problems that current MS-based proteomics is facing. A recent review discusses major concerns in bottom-up proteomics [40] where a major problem is the loss of linkage information between peptides that occurs as a result of tryptic digestion. Inferring the identity and modification status of a protein from a collection of non overlapping peptides is unfeasible. Protein inference becomes complicated because a single MS/MS spectrum can point to several peptides with the same score and a single peptide can point to several proteins mainly due to protein homology. Most search engines present a minimal solution that can explain the observed MS/MS data, but this minimal solution will almost for certain not accurately reflect reality of what is present in the sample. An alternative approach is to use tools that visually present the complexity of the spectrum, peptide and protein inference networks. Such approach is not fully automated. However, useful for manual validation.

As mentioned above the linkage loss between peptides also causes problems for inferring the modification status of a protein. For example, a protein with the two peptides: ELVEMSNHGLK and VENSMNLR that both can be phosphorylated on the serine residue will have four possible states (S,S; pS,S; S,pS; pS,pS). It is likely that these four protein states do not have exactly the same function, role or activity in the cell and with current MS-based bottom approaches we does not generate enough information to obtain the distribution of the possible protein states [40, 133].

The goal of proteomics to quantitatively measure proteins and their modifications in different cells over time remains challenging. Improvement of MS instruments capability to analyze larger peptide fragments will facilitate such holistic analysis. Alternatively, the increased sensitivity gain obtained in recent MS instrument might allow better sampling of overlapping peptides which will facilitate downstream data analysis. Accuracy of quantitation will likely improve in the future by better parent ion separation in the survey scans. A better understanding of peptide fragmentation allows developments of better scoring function for matching experimental against theoretical spectra.

Furthermore, common to all data base search programs is the lack of fully accounting for all the data provided in the MS or MS/MS spectra. The reason is that the MS/MS spectra are complex, containing many mass peaks of known and unknown origin. Several major problems need attention. The fragmentation chemistry might be improved or become more comprehensive by including complementary fragmentation data as CID/ETD for each precursor ion. The interpretation algorithms might be improved to recognize and predict sequence specific fragmentation patterns. Interpretation algorithms might also improve by including the intensity of the fine isotopic distribution during the interpretation [134] or other ion features. Much data is also lost because of the rich chemistry of peptides and the many post translational modifications are not considered when the data is searched against protein databases.

8   Notes

1. Mass is typically stated in unified mass unit (u) or Dalton (Da). 1 u is defined as $1/12$ mass of carbon-12 ($1\ u = 1/12 * m$ (^{12}C)). The x-axis of mass spectra is commonly labelled the symbol m/z serving as an abbreviation for mass-to-charge ratio. m/z is a dimensionless quantity obtained by dividing the mass of an ion by its charge number. Thomson (Th) is defined as $1\ u/e$ [135]. Th serves as a unit of mass-to-charge ratio (*see* Wikipedia for more details). Nominal mass is defined as the count of elements in the ion times the integer mass of the

most abundant isotope for each element. Monoisotopic mass is the exact mass calculated from the most abundant isotope for each element. Average mass is the mass weighted by isotope peak intensity.

2. In most MS-based proteomics applications the instrument is run in positive mode where positive charged ions are injected into the instrument. The positive mode has several advantages for bottom-up proteomics where tryptic peptides are typically analyzed under acidic conditions. This means that most tryptic peptides have a charge state of +2 or higher since most peptides will have a basic N terminal and lysine or arginine in the C-terminal. The negative mode also has useful applications for example when precursor ion scanning is used to identify negative charged phosphopeptides [136].

3. MS/MS spectra are obtained by data dependent acquisition (DDA) or data-independent acquisition (DIA) [137].

In DDA the MS scan or survey scan is analyzed on the fly and the instrument selects ions for fragmentation to generate the MS/MS spectra.

DIA is a method of molecular structure determination in which all ions within a selected m/z range are fragmented and analyzed in a second stage of tandem mass spectrometry. MS^E [138] and Sequential Windowed Acquisition of All Theoretical Fragment Ion Mass Spectra (SWATH-MS) [139] are examples of DIA methodologies. In MS^E the instrument shifts between low-energy CID and high-energy CID. The low-energy CID is to obtain precursor ion mass spectra whereas the high-energy CID is for product ion information by tandem mass spectrometry. No selection of ions is made and therefore all ions in the m/z range are fragmented. The obtained MS/MS spectra contain fragmentations from all parent ions eluting from the chromatography that is within the mass range of the first mass analyzer. To analyze these MS/MS spectra the parent ion intensity in the MS spectra over the elution profile of the peptide is correlated with the ion intensity of the fragment ions and in this way a link between parent ions and fragment ions is established. The advantage of MS^E is that it is more sensitive since no ions filtered away.

In MS^3 a fragment ion in MS/MS is selected for another round of isolation and fragmentation and the spectrum of the new fragments is the MS^3 spectrum. MS^3 can generate more information about the peptide, which is especially relevant for accurate identification of modified peptides. For example, for many phosphopeptides most of the absorbed energy from collision with the inert gas is lost with the loss of phosphoric acid (H_3PO_4) with the result that the peptide backbone will not fragment further. In MS^3 additional energy is added for the fragmentation of the peptide backbone.

4. Annotating fragment ions in MSMS are frequently performed computationally by considering mass accuracy and charge state of the observed ions. However, ions can sometimes have several possible explanations. The annotation depicted in this chapter selects the highest intensity ion within a selected mass accuracy interval that match the charge state. If two theoretical ions match a peak, then the peak with highest mass accuracy is selected and if the mass accuracy is the same then both annotations are maintained. Another issue is that some ion annotations are less likely from a chemical point of view. To the best of our knowledge such chemical considerations are not currently considered by MSMS annotations programs. For example, in Fig. 3b the ion 129 m/z theoretically match b1, b2++, and y1–18. However, b1 fragment is not expected for this sequence, and the b2++ ion is doubtful because of two charges in close proximity. The water loss from y1 is possible, but due to doubts this observed ion was not annotated in Fig. 3a.

5. LC-MS techniques frequently suffer from ion suppression effects, which are caused by known and unknown compounds. These can derive from a multitude of sources including the analyzed sample, ambient air, contaminants from LC gradient buffers or from sample preparation procedures. Ion suppression negatively affects ion intensity, precision, and accuracy. The origin of ion suppression is not fully understood but sample clean up and changes in chromatography conditions can improve the analysis of specific analytes.

6. The MaxQuant homepage contains a useful list of common protein contaminants that can be searched together with the protein database of target proteins of interest. The list of protein contaminants are proteases from digestion protocols, proteins from cell culture medium, human and sheep hair keratins, and protein tags such as green fluorescent protein.

7. The rule of thumb is that the enzyme trypsin cleaves after Lys—and Arg if not followed by Pro. However, in LC-MS/MS runs, MS/MS spectra are often observed for peptides where the tryptic cleavage of Lys, or Arg happened even though the cleavage site is followed by Pro. In these cases, the MS/MS spectra of the peptide corresponding to no cleavage is often also observed and with much higher intensity if it is not outside the mass range of the mass spectrometer. In general, one can also observe missed cleavage sites, which often occur when Lys or Arg is neighboring to Asp, Glu, Lys, or Arg or if there is a phosphorylated residue nearby the cleavage site.

Acknowledgments

R.M. is supported by Fundação para a Ciência e a Tecnologia (CEEC position, 2019-2025 investigator). iNOVA4Health— UID/Multi/04462/2013, a program financially supported by Fundação para a Ciência e Tecnologia/Ministério da Educação e Ciência, through national funds and cofunded by FEDER under the PT2020 Partnership Agreement. This work is also funded by FEDER funds through the COMPETE 2020 Programme and National Funds through FCT – Portuguese Foundation for Science and Technology under the projects number PTDC/BTM-TEC/30087/2017 and PTDC/BTM-TEC/30088/2017.

Abbreviations

2D-GE	Two-dimensional gel electrophoresis
2D-PAGE	Two-dimensional polyacrylamide gel electrophoresis
AbI	Abundance index
AC	Alternating current
ACN	Acetonitrile
AMT	Accurate mass and time tag
ANN	Artificial neural networks
AUROC	Area under the ROC curve
BabA	Blood group antigen-binding adhesin
BIRD	Blackbody infrared dissociation
BMI	Body mass index
BPC	Base peak chromatogram
BSA	Bovine serum albumin
CAD	Collision-activated dissociation
CCDS	Consensus coding sequence
CE	Capillary electrophoresis
CID	Collision-induced dissociation
CML	Chronic myelogenous leukemia
CNBr	Cyanogen bromide
COFRADIC	Combined fractional diagonal chromatography
CRAN	Comprehensive R Archive Network
C-TAILS	Carboxy-terminal amine-based isotope labeling of substrates
CV	Coefficient of variation
CV	Controlled vocabulary
DAC	Data Access Compliance
DACO	Data Access Compliance Office
DDA	Data-dependent acquisition
DE	Delayed ion extraction
DEHP	Bis(2-ethylhexyl)(Phthalate)

DHB	2,5-Dihydroxybenzoic acid
DMSO	Dimethylsulfoxide
DTT	Dithiothreitol
DUB	Deubiquitinase
ECD	Electron capture dissociation
eGFR	Estimated glomerular filtration rate
EHR	Electronic health records
EM	Expectation maximization
EOS	Empirical observability score
ERLIC	Electrostatic repulsion–hydrophilic interaction chromatography
ESI	Electrospray ionization
ESS	Empirical suitability score
ESSI	Electrosonic spray ionization
ESTs	Expressed sequence tags
ETD	Electron transfer dissociation
FAB	Fast-atom bombardment
FAK	Focal adhesion kinase
FASP	Filter-aided sample preparation
FAT	Focal adhesion targeting
FDR	False discovery rate
FFT	Fast Fourier transform
FTICR	Fourier transform ion cyclotron resonance
FWHM	Full-width at half maximum
GA	Genetic algorithm
GlcNAc	N-Acetylglucosamines
GO	Gene ontology
HCCA	α-Cyano-4-hydroxy cinnamic acid
HCD	High-energy collisional dissociation/higher-energy C-trap dissociation
HDX-MS	Hydrogen exchange mass spectrometry
HED	High-energy dynode detector
HILIC	Hydrophilic interaction liquid chromatography
HPLC	High-performance liquid chromatography
HPRD	Human Proteome Reference Database
HX-MS	Hydrogen exchange mass spectrometry
IAA	Iodoacetamide
ICAT	Isotope-coded affinity tag
IEF	Isoelectric focusing
IgG	Immunoglobulin
IMAC	Immobilized metal affinity chromatography
IRMPD	Infrared multiphoton dissociation
IT	Ion trap
iTRAQ	Isobaric tags for relative and absolute quantitation
KNN	K-nearest neighbor classifier
KS	Kolmogorov–Smirnov
LC	Liquid chromatography
LDA	Linear discriminant analysis
LDI	Laser desorption ionization

LFQ	Label-free quantitation
LIMS	Laboratory information management systems
LOOCV	Leave-one-out cross-validation
m/z	Mass over charge
MAD	Median absolute deviation
MALDI	Matrix-assisted laser desorption/ionization
MCP	Microchannel plate detector
mgf	Mascot generic files
MI	Molecular interactions
MIAPE	Minimal information about a proteomics experiment
MRM	Multiple reaction monitoring
MS	Mass spectrometry
MS	Mass spectrum/spectra
MSMS	Tandem mass spectrometry
MudPIT	Multidimensional protein identification technology
mziXIC	Multiple charge state and isotope envelope dependent XIC
NeuAc	Neuraminic acid
NGQ	Next-generation quantification
NMR	Nuclear magnetic resonance
N-TAILS	N-terminal amine isotopic labeling of substrates
OOB	Out-of-bag
PCA	Principal component analysis
PCM	Polycyclodimethylsiloxane
PD	^{252}Cf plasma desorption
PEG	Polyethylene glycol
PLS	Partial least squares
PMF	Peptide mass fingerprinting
PMM	Peptide mass mapping
PNGase	F Peptide-N-glycosidase F
PPC	Peak probability contrasts
ppm	Parts per million
PRIDE	PRoteomics IDEntifications
PRM	Parallel reaction monitoring
PSA	Peptide spectra assignments
PSA	Prostate-specific antigen
PSAQ™	Protein Standard Absolute Quantification
PSD	Post-source decay
PSI	Proteomics Standards Initiative
PSICQUIC	Proteomics standard initiative common query interface
PSO	Particle swarm optimization
PSSM	Position specific scoring matrices
PTM	Post-translational modifications
QA	Quality assessment
QC	Quality control
QDA	Quadratic discriminant analysis
RF	Random forest
ROC	Receiver operating characteristic

RP	Reversed-phase
RPPA	Reverse-phase protein microarrays
R-SVM	Recursive support vector machine
RT	Retention time
SA	Sinapinic acid
SabA	Sialic-acid binding adhesin
SAP	Single amino acid polymorphism
SAX	Strong anion exchange chromatography
SBS	Sequential backward selection
SCX	Strong cation exchanger
SDS	Sodium dodecyl sulfate
SDS-PAGE	Sodium dodecyl sulfate–polyacrylamide gel electrophoresis
SE	Spectral error
SELDI	Surface enhanced laser desorption/ionization
SEM	Standard error of the mean
SFS	Sequential forward selection
SID	Surface-induced dissociation
SIL	Stable isotope labeling
SILAC	Stable isotope labeling by amino acids in cell culture
SISCAPA	Stable isotope standards with capture by anti-peptide antibody
SMIRP	Subtle modification of isotope ratio proteomics
SNP	Single nucleotide polymorphism
SNR	Signal-to-noise ratio
SOM	Self-organizing maps
SOPs	Standard operating procedures
SPC	Statistical process control
SpI	Spectral index
SRM	Selected reaction monitoring
SVM	Support vector machine
SVM-RFE	Support vector machine recursive feature elimination
SVR	Support vector regressor
TAP	Third-party annotation
TFA	Trifluoroacetic acid
THAP	2,4,6-Trihydroxyacetophenone
TIC	Total ion chromatogram
TiO_2	Titanium dioxide
TMT	Tandem mass tags
TOF	Time-of-flight
TPCK	N-Tosyl-L-phenyl chloromethyl ketone
TPP	Trans-Proteomic Pipeline
Ub	Ubiquitin
UbL	Ubiquitin-like
UND	Undeuterated control
WMS	Workflow management system
XIC	Extract ion chromatograms
ziXIC	Charge state and isotope envelope-dependent XIC

Appendix Masses of Matrix Clusters

The masses of matrix clusters can, for example, be used as a standard for internal calibration. Possible masses of HCCA matrix clusters were calculated using Eqs. 2–4. The number of possible combination for a given n is given by $(n + 2) * \frac{n+3}{2} - 1$.

Mass (Da)	n	x	y	z
212.03	1	0	0	1
228.01	1	0	1	0
234.01	1	1	0	2
249.99	1	1	1	1
265.96	1	1	2	0
401.07	2	0	0	1
417.05	2	0	1	0
423.06	2	1	0	2
439.03	2	1	1	1
445.04	2	2	0	3
455.00	2	1	2	0
461.01	2	2	1	2
476.99	2	2	2	1
492.96	2	2	3	0
590.12	3	0	0	1
606.09	3	0	1	0
612.10	3	1	0	2
628.07	3	1	1	1
634.08	3	2	0	3
644.05	3	1	2	0
650.06	3	2	1	2
656.06	3	3	0	4
666.03	3	2	2	1
672.04	3	3	1	3
682.00	3	2	3	0
688.01	3	3	2	2
703.99	3	3	3	1
719.96	3	3	4	0

(continued)

Mass (Da)	n	x	y	z
779.16	4	0	0	1
795.13	4	0	1	0
801.14	4	1	0	2
817.12	4	1	1	1
823.12	4	2	0	3
833.09	4	1	2	0
839.10	4	2	1	2
845.11	4	3	0	4
855.07	4	2	2	1
861.08	4	3	1	3
867.09	4	4	0	5
871.05	4	2	3	0
877.05	4	3	2	2
883.06	4	4	1	4
893.03	4	3	3	1
899.04	4	4	2	3
909.00	4	3	4	0
915.01	4	4	3	2
930.98	4	4	4	1
946.96	4	4	5	0

Table of Amino Acid Residue Masses and Some Elemental Masses

Compound	Monoisotopic mass (Da)
E	0.000549
H^+	1.00728
H	1.00783
O	15.9949146
H_2O	18.01056
Glycine (G)	57.02146
Alanine (A)	71.03711
Serine (S)	87.03203

(continued)

Compound	Monoisotopic mass (Da)
Proline (P)	97.05276
Valine (V)	99.06841
Homoserine lactone (HSL)	100.03985
Threonine (T)	101.04768
Cysteine (C)	103.00919
Isoleucine (I)	113.08406
Leucine (L)	113.08406
Asparagine (N)	114.04293
Aspartic acid (D)	115.02694
Glutamine (Q)	128.05858
Lysine (K)	128.09496
Glutamic acid (E)	129.04259
Methionine (M)	131.04049
Histidine (H)	137.05891
Methionine sulfoxide (MSO)	147.0354
Phenylalanine (F)	147.06841
Cysteic acid	150.993935
Arginine (R)	156.10111
Carboxyamidomethyl cysteine (Cys_CAM)	160.03065
Carboxymethyl cysteine (Cys_CM)	161.01466
Methionine sulfone	163.03032
Tyrosine (Y)	163.06333
Propionamide cysteine (Cys_PAM)	174.04631
Tryptophan (W)	186.07931
N-Formylkynurenine	186.07931
Pyridyl-ethyl cysteine (Cys_PE)	208.067039

Adapted from [89, 140], and ExPASy http://www.expasy.org/tools/findmod/findmod_masses.html

Tables of Immonium Ion and Neutral Loss Masses

Immonium ions	Monoisotopic mass (Da)	
Glycine (G)	30.032362	
Alanine (A)	44.052362	
Serine (S)	60.042362	−
Proline (P)	70.062362	+
Valine (V)	72.082362	+
Threonine (T)	74.062362	−
Cysteine (C)	76.022362	
Isoleucine (I)	86.092362	+
Leucine (L)	86.092362	+
Asparagine (N)	87.052362	
Aspartic acid (D)	88.042362	
Glutamine (Q)	101.072362	
Lysine (K)	101.102362	−
Glutamic acid (E)	102.052362	
Methionine (M)	104.052362	
Histidine (H)	110.072362	+
Phenylalanine (F)	120.052362	+
Arginine (R)	129.112362	−
Tyrosine (Y)	136.072362	+
Cysteic acid	152.001211	
Tryptophan (W)	159.092362	+
Methionine sulfone	164.037596	
N-Formylkynurenine	219.076416	

The masses were calculated with MassCal part of the VEMS v2.0 application. (+) and (−) indicates how well the amino acids are in forming the corresponding immonium ion

References

1. Yamashita M, Fenn JB (1984) Electrospray ion source. Another variation on the free-jet theme. J Phys Chem 88(20):4451–4459
2. Tanaka K, Waki H, Ido Y, Akita S, Yoshida Y, Yoshida T, Matsuo T (1988) Protein and polymer analyses up to m/z 100 000 by laser ionization time-of flight mass spectrometry. Rapid Commun Mass Spectrom 2 (20):151–153

3. Patterson SD, Aebersold R (1995) Mass spectrometric approaches for the identification of gel-separated proteins. Electrophoresis 16 (10):1791–1814

4. Hu Q, Noll RJ, Li H, Makarov A, Hardman M, Graham Cooks R (2005) The Orbitrap: a new mass spectrometer. J Mass Spectrom 40(4):430–443. https://doi.org/10.1002/jms.856

5. Fenn JB, Mann M, Meng CK, Wong SF, Whitehouse CM (1989) Electrospray ionization for mass spectrometry of large biomolecules. Science 246(4926):64–71

6. Jonscher KR, Yates JR 3rd (1997) The quadrupole ion trap mass spectrometer—a small solution to a big challenge. Anal Biochem 244(1):1–15. https://doi.org/10.1006/abio.1996.9877. S0003-2697(96)99877-2

7. Hadden WF, McLafferty FW (1968) Metastable ion characteristics. VII. Collision-induced metastables. J Am Chem Soc 90:4745–4746

8. Siuzak G (1996) Mass spectrometry in biotechnology. Academic Press, San Diego

9. Karas M, Hillenkamp F (1988) Laser desorption ionization of proteins with molecular masses exceeding 10,000 daltons. Anal Chem 60(20):2299–2301

10. Olsen JV, Macek B, Lange O, Makarov A, Horning S, Mann M (2007) Higher-energy C-trap dissociation for peptide modification analysis. Nat Methods 4(9):709–712. https://doi.org/10.1038/nmeth1060. nmeth1060

11. Laskin J, Futrell JH (2003) Collisional activation of peptide ions in FT-ICR mass spectrometry. Mass Spectrom Rev 22 (3):158–181. https://doi.org/10.1002/mas.10041

12. Kuwata H, Yip TT, Yip CL, Tomita M, Hutchens TW (1998) Bactericidal domain of lactoferrin: detection, quantitation, and characterization of lactoferricin in serum by SELDI affinity mass spectrometry. Biochem Biophys Res Commun 245(3):764–773. https://doi.org/10.1006/bbrc.1998.8466. S0006-291X(98)98466-2

13. Syka JE, Coon JJ, Schroeder MJ, Shabanowitz J, Hunt DF (2004) Peptide and protein sequence analysis by electron transfer dissociation mass spectrometry. Proc Natl Acad Sci U S A 101(26):9528–9533. https://doi.org/10.1073/pnas.0402700101. 0402700101

14. Burlingame AL, Boyd RK, Gaskell SJ (1994) Mass spectrometry. Anal Chem 66:634R–683R

15. Zubarev RA, Horn DM, Fridriksson EK, Kelleher NL, Kruger NA, Lewis MA, Carpenter BK, McLafferty FW (2000) Electron capture dissociation for structural characterization of multiply charged protein cations. Anal Chem 72(3):563–573

16. James H, Barnes IV, Hieftje GM (2004) Recent advances in detector-array technology for mass spectrometry. Int J Mass Spectrom 238:33–46

17. Takats Z, Wiseman JM, Gologan B, Cooks RG (2004) Electrosonic spray ionization. A gentle technique for generating folded proteins and protein complexes in the gas phase and for studying ion-molecule reactions at atmospheric pressure. Anal Chem 76 (14):4050–4058. https://doi.org/10.1021/ac049848m

18. Todd JFJ, March RE (1999) A retrospective review of the development and application of the quadrupole ion trap prior to the appearance of commercial instruments. Int J Mass Spectrom 190(191):9–35

19. Spengler B, Kirsch D, Kaufmann R (1991) Metastable decay of peptides and proteins in matrix assisted laser desorption mass spectrometry. Rapid Commun Mass Spectrom 5:198–202

20. Thorne LR, Beauchamp JL (1984) In: Bowers MT (ed) Gas phase ion chemistry. Academic Press, London, p 41

21. James P (2001) Proteome research: mass spectrometry. Springer-Verlag, Berlin, Heidelberg, New York, p 35

22. Price WD, Schnier PD, Williams ER (1996) Tandem mass spectrometry of large biomolecule ions by blackbody infrared radiative dissociation. Anal Chem 68(5):859–866. https://doi.org/10.1021/ac951038a

23. Barefoot RR (2004) Determination of platinum group elements and gold in geological materials: a review of recent magnetic sector and laser ablation applications. Anal Chim Acta 509:119–125

24. Cooks RG, Ast T, Beynon JH (1975) Anomalous metastable peaks. Int J Mass Spectrom Ion Phys 16:55

25. Chait BT (2006) Chemistry. Mass spectrometry: bottom-up or top-down? Science 314 (5796):65–66. https://doi.org/10.1126/science.1133987. 314/5796/65

26. Wu C, Tran JC, Zamdborg L, Durbin KR, Li M, Ahlf DR, Early BP, Thomas PM, Sweedler JV, Kelleher NL (2012) A protease for 'middle-down' proteomics. Nat Methods 9 (8):822–824. https://doi.org/10.1038/nmeth.2074

27. Kelleher NL (2004) Top-down proteomics. Anal Chem 76(11):197A–203A

28. Perry RH, Cooks RG, Noll RJ (2008) Orbitrap mass spectrometry: instrumentation, ion motion and applications. Mass Spectrom Rev 27(6):661–699. https://doi.org/10.1002/mas.20186

29. Armirotti A, Damonte G (2010) Achievements and perspectives of top-down proteomics. Proteomics 10(20):3566–3576. https://doi.org/10.1002/pmic.201000245

30. Washburn MP, Wolters D, Yates JR 3rd (2001) Large-scale analysis of the yeast proteome by multidimensional protein identification technology. Nat Biotechnol 19(3):242–247. https://doi.org/10.1038/85686. 85686

31. Doucet A, Kleifeld O, Kizhakkedathu JN, Overall CM (2011) Identification of proteolytic products and natural protein N-termini by Terminal Amine Isotopic Labeling of Substrates (TAILS). Methods Mol Biol 753:273–287. https://doi.org/10.1007/978-1-61779-148-2_18

32. Schilling O, Huesgen PF, Barre O, Overall CM (2011) Identification and relative quantification of native and proteolytically generated protein C-termini from complex proteomes: C-terminome analysis. Methods Mol Biol 781:59–69. https://doi.org/10.1007/978-1-61779-276-2_4

33. Prudova A, auf dem Keller U, Butler GS, Overall CM (2010) Multiplex N-terminome analysis of MMP-2 and MMP-9 substrate degradomes by iTRAQ-TAILS quantitative proteomics. Mol Cell Proteomics 9(5):894–911. https://doi.org/10.1074/mcp.M000050-MCP201. M000050-MCP201 [pii]

34. Van Damme P, Martens L, Van Damme J, Hugelier K, Staes A, Vandekerckhove J, Gevaert K (2005) Caspase-specific and non-specific in vivo protein processing during Fas-induced apoptosis. Nat Methods 2(10):771–777. https://doi.org/10.1038/nmeth792. nmeth792 [pii]

35. Timmer JC, Enoksson M, Wildfang E, Zhu W, Igarashi Y, Denault JB, Ma Y, Dummitt B, Chang YH, Mast AE, Eroshkin A, Smith JW, Tao WA, Salvesen GS (2007) Profiling constitutive proteolytic events in vivo. Biochem J 407(1):41–48. https://doi.org/10.1042/BJ20070775. BJ20070775

36. La Scola B (2011) Intact cell MALDI-TOF mass spectrometry-based approaches for the diagnosis of bloodstream infections. Expert Rev Mol Diagn 11(3):287–298. https://doi.org/10.1586/erm.11.12

37. Zhang Q, Willison LN, Tripathi P, Sathe SK, Roux KH, Emmett MR, Blakney GT, Zhang HM, Marshall AG (2011) Epitope mapping of a 95 kDa antigen in complex with antibody by solution-phase amide backbone hydrogen/deuterium exchange monitored by Fourier transform ion cyclotron resonance mass spectrometry. Anal Chem 83(18):7129–7136. https://doi.org/10.1021/ac201501z

38. Margolis J, Kenrick KG (1969) 2-Dimensional resolution of plasma proteins by combination of polyacrylamide disc and gradient gel electrophoresis. Nature 221(5185):1056–1057

39. Klose J, Kobalz U (1995) Two-dimensional electrophoresis of proteins: an updated protocol and implications for a functional analysis of the genome. Electrophoresis 16(6):1034–1059

40. Matthiesen R, Azevedo L, Amorim A, Carvalho AS (2011) Discussion on common data analysis strategies used in MS-based proteomics. Proteomics 11(4):604–619. https://doi.org/10.1002/pmic.201000404

41. Inagaki N, Katsuta K (2004) Large gel two-dimensional electrophoresis: improving recovery of cellular proteome. Curr Proteomics 1:35–39

42. Gygi SP, Rochon Y, Franza BR, Aebersold R (1999) Correlation between protein and mRNA abundance in yeast. Mol Cell Biol 19:1720–1730

43. Patton WF, Schulenberg B, Steinberg TH (2002) Two-dimensional gel electrophoresis; better than a poke in the ICAT? Curr Opin Biotechnol 13:321–328

44. Rosenfeld J, Capdevielle J, Guillemot JC, Ferrara P (1992) In-gel digestion of proteins for internal sequence analysis after one- or two-dimensional gel electrophoresis. Anal Biochem 203(1):173–179. 0003-2697(92)90061-B [pii]

45. Hellman U, Wernstedt C, Gonez J, Heldin CH (1995) Improvement of an "In-Gel" digestion procedure for the micropreparation of internal protein fragments for amino acid sequencing. Anal Biochem 224(1):451–455. https://doi.org/10.1006/abio.1995.1070. S0003-2697(85)71070-6 [pii]

46. Bienvenut WV, Sanchez JC, Karmime A, Rouge V, Rose K, Binz PA, Hochstrasser DF (1999) Toward a clinical molecular scanner for proteome research: parallel protein chemical processing before and during western blot. Anal Chem 71(21):4800–4807

47. Aebersold R, Mann M (2003) Mass spectrometry-based proteomics. Nature 422 (6928):198–207. https://doi.org/10.1038/nature01511. nature01511

48. Bunkenborg J, Garcia GE, Paz MI, Andersen JS, Molina H (2010) The minotaur proteome: avoiding cross-species identifications deriving from bovine serum in cell culture models. Proteomics 10(16):3040–3044. https://doi.org/10.1002/pmic.201000103

49. Zhao C, O'Connor PB (2007) Removal of polyethylene glycols from protein samples using titanium dioxide. Anal Biochem 365 (2):283–285. https://doi.org/10.1016/j.ab.2007.03.024. S0003-2697(07)00185-6 [pii]

50. Williams S (2004) Ghost peaks in reversed-phase gradient HPLC: a review and update. J Chromatorgr A 1052:1–11

51. Wisniewski JR, Zougman A, Mann M (2009) Combination of FASP and StageTip-based fractionation allows in-depth analysis of the hippocampal membrane proteome. J Proteome Res 8(12):5674–5678. https://doi.org/10.1021/pr900748n

52. Matthiesen R, Bauw G, Welinder KG (2004) Use of performic acid oxidation to expand the mass distribution of tryptic peptides. Anal Chem 76:6848–6852

53. Gilar M, Olivova P, Daly AE, Gebler JC (2005) Orthogonality of separation in two-dimensional liquid chromatography. Anal Chem 77(19):6426–6434. https://doi.org/10.1021/ac050923i

54. Batth TS, Francavilla C, Olsen JV (2014) Off-line high-pH reversed-phase fractionation for in-depth phosphoproteomics. J Proteome Res 13(12):6176–6186. https://doi.org/10.1021/pr500893m

55. Bunkenborg J, Espadas G, Molina H (2013) Cutting edge proteomics: benchmarking of six commercial trypsins. J Proteome Res 12 (8):3631–3641. https://doi.org/10.1021/pr4001465

56. Smith RL, Shaw E (1969) Pseudotrypsin. A modified bovine trypsin produced by limited autodigestion. J Biol Chem 244 (17):4704–4712

57. Wilkins MR, Gasteiger E, Bairoch A, Sanchez JC, Williams KL, Appel RD, Hochstrasser DF (1999) Protein identification and analysis tools in the ExPASy server. Methods Mol Biol 112:531–552

58. S. G (2017) cleaver: cleavage of polypeptide sequences. R package version 1.18.0. https://github.com/sgibb/cleaver/

59. Willett WS, Gillmor SA, Perona JJ, Fletterick RJ, Craik CS (1995) Engineered metal regulation of trypsin specificity. Biochemistry 34 (7):2172–2180

60. Tran JC, Zamdborg L, Ahlf DR, Lee JE, Catherman AD, Durbin KR, Tipton JD, Vellaichamy A, Kellie JF, Li M, Wu C, Sweet SM, Early BP, Siuti N, LeDuc RD, Compton PD, Thomas PM, Kelleher NL (2011) Mapping intact protein isoforms in discovery mode using top-down proteomics. Nature 480(7376):254–258. https://doi.org/10.1038/nature10575. nature10575

61. Larsen MR, Cordwell SJ, Roepstorff P (2002) Graphite powder as an alternative or supplement to reversed-phase material for desalting and concentration of peptide mixtures prior to matrix-assisted laser desorption/ionization-mass spectrometry. Proteomics 2:1277–1287

62. Hagglund P, Bunkenborg J, Elortza F, Jensen ON, Roepstorff P (2004) A new strategy for identification of N-glycosylated proteins and unambiguous assignment of their glycosylation sites using HILIC enrichment and partial deglycosylation. J Proteome Res 3 (3):556–566

63. Pinkse MWH, Uitto PM, Hilhorst MJ, Ooms B, Heck AJR (2004) Selective isolation at the femtomole level of phosphopeptides from proteolytic digests using 2D-NanoLC-ESI-MS/MS and titanium oxide precolumns. Anal Chem 76:3935–3943

64. Larsen MR, Thingholm TE, Jensen ON, Roepstorff P, Jorgensen TJ (2005) Highly selective enrichment of phosphorylated peptides from peptide mixtures using titanium dioxide microcolumns. Mol Cell Proteomics 4(7):873–886. https://doi.org/10.1074/mcp.T500007-MCP200. T500007-MCP200 [pii]

65. Gerber SA, Rush J, Stemman O, Kirschner MW, Gygi SP (2003) Absolute quantification of proteins and phosphoproteins from cell lysates by tandem MS. Proc Natl Acad Sci U S A 100(12):6940–6945. https://doi.org/10.1073/pnas.0832254100. 0832254100

66. Pratt JM, Simpson DM, Doherty MK, Rivers J, Gaskell SJ, Beynon RJ (2006) Multiplexed absolute quantification for proteomics using concatenated signature peptides encoded by QconCAT genes. Nat Protoc 1 (2):1029–1043. https://doi.org/10.1038/nprot.2006.129. nprot.2006.129 [pii]

67. Brun V, Dupuis A, Adrait A, Marcellin M, Thomas D, Court M, Vandenesch F, Garin J (2007) Isotope-labeled protein standards: toward absolute quantitative proteomics.

Mol Cell Proteomics 6(12):2139–2149. https://doi.org/10.1074/mcp.M700163-MCP200. M700163-MCP200 [pii]

68. Brownridge P, Beynon RJ (2011) The importance of the digest: proteolysis and absolute quantification in proteomics. Methods 54 (4):351–360. https://doi.org/10.1016/j.ymeth.2011.05.005. S1046-2023(11)00097-1

69. Sachs AN, Pisitkun T, Hoffert JD, Yu MJ, Knepper MA (2008) LC-MS/MS analysis of differential centrifugation fractions from native inner medullary collecting duct of rat. Am J Physiol Renal Physiol 295(6):F1799–F1806. https://doi.org/10.1152/ajprenal.90510.2008. 90510.2008 [pii]

70. Seib FP, Jones AT, Duncan R (2006) Establishment of subcellular fractionation techniques to monitor the intracellular fate of polymer therapeutics I. Differential centrifugation fractionation B16F10 cells and use to study the intracellular fate of HPMA copolymer – doxorubicin. J Drug Target 14 (6):375–390. https://doi.org/10.1080/10611860600833955. Q2801G62J27V1H10 [pii]

71. Vogelmann R, Nelson WJ (2007) Separation of cell-cell adhesion complexes by differential centrifugation. Methods Mol Biol 370:11–22. https://doi.org/10.1007/978-1-59745-353-0_2. 1-59745-353-6:11 [pii]

72. Simpson RJ (2006) Fractional precipitation of proteins by ammonium sulfate. CSH Protoc 2006(1). https://doi.org/10.1101/pdb.prot4309. 2006/1/pdb.prot4309

73. Wingfield P (2001) Protein precipitation using ammonium sulfate. Curr Protoc Protein Sci Appendix 3:Appendix 3F. https://doi.org/10.1002/0471140864.psa03fs13

74. Lopitz-Otsoa F, Rodriguez-Suarez E, Aillet F, Casado-Vela J, Lang V, Matthiesen R, Elortza F, Rodriguez MS (2011) Integrative analysis of the ubiquitin proteome isolated using Tandem Ubiquitin Binding Entities (TUBEs). J Proteomics. https://doi.org/10.1016/j.jprot.2011.12.001. S1874-3919(11)00668-3

75. Dai Z, Zhou J, Qiu SJ, Liu YK, Fan J (2009) Lectin-based glycoproteomics to explore and analyze hepatocellular carcinoma-related glycoprotein markers. Electrophoresis 30 (17):2957–2966. https://doi.org/10.1002/elps.200900064

76. Fung KY, Cursaro C, Lewanowitsch T, Brierley GV, McColl SR, Lockett T, Head R, Hoffmann P, Cosgrove L (2011) A combined free-flow electrophoresis and DIGE approach to identify proteins regulated by butyrate in HT29 cells. Proteomics 11(5):964–971. https://doi.org/10.1002/pmic.201000429

77. Kim KH, Kim JY, Kim MO, Moon MH (2012) Two dimensional (pI & d(s)) separation of phosphorylated proteins by isoelectric focusing/asymmetrical flow field-flow fractionation: Application to prostatic cancer cell line. J Proteomics. https://doi.org/10.1016/j.jprot.2012.01.034. S1874-3919(12)00074-7

78. Bökelmann V, Spengler B, Kaufmann R (1995) Dynamical parameters of ion ejection and ion formation in matrix-assisted laser desorption/ionization. Eur Mass Spectrom 27:156–158

79. Beavis RC, Chait BT (1990) Rapid, sensitive analysis of protein mixtures by mass spectrometry. Proc Natl Acad Sci U S A 87 (17):6873–6877

80. Cohen SL, Chait BT (1996) Influence of matrix solution conditions on the MALDI-MS analysis of peptides and proteins. Anal Chem 68(1):31–37

81. Vorm O, Roepstorff P, Mann M (1994) Improved resolution and very high sensitivity in MALDI TOF of matrix surfaces made by fast evaporation. Anal Chem 66:3281–3287

82. Kussmann M, Nordhoff E, Nielsen RB, Hábel S, Larsen MR, Jakobsen L, Gobom J, Mirgorodskaya E, Kristensen AK, Palm L, Roepstorff P (1997) Matrix-assisted laser desorption/ionization mass spectrometry sample preparation techniques designed for various peptide and protein analytes. J Mass Spectrom 32:593–601

83. Guilhaus M (1995) Principles and instrumentation in time-of-flight mass spectrometry. J Mass Spectrom 30:1519–1552

84. Matthiesen R (2007) Methods, algorithms and tools in computational proteomics: a practical point of view. Proteomics 7 (16):2815–2832. https://doi.org/10.1002/pmic.200700116

85. Vestal ML, Juhasz P, Martin SA (1995) Delayed extraction matrix-assisted laser desorption time-of-flight mass spectrometry. Rapid Commun Mass Spectrom 9:1044–1050

86. Patterson SD (1995) Matrix-assisted laser-desorption/ionization mass spectrometric approaches for the identification of gel-separated proteins in the 5–50 pmol range. Electrophoresis 16(7):1104–1114

87. Keller BO, Li L (2000) Discerning matrix-cluster peaks in matrix-assisted laser desorption/ionization time-of-flight mass spectra of

dilute peptide mixtures. J Am Soc Mass Spectrom 11:88–93

88. Guo Z, Zhang Q, Zou H, Guo B, Ni J (2002) A method for the analysis of low-mass molecules by MALDI-TOF mass spectrometry. Anal Chem 74:1637–1641

89. Coursey JS, Schwab DJ, Dragoset RA (2001) Atomic weights and isotopic compositions (version 2.3.1). Available: http://physics.nist.gov/Comp. 7 July 2003. National Institute of Standards and Technology, Gaithersburg, MD

90. Snyder AP (2000) Interpreting protein mass spectra, a comprehensive resource. Oxford University Press, Oxford

91. Zheng PP, Luider TM, Pieters R, Avezaat CJ, van den Bent MJ, Smitt SPA, Kros JM (2003) Identification of tumor-related proteins by proteomic analysis of cerebrospinal fluid from patients with primary brain tumors. J Neuropathol Exp Neurol 62:855–862

92. Jensen ON, Podtelejnikov AV, Mann M (1997) Identification of the components of simple protein mixtures by high-accuracy peptide mass mapping and database searching. Anal Chem 69(23):4741–4750

93. Zhang N, Aebersold R, Schwikowski B (2002) ProbID: a probabilistic algorithm to identify peptides through sequence database searching using tandem mass spectral data. Proteomics 2:1406–1412

94. Smith RD, Loo JA, Edmonds CG, Barinaga CJ, Udseth HR (1990) New developments in biochemical mass spectrometry: electrospray ionization. Anal Chem 62(9):882–899

95. Ikonomou MG, Blades AT, Kebarle P (1991) Electrospray-ion spray: a comparison of mechanisms and performance. Anal Chem 63:1989–1998

96. Covey TR, Bonner RF, Shushan BI, Henion J (1988) The determination of protein, oligonucleotide and peptide molecular weights by ion-spray mass spectrometry. Rapid Commun Mass Spectrom 2:249–256

97. Mirza UA, Chait BT (1994) Effects of anions on the positive ion electrospray ionization mass spectra of peptides and proteins. Anal Chem 66(18):2898–2904

98. Köcher T, Allmaier G, Wilm M (2003) Nanoelectrospray-based detection and sequencing of substoichiometric amounts of phosphopeptides in complex mixtures. J Mass Spectrom 38:131–137

99. Gatlin CL, Tureček F (1994) Acidity determination in droplets formed by electrospraying methanol-water solutions. Anal Chem 66:712–718

100. Wilm M, Mann M (1996) Analytical properties of the nanoelectrospray ion source. Anal Chem 68(1):1–8

101. Choudhary G, Apffel A, Yin H, Hancock W (2000) Use of on-line mass spectrometric detection in capillary electrochromatography. J Chromatogr A 887:85–101

102. Urvoas A, Amekraz B, Moulin C, Clainche LL, Stocklin R, Moutiez M (2003) Analysis of the metal-binding selectivity of the metallochaperone CopZ from Enterococcus hirae by electrospray ionization mass spectrometry. Rapid Commun Mass Spectrom 17:1889–1896

103. Wilm M, Shevchenko A, Houthaeve T, Breit S, Schweigerer L, Fotsis T, Mann M (1996) Femtomole sequencing of proteins from polyacrylamide gels by nanoelectrospray mass spectrometry. Nature 379 (6564):466–469. https://doi.org/10.1038/379466a0

104. Chernushevich IV, Loboda AV, Thomson BA (2001) An introduction to quadrupole–time-of-flight mass spectrometry. J Mass Spectrom 36:849–865

105. Lammert SA, Rockwood AA, Wang M, Lee ML, Lee ED, Tolley SE, Oliphant JR, Jones JL, Waite RW (2006) Miniature toroidal radio frequency ion trap mass analyzer. J Am Soc Mass Spectrom 17(7):916–922. https://doi.org/10.1016/j.jasms.2006.02.009.S1044-0305(06)00125-5

106. Michalski A, Damoc E, Hauschild JP, Lange O, Wieghaus A, Makarov A, Nagaraj N, Cox J, Mann M, Horning S (2011) Mass spectrometry-based proteomics using Q Exactive, a high-performance benchtop quadrupole Orbitrap mass spectrometer. Mol Cell Proteomics 10(9):M111.011015. https://doi.org/10.1074/mcp.M111.011015

107. Hagglund P, Matthiesen R, Elortza F, Hojrup P, Roepstorff P, Jensen ON, Bunkenborg J (2007) An enzymatic deglycosylation scheme enabling identification of core fucosylated N-glycans and O-glycosylation site mapping of human plasma proteins. J Proteome Res 6(8):3021–3031. https://doi.org/10.1021/pr0700605

108. Carr SA, Huddleston MJ, Annan RS (1996) Selective detection and sequencing of phosphopeptides at the femtomole level by mass spectrometry. Anal Biochem 239 (2):180–192. https://doi.org/10.1006/abio.1996.0313. S0003-2697(96)90313-9

109. Cham Mead JA, Bianco L, Bessant C (2010) Free computational resources for designing selected reaction monitoring transitions.

Proteomics 10(6):1106–1126. https://doi.org/10.1002/pmic.200900396

110. Sherman J, McKay MJ, Ashman K, Molloy MP (2009) How specific is my SRM?: the issue of precursor and product ion redundancy. Proteomics 9(5):1120–1123. https://doi.org/10.1002/pmic.200800577

111. Kinter M, Sherman NE (2000) Protein sequencing and identification using tandem mass spectrometry. John Wiley & Sons, New York

112. Creasy DM, Cottrell JS (2002) Error tolerant searching of uninterpreted tandem mass spectrometry data. Proteomics 2(10):1426–1434. https://doi.org/10.1002/1615-9861(200210)2:10<1426::AID-PROT1426>3.0. CO;2-5

113. Eng J, McCormack A, Yates J (1994) An approach to correlate tandem mass spectral data of peptides with amino acid sequences in a protein database. J Am Soc Mass Spectrom 5(11):976–989

114. Colinge J, Masselot A, Giron M, Dessingy T, Magnin J (2003) OLAV: towards high-throughput tandem mass spectrometry data identification. Proteomics 3(8):1454–1463. https://doi.org/10.1002/pmic.200300485

115. Matthiesen R (2007) Virtual Expert Mass Spectrometrist v3.0: an integrated tool for proteome analysis. Methods Mol Biol 367:121–138. https://doi.org/10.1385/1-59745-275-0:121

116. Bjornson RD, Carriero NJ, Colangelo C, Shifman M, Cheung KH, Miller PL, Williams K (2008) X!!Tandem, an improved method for running X!tandem in parallel on collections of commodity computers. J Proteome Res 7(1):293–299. https://doi.org/10.1021/pr0701198

117. Cox J, Neuhauser N, Michalski A, Scheltema RA, Olsen JV, Mann M (2011) Andromeda: a peptide search engine integrated into the MaxQuant environment. J Proteome Res 10 (4):1794–1805. https://doi.org/10.1021/pr101065j

118. Tharakan R, Martens L, Van Eyk JE, Graham DR (2008) OMSSAGUI: an open-source user interface component to configure and run the OMSSA search engine. Proteomics 8 (12):2376–2378. https://doi.org/10.1002/pmic.200701126

119. Wysocki VH, Tsaprailis G, Smith LL, Breci LA (2000) Mobile and localized protons: a framework for understanding peptide dissociation. J Mass Spectrom 35:1399–1406

120. Paizs B, Suhai S (2002) Towards understanding some ion intensity relationships for the tandem mass spectra of protonated peptides. Rapid Commun Mass Spectrom 16:1699–1702

121. Roepstorff P, Fohlman J (1984) Proposal for a common nomenclature for sequence ions in mass spectra of peptides. Biomed Mass Spectrom 11(11):601. https://doi.org/10.1002/bms.1200111109

122. Havilio M, Haddad Y, Smilansky Z (2003) Intensity-based statistical scorer for tandem mass spectrometry. Anal Chem 75 (3):435–444

123. Summerfield SG, Bolgar MS, Gaskell SJ (1997) Promotion and stabilization of ions in peptide b1 phenythiocarbamoyl derivatives: analogies with condensed-phase chemistry. J Mass Spectrom 32:225–231

124. Schlosser A, Lehmann WD (2002) Patchwork peptide sequencing: extraction of sequence information from accurate mass data of peptide tandem mass spectra recorded at high resolution. Proteomics 2(5):524–533. https://doi.org/10.1002/1615-9861(200205)2:5<524::AID-PROT524>3.0. CO;2-O

125. Harrison AG, Csizmadia IG, Tang TH, Tu YP (2000) Reaction competition in the fragmentation of protonated dipeptides. J Mass Spectrom 35:683–688

126. Sleno L, Volmer DA (2004) Ion activation methods for tandem mass spectrometry. J Mass Spectrom 39(10):1091–1112. https://doi.org/10.1002/jms.703

127. Csonka IP, Paizs B, Lendvay G, Suhai S (2000) Proton mobility in protonated peptides: a joint molecular orbital and RRKM study. Rapid Commun Mass Spectrom 14 (6):417–431. https://doi.org/10.1002/(SICI)1097-0231(20000331)14:6<417:: AID-RCM885>3.0.CO;2-J

128. Lee SA, Jiao CQ, Huang Y, Freiser BS (1993) Multiple excitation collisional activation in Fourier-transform mass spectrometry. Rapid Commun Mass Spectrom 7:819–821

129. Jedrychowski MP, Huttlin EL, Haas W, Sowa ME, Rad R, Gygi SP Evaluation of HCD- and CID-type fragmentation within their respective detection platforms for murine phosphoproteomics. Mol Cell Proteomics 10(12). https://doi.org/10.1074/mcp.M111. 009910. M111.009910 [pii]

130. Cooper HJ, Hudgins RR, Håkansson K, Marshall AG (2002) Characterization of amino acid side chain losses in electron capture dissociation. J Am Soc Mass Spectrom 13:241–249

131. Zubarev RA, Kruger NA, Fridriksson EK, Lewis MA, Horn DM, Carpenter BK, McLafferty FW (1999) Electron capture dissociation of gaseous multiply-charged proteins is favoured at disulphide bonds and other sites of high hydrogen atom affinity. J Am Chem Soc 121:2857–2862

132. Bakhtiar R, Guan Z (2005) Electron capture dissociation mass spectrometry in characterization of post-translational modifications. Biochem Biophys Res Commun 334:1–8

133. Carvalho AS, Penque D, Matthiesen R (2015) Bottom up proteomics data analysis strategies to explore protein modifications and genomic variants. Proteomics 15 (11):1789–1792. https://doi.org/10.1002/pmic.201400186

134. Cannon WR, Jarman KD (2003) Improved peptide sequencing using isotope information inherent in tandem mass spectra. Rapid Commun Mass Spectrom 17:1793–1801

135. Cooks RG, Rockwood AL (1991) Rapid Commun Mass Spectrom 5, 93

136. Busman M, Schey KL, Oatis JE, Knapp DR (1996) Identification of phosphorylation sites in phosphopeptides by positive and negative mode electrospray ionization-tandem mass spectrometry. J Am Soc Mass Spectrom 7 (3):243–249. https://doi.org/10.1016/1044-0305(95)00675-3

137. Venable JD, Dong MQ, Wohlschlegel J, Dillin A, Yates JR (2004) Automated approach for quantitative analysis of complex peptide mixtures from tandem mass spectra. Nat Methods 1(1):39–45. https://doi.org/10.1038/nmeth705

138. Plumb RS, Johnson KA, Rainville P, Smith BW, Wilson ID, Castro-Perez JM, Nicholson JK (2006) UPLC/MSE; a new approach for generating molecular fragment information for biomarker structure elucidation. Rapid Commun Mass Spectrom 20(13):1989–1994

139. Aebersold R, Bensimon A, Collins BC, Ludwig C, Sabido E (2016) Applications and developments in targeted proteomics: from SRM to DIA/SWATH. Proteomics 16 (15–16):2065–2067. https://doi.org/10.1002/pmic.201600203

140. Lide DR (1992–1993) Handbook of chemistry and physics, CRC Press, 73rd edn, p 11.2

Chapter 2

LC-MS Spectra Processing

Rune Matthiesen

Abstract

Peak extraction from raw data is the first step in LC-MS data analysis. The quality of this procedure can have dramatic effects on the quality and accuracy of all subsequent data analysis steps such as database searches and peak quantitation. The most important and most accurately measured physical entity provided by mass spectrometers is m/z. Peak processing algorithms must extract m/z values unaffected from overlapping peaks to avoid confusing downstream algorithms. The aim of this chapter is to provide a discussion of peak processing methods and furthermore discuss some of the yet unresolved or neglected issues. The chapter mainly discusses possible software developed in R for spectra processing and free software to generate Mascot generic files (mgf—*see* Chapter 1).

Key words Noise filtering, Peak extraction, Deisotoping, Decharging

1 Introduction

Liquid chromatography—electrospray ionization—mass spectrometry (LC-ESI-MS) of peptides produces a wealth of information in the form of peptide masses, peptide fragment masses and peptide retention time(s) on the liquid chromatography column. In MS-based proteomics the liquid chromatography system is in general a single hydrophobic reverse phase column (one dimensional separation) or anionic/cationic column followed by a hydrophobic reverse phase column (multiple dimensional separation) [1]. The electrospray ion source is responsible for formation of charged peptides in the gas phase resulting in tryptic peptides with charge states ranging typically between +1–4/5 (depending on the instrument and buffer conditions). A single peptide can appear with different charge states depending on the number of proton accepting groups in the peptide [2]. The mass spectrometer used in LC-MS-based proteomics is most often a tandem mass spectrometer which produces MS or both MS and MS/MS data (or even higher order fragmentation data, *see* Chapter 1). The raw data obtained from these experiments contains, in general, distorted signals of the

Rune Matthiesen (ed.), *Mass Spectrometry Data Analysis in Proteomics*, Methods in Molecular Biology, vol. 2051, https://doi.org/10.1007/978-1-4939-9744-2_2, © Springer Science+Business Media, LLC, part of Springer Nature 2020

ideal physical quantity of interest which are the masses of the intact tryptic peptides, the peptide fragment masses of selected peptide ions and the retention time. The conversion of raw data to peak list, used downstream for database dependent searches, de novo sequencing, and quantitation might consists of the following steps: (1) removal of instrument introduced noise and bias from the spectra, (2) peak detection, (3) extraction of the monoisotopic single charged mass by decharging and deisotoping the peaks in the survey scans (MS scans), (4) extraction of the retention time(s) for the peptides, and (5) extraction of the MS/MS peaks.

Both the survey scans and the MS/MS scans are used to generate peak lists (e.g., *see* Chapter 1 for details on the mgf format) for subsequent peptide and protein identifications. Spectra processing is also essential for quantitation and indispensable for optimal peptide and protein identification. This chapter provides a theoretical review and a practical protocol for MS-based proteomics data processing.

1.1 Noise Filtering

The quality of the signal is often expressed as the true signal divided by the standard deviation of the noise also referred as signal to noise ratio (SNR). If SNR is higher than 1 then there is more signal than noise. SNR is defined as the ratio of the power of a signal (targeted signal) to the power of background noise (untargeted signal) or $SNR = P_{signal}/P_{noise}$. There are other noise definitions, for example in imaging the reciprocal of coefficient of variance is often used [3]. The part of the signal defined as signal versus the part defined as noise depends on the objective of the analysis. The signal is estimated by filtering out the noise as described below. Then the noise is estimated by subtracting the estimated signal from the observed signal.

Two types of errors are present in experimental data which are systematic and random errors. Systematic errors are often removed by calibration. Random errors are also called noise or short-term variations. Noise is typically divided into low and high frequency noise. Low-frequency noise or base line is often caused by electrospray or thermospray are inefficient at selectively ionize only the analyte of interest. Buffer related clusters (*see* Chapter 1) and bleeding columns often result in chromatograms with a high level of noise. High frequency noise arises typically from the ion source interface between LC and MS parts and the MS ion optics [4].

The signal is reproducible in contrast to the noise. The signal reproducibility enables averaging over many spectra to improve signal to noise ratio. Data noise filtering ideally gives the true signal although this is not fully accomplished for many filtering algorithms such as for example Savitzky–Golay filters as it provides solely an estimate of the true signal [5]. This also means that filters such as Savitzky–Golay filter are mainly used (1) for data visualization, (2) for peak maxima localization, or (3) to obtain preliminary parameter estimates that are used to fit models against the raw data to obtain more accurate parameters.

Many methods for noise removal exists such as linear filters [6, 7], penalized least square [8], Fourier transform [9], and wavelets [10]. This section focus on linear filters which are simple to implement, computational fast, widely used with a satisfactory performance for MS-based proteomics data. In data from the latest instruments, it is typically unnecessary to filter out noise in the raw mass scans. However, often the extracted ion chromatograms (XIC) require noise filtering for peak detection and quantitation. Linear filters are useful for noise removal of XIC. In general, linear filters convert a series to another by a linear operation. A general mathematical expression of linear filter is given by [7]

$$y_t = \sum_{r=-q}^{+s} a_r x_{t+r} \qquad (1)$$

where y_t is the smoothed signal, x_t is the current data point, and r iterates over neighboring data points. The smooth width m is equal to $q + s + 1$. a_r represent constant numbers (weights) and are dependent on the filter type. For example, for a simple unweighted sliding-average smooth $a_r = 1/m$ for all r. Savitzky–Savitzky–Golay filter is frequently used in mass spectrometry [6] which weights result in a smoothed signal that corresponds to estimating a low order polynomial to all smooth intervals (*see* Fig. 1). The data in Fig. 1 was obtained from PRIDE data set PXD000380 [11].

The code for Savitzky–Golay filters is provided in "Numerical recipes in C" by Cambridge University Press. Additionally, it provides code for generations of arbitrary filters of different window length, order of derivative, smoothing polynomial and highest conserved moment [5]. Alternatively, functions and code Savitzky–Golay filters is also available in the R package "signal". The symmetrical filter properties of Savitzky–Golay gives raise to start and end effect problems (*see* **Note 1**). Combining symmetric and asymmetric filters overcome this problem for mass spectrometry data. Savitzky–Golay filters are most appropriated for large window sizes since smaller windows only have a relative small smoothing effect [5]. Then, how to choose the optimal filter parameters for Savitzky–Golay filters? Previous studies addressed the problem of defining the optimal parameters for Savitzky–Golay filtering for NMR, LC, and mass spectrometry data [12]. This proposed method includes the following three steps: (1) the lag-one autocorrelation value of the noise of the instrument is computed through the study of a blank signal, (2) the Savitzky–Golay algorithm is applied to filter the signal using different window sizes and (3) The window size that yields an autocorrelation of the residuals closest to the autocorrelation of the noise of the instrument is considered optimal. The drawback of this method is that a blank signal is

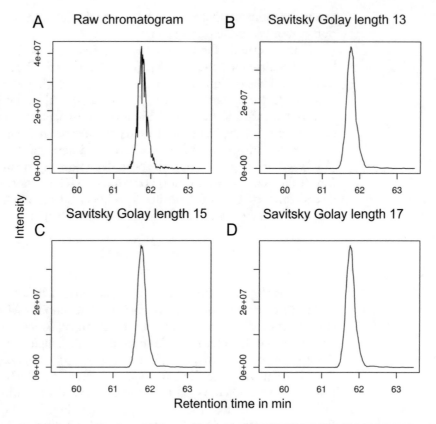

Fig. 1 Savitzky–Golay noise filtering of $XIC_{0.005}$ of peptide NMGGPYGGGNYGPGGSGGSGGYGGR with charge state 2. (**a**) Raw MS data. MS data from (**a**) after applying the function "sgolayfil" from the R package signal using a filter length of 13 (**b**), 15 (**c**), and 17 points (**d**)

needed to extract the estimates of instrument noise. Blank signals (data acquired on sample with no analyte of interest) is not always available.

For mass spectrometry data this is mainly an issue for XIC where noise level, peak shape and resolution tend to vary from XIC to XIC (*see* Fig. 2). Automatical evaluation of the quality of the smooth is possible by estimating the lack of fit, the roughness of the data, by maximum entropy, spike frequency or the final number of estimated peaks (*see* **Note 2**).

In MS scans the peak shape and resolution is more constant and fixed window length is often used for MS peaks across data sets obtained with the same or similar experimental settings and instrument. Although, the peak broadness increase for higher m/z values in data obtained from TOF [13] and Orbitrap instruments (*see* Fig. 3). This increasing peak broadness still requires careful considerations.

The full-width at half maximum (FWHM) values have a small quadratic increase as a function of m/z. The data points far from

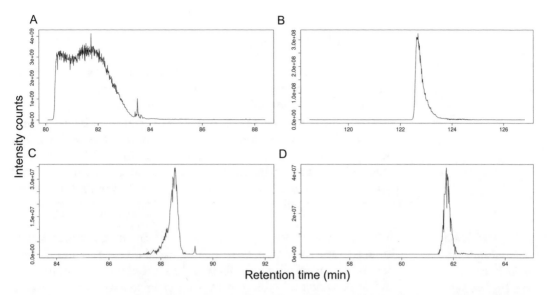

Fig. 2 XIC$_{0.005, z = +2}$ of four different peptides LLLPGELAK (**a**), MTDQEAIQDLWQWR (**b**), VGINYQPPTVVPGG DLAK (**c**) and NMGGPYGGGNYGPGGSGGSGGYGGR (**d**) with different noise level, peak resolutions and decreasing overall signal. The retention time intervals on the X-axis are all 8 min to facilitate comparison of peak broadness

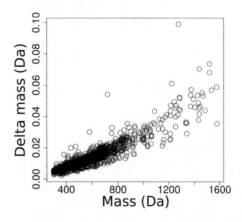

Fig. 3 FWHM as a function of *m*/*z* values estimated from Q-Exactive survey scans

the main trend lines are likely FWHM estimation artefacts caused by for example peak overlaps. This claim on outliers is supported by the fact that all outliers are overestimates of the FWHM and we have further validated some cases.

Alternatively, a geometric mean filter with window size $2m + 1$ can be used in combination with a Savitzky–Golay filter, where m is the length of the upstream or downstream data points from the center of the window.

$$y_t = \left(\prod_i x(t + i\Delta t) \right)^{1/(2m+1)} \quad i = 0, \pm 1, \pm, 2, \ldots, \pm m \quad (2)$$

The geometric mean filter can remove spikes because of the requirement of nonzero neighboring data point for maintaining a signal. A combination of geometric mean and Savitzky–Golay filter in which the window size is automatically optimized by evaluating the spike frequency for a specific XIC is a simple way to automate peak processing of peaks with different profiles. The spike frequency is defined as the ratio between the sum of data points for which the two neighboring data points is lower and the total number of data points.

1.2 Peak Picking, Calibration, Deisotoping, and Decharging LC-MS/MS Data

1.2.1 Peak Picking

Neither raw data nor filtered (smoothed) raw data are practical as input to a search engine. Peak picking based on the raw and filtered raw data is a possibility. The filtered data is useful to locate the peak and to provide preliminary parameter estimates for peak fitting. A simple peak picking algorithm locates all data points in the smoothed data where a predefined number of n neighboring data points are lower (i.e., local maximum search). Success of this method depends on the degree of smoothing, which often needs optimization with regards to specific resolution, peak shape and noise. As mentioned above in some cases the MS scans do not require noise filtering before peak picking. An alternative method for peak picking is to apply Savitzky–Golay first derivative filter directly on the raw data (*see* Fig. 4a).

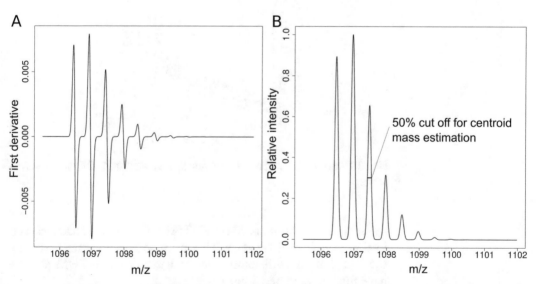

Fig. 4 Converting profile data to peak lists. (**a**) First derivative method. The first derivative of the signal is calculated, and preliminary estimates of the peak masses are determined at the mass points where the first derivative is cutting the *X*-axis. (**b**) 50% centroid method. 50% of the resolved part of the peak is used for determining the mass. The mass is calculated by an intensity weighted average of the masses in the peak that are above the threshold. This is equivalent to finding the vertical line passing through the center of gravity of the peak

After peak picking or localization, an accurate mass estimate is calculated for all the peaks. A method frequently used calculates the centroid mass at a specific percentage from the raw data points. This method is simple although peak overlaps might give problems (*see* Fig. 4b and **Note 3**). In our hands fitting Gaussian (mainly used for MS spectra) or mixed Guassian/Lorentzian (mainly used for XIC) using all data points of the peak in the raw data (*see* **Note 4**) slightly improves in mass accuracy. However, fitting procedures are computationally slower and less robust than the centroid method described above. Therefore the method used by MaxQuant, which consists of three different approaches depending on the number of peak data points available for mass estimations, is practical and also a more robust computational solution [15]. The following three rules define how the mass is estimated: (1) If the peak consists of one raw data point solely, that *m/z* value is taken as the centroid position, (2) if there are two raw data points in a peak, then the centroid position is defined as the average of the two raw *m/z* values, weighted by the raw intensities and (3) if at least three points exists then it is determined by the three-point Gaussian estimation given by the equation below.

$$m = \frac{1}{2} \times \frac{(L_0 - L_1)m_{-1}^2 + (L_1 - L_{-1})m_0^2 + (L_{-1} - L_0)m_1^2}{(L_0 - L_1)m_{-1} + (L_1 - L_{-1})m_0 + (L_{-1} - L_0)m_1}, \quad (3)$$

where $m_{-1,0,1}$ are the *m/z* positions of the three central raw data points, and $L_{-1,0,1}$ are the natural logarithms of the corresponding raw intensities [16]. The three-point Gaussian estimation also facilitates calculation of other parameters and properties of the Gaussian curve [16]. The standard deviation is given by

$$\sigma = \sqrt{\frac{1}{2} \times \frac{(m_{-1} - m_1)*(m_0 - m_{-1})*(m_1 - m_0)}{(L_0 - L_1)m_{-1} + (L_1 - L_{-1})m_0 + (L_{-1} - L_0)m_1}} \quad (4)$$

FWHM is now calculated as

$$\text{FWHM} = 2(2\ln 2)^{1/2}\sigma = 2.355\sigma \quad (5)$$

The area is now given by

$$A = \sqrt{2\pi\sigma^2}(I_{-1} \times I_0 \times I_1)^{1/3}$$
$$\times \exp\left[\frac{(m_{-1} - \mu)^2 + (m_0 - \mu)^2 + (m_1 - \mu)^2}{6\sigma^2}\right] \quad (6)$$

where $I_{-1,0,1}$ are the raw intensities. Note that the above Eqs. 3–6 though practical ignore potential binning bias. These equations are inaccurate if there are peak overlaps.

1.2.2 Calibration

A large number of calibration methods for MS data exists: (1) lockspray calibration, (2) calibration using contaminants (bovine peptides from culture media, keratins from hair, skin, nails, protease autodigested peptides), (3) Calibration using polycyclodimethylsiloxane (PCMs—outgassed material from semiconductors) and bis (2-ethylhexyl) phthalate (DEHP—from plastic) present in the ambient air [17, 18], (4) calibration using all theoretical fragment ions from all theoretical peptides during a database dependent search [19], (5) calibration using charge state pairs [20], (6) calibration using local Gaussian mass distribution of canonical trypic peptides and, (7) Calibration using peptides identified in a first database dependent search using low mass accuracy [21]. It has recently been reported that Orbitrap instrument data show nonlinear m/z errors over retention time [18]. Figure 5a shows delta m/z values of PCMs and DEHP extracted from Q Exactive data run (i.e., observed − theoretical m/z values). The nonlinear calibration of the data shown in Fig. 5 involves the following steps: (1) calculation of the mean or median for each scan number, (2) use of interpolation for scans where no data points exists and (3) use of moving average filter with window size of 500 scans to smooth the calibration curve. It is reassuring that two different contaminants outline the same overall calibration trend (Fig. 5a).

1.2.3 Deisotoping and Decharging

Once the above estimation of peak masses are calculated, the next step estimates the monoisotopic peak and the charge state of the parent. A peak list containing the monoisotopic peak of the parent

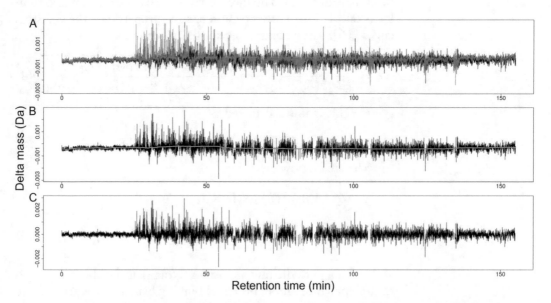

Fig. 5 Calibration of nonlinear *m/z* errors over retention time. (**a**) *m/z* errors of two frequent occurring contaminants observed in the MS scans depicted with black and red. (**b**) Moving average filter with window size of 500 scans to estimate a calibration curve (depicted in green). (**c**) The result after nonlinear calibration of the data in "**a**" by subtracting the estimated curve in "**b**"

ion and its charge state followed by the fragment ions is often used as input to database dependent search engines. For complex samples with many peak overlaps this constitutes a difficult problem and a number of approaches have been proposed in the literature [15, 22–25]. MaxQuant generates an undirected graph with the 3D peaks as vertices. An edge is put between two peaks, whenever there is a possibility that they are neighbors in an isotope pattern [15]. These peak connections are based on mass differences between peaks and the uncentered Pearson correlation >0.6 in retention time intensity profiles. MaxQuant further checks the correlation of the measured isotope pattern for a correlation of at least 0.6 with the averagine [26] isotope pattern of the same mass (*see* **Note 5**).

A novel strategy consists of calculating the area under the multiple charge state and isotope envelope dependent XIC (mziXIC) curve (*see* Fig. 6) for different potential charge states (typically $z = +1–5$) and potential alternative monoisotopic peaks (typically set to a maximum of two or three) which increases SNR [27].

Figure 6b depicts the mziXIC curve for $z = +2–3$ and for two potential monoisotopic peaks. mziXIC is simple, fast to calculate, and easy to visualize. mziXIC increases the signal to noise ratio and is less sensitive to overlapping peaks (*see* Fig. 6b). mziXIC is useful for semi quantitative estimation of peptide abundance and to determine peptide charge states.

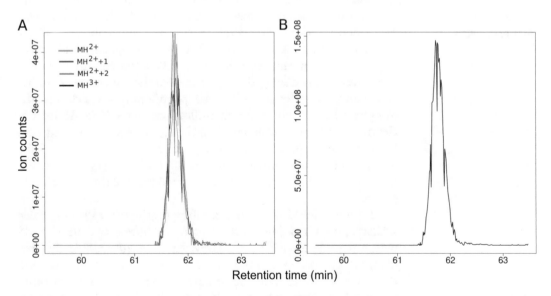

Fig. 6 Different types of extract ion chromatograms (XIC). (**a**) $XIC_{0.005}$ of individual masses and charge states in the isotopic distribution. (**b**) $mziXIC_{0.005,\ z\ =\ +2–3}$ which considers isotope distribution patterns, local mass shifts and correlation of intensity at different m/z in the isotope envelope

1.3   Extraction of MS/MS Peaks— Processing of MS/MS Spectra

The way the fragment ions are represented varies depending on the purpose of the collected data and on the type of analysis performed. If complex modifications are searched then it is an advantage to keep all peaks. However, most search engines will not consider the extra peaks resulting from the complex modification. For most standard database dependent searches picking 6–10 of the most intense peaks per 100 m/z results in a adequate performance. Decharging and deisotoping the fragment ions further shorten the list of fragment ion peaks. This is however, not strictly needed since most database search engines allows to score double charged or even triple charged fragment ions as well. If the processing algorithm available is of poor quality or the data is of low resolution so that optimal decharging of the fragment ions are difficult to obtain, then it is preferable to specify that the database dependent search engine should search for fragment ions of higher charge states.

If using iTRAQ quantitation, then downstream data analysis is facilitated if exception ions are defined to ensure that the iTRAQ reporter ions for quantitation are retained in the MS/MS spectra [28]. For analysis of specific posttranslational modifications defining diagnostic ions characteristic of a specific modification as exception ions is also beneficial. For de novo sequencing, often the full list of fragment ions is parsed to de novo sequencing program.

1.4   Extraction of Retention Time and Processing for Quantitation

There are numerous scientific papers stating that XIC is used for quantification of peptides. This description is both insufficient and inaccurate. Firstly, the mass interval used for calculating XIC needs specification. We therefore here use subscript to indicate the mass interval used when referring to a specific XIC curve (e.g., $XIC_{0.005}$ means ± 0.005 m/z). Changing the mass interval can considerably change the appearance of the XIC curve. Another issue concerning the use of XIC for quantitation is reflected on the number of cases where the identifications from MS/MS do not clearly identify an elution curve in the survey scan associated with the identified peptide. Take for instance the example in Fig. 7 the $XIC_{0.005}$ curve indicates a main elution profile but several MS/MS spectra of the same peptides were detected outside the main elution profile.

If there is no MS/MS identification for the main elution profile available, then it is difficult to associate the intensity counts in MS with a peptide identification. However, it is still feasible that the identification occured in the tail of the peak or in another LC-MS run. This can for example occur because the elution of many other alternative peptides occurs in the same time interval. In this case, the automated quantitation might base the quantitation on a second elution profile at later time points where ion suppression is less or the tail of the main elution profile which will lead to unreliable quantitation. For stable isotope labeled samples this problem

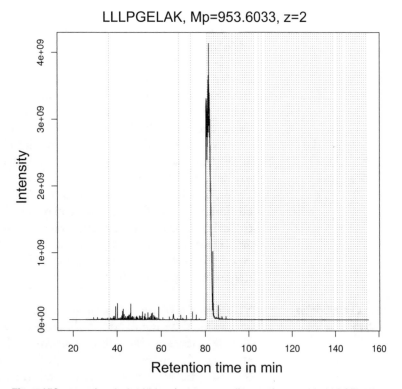

Fig. 7 $XIC_{0.005}$ using 953.6033 *m/z* corresponding to the peptide LLPGELAK

mainly affect accuracy but for label free the result will most likely be an artefact. It frequently occurs that peptides are identified based on a single MS/MS event and often these events are in the high retention time tail of the peptide elution profile. Notice that the majority of the identifications in Fig. 7 are in the right tail of the elution profile. In these cases, the association between the main elution profile and the identified peptide is uncertain. Nevertheless, if no other peptide identification is associated to the main elution peak then the most likely elution profile is the one left to the identification. The following rules to determine the main elution peak of a peptide is one possibility. In a noise filtered mziXIC curve (*see* Subheading 1.1) the median retention time of all MS/MS events that identify the specific peptide across all acquired runs is used as a starting point. If there are more than two MS/MS events that identify the same peptide, then they tend to cluster together around the main elution peak. This means that walking along the curve from the median towards the side with the highest slope inclination, until a local maximum is reached, tend to locate the main elution peak. If the inclination is equal in both directions, then the walk is along the curve towards lower retention time. This choice of walk direction along retention time, when the inclinations in both directions are equal, is supported based on the observation that basically all elution profiles are right skewed. If exactly two MS/MS events for a peptide exist, then a walk is performed from both time points. The walk that identifies an elution peak maximum that is closest to the average elution time of the two events is chosen. If only one event exists, then the walk is performed from that time point. The elution maximum identified by this method is finally checked to validate that no other peptide identification is associated with the identified elution profile. If there is another peptide identifications associated to the elution profile then, the algorithms maintains only the most confident peptide identification for the quantitation.

In conclusion small fluctuation in intensity may appear as local elution points, as for the case displayed in Fig. 7, but is just part of the tail of a main elution peak profile. In case no MS/MS events are associated with the main elution peak this problem can be resolved by targeting the main peak by using inclusion list and/or alternative fragmentation methods.

1.5 Visualization of Data

Visualization of the MS data is important for data quality checks and to validate important findings. XIC, ziXIC, and mziXIC are already discussed above in connection with parent ion annotation and peptide quantitation. Base peak chromatogram (BPC, the highest intensity per scan) and total ion chromatogram (TIC) is often used for quality control. BPC and TIC can for example reveal instability of electrospray and chromatography. BPC is often clearer to evaluate than TIC since there is less background in BPC than TIC.

2   Data and Software

2.1 Data

A large number of data sets are available in the PRIDE database [29, 30]. The data used in this protocol was extracted from PRIDE data set PXD000380 [11]. The data is only available in a vendor specific format and therefore needs conversion by one of the software listed in Subheading 2.2.

2.2 Software

1. MSconvert [31, 32] can convert most of the vendor formats into XML-based formats such as mzXML and MGF, the commonly used format for MSMS database searches. MSconvert is open source; the only drawback is that it only runs on Windows operating systems.

2. RawConvert [33] converts raw data from Thermo Fisher instruments into formats such as mzXML and MGF. It requires Thermo MSFileReader installed as well as described in the installation instructions.

3. R packages: MsnBase [34], the function is to extract data from the formats such as mzXML and mgf. The R packages enviPat [35], protViz [36], and IsoSpecR [37] can calculate estimated isotopic distributions for a given peptide or parent ion mass. The package signal contains useful filtering functions which was used in this chapter together with MsnBase.

3 Methods

3.1 Raw Conversion to mgf Using MSconvert

1. Open MSconvert and press the button "Browse" to select the raw files downloaded from PRIDE database. Then press the button "add."

2. Select output directory and output format mgf.

3. Select "Threshold peak filter" and set it to 50 and press the "add" button to add it to the list of applied filters. Currently, the charge state predictor just outputs several possible charge states and is therefore not that useful.

4. Then press the "start" button.

5. Rename the mgf file in the output and repeat **steps 3–4** with 100, 200, and 1000.

The outputted mgf files have the scan number in the title field and some search engines such rTANDEM needs the scan numbers as a separate field to parse it to the output.

3.2 Raw Conversion to mgf Using rawConvert

1. Press the "+" button to add raw files from Thermo Fisher instruments.

2. Choose output directory and set output format mgf.

3. In the menu set "Options → DDA → charge state" to +1, +2, +3, +4 and +5.

4. Press the button "Go".

3.3 mzXML to mgf Using MsnBase

1. Use the function MsnBase::readMSData to read an mzXML file generated by rawConvert.

2. Use the MsnBase::writeMgfData to generate an mgf file.

The above generated mgf files were compared by searching using rTANDEM against UniProt proteome "3AUP000005640". Using "n_proAcetyl", "M_Oxidation", and "NQ_Deamidation" as variable modifications. Fixed modification was set to carboamido-methylation. 5 ppm for parent ion and 0.01 Da for fragment ion accuracy. MaxPeaks were set to 1000. The search output was compared using 1% FDR threshold of identified proteins and peptides (*see* Table 1). We see that MsnBase and MSconvert needs to output a large number of MSMS fragment ions (large mgf files) to obtained similar identification rate as rawConvert (smaller mgf files).

Table 1
Comparison of the number of identified proteins and peptides using a 1% FDR threshold and mgf files generated by different free software

Program and settings	Isoforms	Encoding genes	Peptides
RawConvert	9586	2941	19,753
Msconvert no prediction 50	7632	2346	12,926
Msconvert no prediction 100	9089	2765	18,389
Msconvert no prediction 200	9535	2915	19,991
Msconvert no prediction 1000	9535	2917	19,985
MsnBase	9658	2949	20,497

4   Notes

1. For symmetric filters $q = s$ in Eq. 1. Symmetric filters provides no estimates in the start and end of the spectrum which correspond to the q first data points and the s last data points in the spectrum, respectively. However, asymmetric filters where q or s equals zero can provide estimates in the start and end of spectrum [7].

2. The quality of function fitting is often evaluated by the lack of fit which is given by $E_{lof} = \sum_{it} (x_t - y_t)^2$. The lack of fit is not the only central measure for the quality of a fit. For example, the roughness of spectrum which is given by $R = \sum_t (y_t - y_{t-1})^2$ is also relevant. A best fit is found by minimizing a weighted sum of E_{lof} and R. Alternatively, peak frequency rather than roughness is used to optimize the level of smoothing. E_{lof} together with maximum entropy of residuals, which is given by $S = -\sum_t p_t \log(p_t)$ where p is residuals at different time points, presents another quality measure. Finally, maximum entropy is very useful for choosing between different models that gives the same E_{lof}.

3. In mass spectrometry the mass of such peaks is often determined by calculating the centroid mass which is more accurate than just taking the mass at the peak maximum. The centroid mass m_c and the corresponding intensity I_c is calculated by the following expressions.

$$m_c = \frac{\sum_{y_i > y_{i,max} x} m_i I_i}{I_c}$$

$$I_c = \sum_{y_i > y_{i,\max} x} I_i$$

where m_i is the mass at a certain mass bin and I_i is the corresponding intensity. x is a specified percentage of the maximum intensity.

4. Peaks obtained from mass spectra or calculated plots based on mass spectra such as XIC can have several different shapes which require different mathematical functions for fitting. The fitting is typically performed using Levenberg-Marquardt method [38]. The source code for Levenberg-Marquardt is available in the free book "Numerical recipes in C" [5]. Alternatively, the function "nls" in R facilitates such optimization problems. Lorentzian (*see* Fig. 8a), Gaussian (*see* Fig. 8a) or a mixture of the two functions often provides accurate peak approximation.

    The equation for the Gaussian function can be formulated as $f(m_i) = A*\exp\text{-}(m_i - m_0)^2/s^2$. Where m_0 is at the center, A is the maximum height at x_0 and s defines the peak width. The width at half-height of a Gaussian peak is given by $s(4 * \ln 2)^{1/2}$, and the area is $As(\pi)^{1/2}$. The equation for a Lorentzian function is given by $f(x_i) = A/(1 + (m_i - m_0)^2/s^2)$ where m_0 is at the midpoint of the peak, and A is the height at the midpoint. The width at half-height of a Lorentzian peak is given by $2s$ and the area is $As\pi$ [39].

5. R functions such as enviPat::isopattern (*see* Fig. 9a) or IsoSpecR::IsoSpecify (*see* Fig. 9b) allows precise estimates of

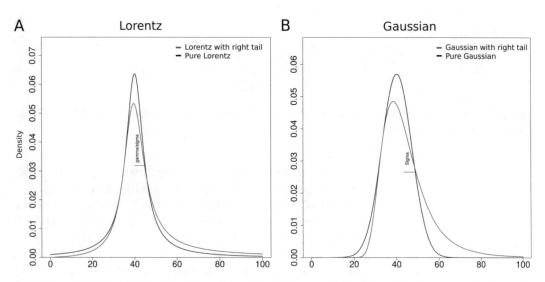

Fig. 8 The peak shape defined by a Lorentzian (**a**) or Gaussian (**b**) equation. The blue curves indicate the Lorentzian or Gaussian with tailing which is obtained by multiplying an increasing function to gamma or sigma, respectively

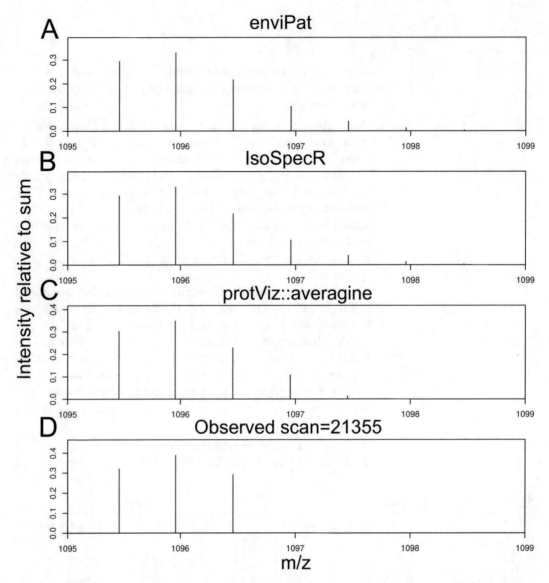

Fig. 9 Theoretical (**a–c**) and observed (**d**) isotopic distribution for the peptide "NMGGPYGGGNYGPGGSGGSG-GYGGR" with charge state +2

isotopic peptide distribution. Approximate isotopic distributions are calculated based on theoretical isotopic distribution of standard tryptic peptides. By having the intensity of the monoisotopic peak the intensity of the following isotopic peaks are approximated by a linear equation [40]. The ratio R between the intensity of the "monoisotopic peak" and the "monoisotopic plus the mass difference between ^{13}C and ^{12}C divided by z" is approximated by $R = 0.0005412 * m - 0.01033$ where m is the mass of the monoisotopic peak. The averaging model [41] is frequently used in deisotoping algorithms and is available in the R package function protViz::averagine (*see* Fig. 9c). It is reassuring that the theoretical isotopic distribution from the different methods are reproducible and provides are reasonable match the observed isotope distribution in Fig. 9d.

For initial deisotoping we do not yet know the corresponding peptide sequence therefore the averaging model are often used to approximate isotopic distributions of all possible combinations of two overlapping isotopic distributions. The combination used is the one which give the best fit on the neighboring peaks. Linear algebra [42] instead of iterating through all reasonable possibilities presents another possibility for deisotoping peak lists.

Acknowledgments

R.M. is supported by Fundação para a Ciência e a Tecnologia (CEEC position, 2019–2025 investigator), iNOVA4Health— UID/Multi/04462/2013, a program financially supported by Fundação para a Ciência e Tecnologia/Ministério da Educação e Ciência, through national funds and is cofunded by FEDER under the PT2020 Partnership Agreement. This work is also funded by FEDER funds through the COMPETE 2020 Programme and National Funds through FCT - Portuguese Foundation for Science and Technology under the projects number PTDC/BTM-TEC/ 30087/2017 and PTDC/BTM-TEC/30088/2017.

References

1. Washburn MP, Wolters D, Yates JR 3rd (2001) Large-scale analysis of the yeast proteome by multidimensional protein identification technology. Nat Biotechnol 19(3):242–247. https://doi.org/10.1038/85686. 85686

2. Fenn JB, Mann M, Meng CK, Wong SF, Whitehouse CM (1989) Electrospray ionization for mass spectrometry of large biomolecules. Science 246(4926):64–71

3. Kieser R, Reynisson P, Mulligan TJ (2005) Definition of signal-to-noise ratio and its critical role in split-beam measurements. ICES J Mar Sci 62(1):123–130

4. Fredriksson M, Petersson P, Jornten-Karlsson-M, Axelsson BO, Bylund D (2007) An objective comparison of pre-processing methods for enhancement of liquid chromatography-mass spectrometry data. J Chromatogr A 1172 (2):135–150. https://doi.org/10.1016/j. chroma.2007.09.077. S0021-9673(07) 01710-4

5. Press WH, Teukolsky SA, Vetterling WT, Flannery BP (1988–1992) Numerical recipes in C: the art of scientific computing. Cambridge University Press, Cambridge

6. Savitzky A, Golay JEM (1964) Smoothing and differentiation of data by simplified least squares procedures. Anal Chem 36:1627–1639

7. Chatfield C (1989) The analysis of time series, an introduction. Chapman & Hall/CRC

8. Eilers PH (2003) A perfect smoother. Anal Chem 75(14):3631–3636

9. Kast J, Gentzel M, Wilm M, Richardson K (2003) Noise filtering techniques for electrospray quadrupole time of flight mass spectra. J Am Soc Mass Spectrom 14(7):766–776. https://doi.org/10.1016/S1044-0305(03) 00264-2. S1044030503002642

10. Morris JS, Coombes KR, Koomen J, Baggerly KA, Kobayashi R (2005) Feature extraction and quantification for mass spectrometry in biomedical applications using the mean

spectrum. Bioinformatics 21(9):1764–1775. https://doi.org/10.1093/bioinformatics/bti254. bti254

11. Carvalho AS, Ribeiro H, Voabil P, Penque D, Jensen ON, Molina H, Matthiesen R (2014) Global mass spectrometry and transcriptomics array based drug profiling provides novel insight into glucosamine induced endoplasmic reticulum stress. Mol Cell Proteomics 13 (12):3294–3307. https://doi.org/10.1074/mcp.M113.034363

12. Vivo-Truyols G, Schoenmakers PJ (2006) Automatic selection of optimal Savitzky-Golay smoothing. Anal Chem 78(13):4598–4608. https://doi.org/10.1021/ac0600196

13. Chernushevich IV, Loboda AV, Thomson BA (2001) An introduction to quadrupole-time-of-flight mass spectrometry. J Mass Spectrom 36(8):849–865. https://doi.org/10.1002/jms.207

14. Bylund D (2001) Chemometrics tools for enhanced performance in liquid chromatography-mass spectrometry. Uppsala University, Sweden, Uppsala

15. Cox J, Mann M (2008) MaxQuant enables high peptide identification rates, individualized p.p.b.-range mass accuracies and proteome-wide protein quantification. Nat Biotechnol 26(12):1367–1372. https://doi.org/10.1038/nbt.1511. nbt.1511

16. Sheppard WF (1898) On the geometrical treatment of the 'normal curve' of statistics, with especial reference to correlation and to the theory of error. Proc Roy Soc 62:170–173

17. Schlosser A, Volkmer-Engert R (2003) Volatile polydimethylcyclosiloxanes in the ambient laboratory air identified as source of extreme background signals in nanoelectrospray mass spectrometry. J Mass Spectrom 38(5):523–525. https://doi.org/10.1002/jms.465

18. Olsen JV, de Godoy LM, Li G, Macek B, Mortensen P, Pesch R, Makarov A, Lange O, Horning S, Mann M (2005) Parts per million mass accuracy on an Orbitrap mass spectrometer via lock mass injection into a C-trap. Mol Cell Proteomics 4(12):2010–2021. https://doi.org/10.1074/mcp.T500030-MCP200. T500030-MCP200

19. Matthiesen R, Trelle MB, Hojrup P, Bunkenborg J, Jensen ON (2005) VEMS 3.0: algorithms and computational tools for tandem mass spectrometry based identification of post-translational modifications in proteins. J Proteome Res 4(6):2338–2347. https://doi.org/10.1021/pr050264q

20. Cox J, Mann M (2009) Computational principles of determining and improving mass

precision and accuracy for proteome measurements in an Orbitrap. J Am Soc Mass Spectrom 20(8):1477–1485. https://doi.org/10.1016/j.jasms.2009.05.007. S1044-0305(09)00378-X

21. Zubarev R, Mann M (2007) On the proper use of mass accuracy in proteomics. Mol Cell Proteomics 6(3):377–381. https://doi.org/10.1074/mcp.M600380-MCP200

22. Wehofsky M, Hoffmann R (2002) Automated deconvolution and deisotoping of electrospray mass spectra. J Mass Spectrom 37(2):223–229. https://doi.org/10.1002/jms.278

23. Zhang Z, Marshall AG (1998) A universal algorithm for fast and automated charge state deconvolution of electrospray mass-to-charge ratio spectra. J Am Soc Mass Spectrom 9 (3):225–233. https://doi.org/10.1016/S1044-0305(97)00284-5. S1044-0305(97)00284-5

24. Senko MW, Beu SC, McLafferty FW (1995) Automated assignment of charge states from resolved isotopic peaks for multiply charged ions. J Am Soc Mass Spectrom 6:52–56

25. Kaur P, O'Connor PB (2006) Algorithms for automatic interpretation of high resolution mass spectra. J Am Soc Mass Spectrom 17 (3):459–468. https://doi.org/10.1016/j.jasms.2005.11.024. S1044-0305(05)00984-0

26. Senko MW, Beru SC, McLafferty FW (1995) Determination of monoisotopic masses and ion populations for large biomolecules from resolved isotopic distributions. J Am Soc Mass Spectrom 6:229–233

27. Matthiesen R (2013) LC-MS spectra processing. Methods Mol Biol 1007:47–63. https://doi.org/10.1007/978-1-62703-392-3_2

28. Rodriguez-Suarez E, Gubb E, Alzueta IF, Falcon-Perez JM, Amorim A, Elortza F, Matthiesen R (2010) Virtual expert mass spectrometrist: iTRAQ tool for database-dependent search, quantitation and result storage. Proteomics 10(8):1545–1556. https://doi.org/10.1002/pmic.200900255

29. Vizcaino JA, Csordas A, Del-Toro N, Dianes JA, Griss J, Lavidas I, Mayer G, Perez-Riverol Y, Reisinger F, Ternent T, Xu QW, Wang R, Hermjakob H (2016) 2016 update of the PRIDE database and its related tools. Nucleic Acids Res 44(22):11033. https://doi.org/10.1093/nar/gkw880

30. Hermjakob H, Apweiler R (2006) The proteomics identifications database (PRIDE) and the ProteomExchange consortium: making proteomics data accessible. Expert Rev Proteomics 3 (1):1–3. https://doi.org/10.1586/14789450.3.1.1

31. Adusumilli R, Mallick P (2017) Data conversion with ProteoWizard msConvert. Methods Mol Biol 1550:339–368. https://doi.org/10.1007/978-1-4939-6747-6_23

32. French WR, Zimmerman LJ, Schilling B, Gibson BW, Miller CA, Townsend RR, Sherrod SD, Goodwin CR, McLean JA, Tabb DL (2015) Wavelet-based peak detection and a new charge inference procedure for MS/MS implemented in ProteoWizard's msConvert. J Proteome Res 14(2):1299–1307. https://doi.org/10.1021/pr500886y

33. He L, Diedrich J, Chu YY, Yates JR 3rd (2015) Extracting accurate precursor information for tandem mass spectra by RawConverter. Anal Chem 87(22):11361–11367. https://doi.org/10.1021/acs.analchem.5b02721

34. Gatto L, Lilley KS (2012) MSnbase-an R/Bioconductor package for isobaric tagged mass spectrometry data visualization, processing and quantitation. Bioinformatics 28(2):288–289. https://doi.org/10.1093/bioinformatics/btr645

35. Loos M, Gerber C, Corona F, Hollender J, Singer H (2015) Accelerated isotope fine structure calculation using pruned transition trees. Anal Chem 87(11):5738–5744

36. Panse C, Grossmann J (2012) protViz: visualizing and analyzing mass spectrometry related data in proteomics. R package

37. Startek MKŁaM (2017) IsoSpecR: the IsoSpec algorithm. R package version 103

38. Levnberg K (1944) A method for the solution of certain non-linear problems in least squares. Q Appl Math 2:164–168

39. Brereton RG (2003) Data analysis for the laboratory and chemical plant. Wiley, Chichester

40. Wehofsky M, Hoffmann R, Hubert M, Spengler B (2001) Isotopic deconvolution of matrix-assisted laser desorption/ionization mass spectra for substances-class specific analysis of complex samples. Eur J Mass Spectrom 7:39–46

41. Horn DM, Zubarev RA, McLafferty FW (2000) Automated reduction and interpretation of high resolution electrospray mass spectra of large molecules. J Am Soc Mass Spectrom 11(4):320–332

42. Meija J, Caruso JA (2004) Deconvolution of isobaric interferences in mass spectra. J Am Soc Mass Spectrom 15(5):654–658. https://doi.org/10.1016/j.jasms.2003.12.016. S1044030504000169

Chapter 3

Isotopic Distributions

Alan L. Rockwood and Magnus Palmblad

Abstract

Isotopic information determined by mass spectrometry can be used in a wide variety of applications. Broadly speaking these could be classified as "passive" applications, meaning that they use naturally occurring isotopic information, and "active" applications, meaning that the isotopic distributions are manipulated in some way. The classic passive application is the determination of chemical composition by comparing observed isotopic patterns of molecules to theoretically calculated isotopic patterns. Active applications include isotope exchange experiments of a variety of types, as well as isotope labeling in tracing studies and to provide references for quantitation. Regardless of the type of application considered, the problem of theoretical calculation of isotopic patterns almost invariably arises. This chapter reviews a number of application examples and computational approaches for isotopic studies in mass spectrometry.

Key words Isotopes, Isotopic distributions, Mass spectrometry

1 Introduction

Isotopes are atoms of the same chemical element having different numbers of neutrons and hence different mass. The presence of isotopes means not all molecules of a chemical compound have the same mass. As a mass spectrometer analyzes molecules individually rather than as a bulk quantity like a laboratory balance, most compounds will show a *distribution* of mass-to-charge ratios rather than a well-defined value. This isotopic distribution in turn depends on the abundances of the isotopes and the elemental composition of the molecule. The abundances can usually be assumed to be close to their average "natural" values. Although isotope abundances are not constant and detailed knowledge about changes or variability in isotope ratios is used in carbon dating [1], geochronology [2], forensics [3], paleoclimatology [4], and other fields, we can usually assume that the isotopic abundances in unknown analytes are well approximated by the natural averages.

Isotopes are either stable or unstable. Unstable isotopes generally decay to stable or unstable isotopes of other elements via

Rune Matthiesen (ed.), *Mass Spectrometry Data Analysis in Proteomics*, Methods in Molecular Biology, vol. 2051, https://doi.org/10.1007/978-1-4939-9744-2_3, © Springer Science+Business Media, LLC, part of Springer Nature 2020

radioactive decay processes. There are many theoretically unstable isotopes for which no radioactive decay has been observed. Among 255 known observationally stable isotopes, only 90 are also theoretically stable [5]. The other 165 are theoretically unstable with respect to decay to an isotope of another element, though their half-lives are so long that no radioactive decay processes have been observed. Some of these are relatively abundant isotopes of biologically important elements such as calcium (^{40}Ca, 97% abundance), iron (^{54}Fe, 6% abundance), and zinc (^{64}Zn, 49% abundance and ^{70}Zn, 0.6% abundance). For mass spectrometry in general, it is the *abundance* that matters, not whether an isotope is stable or not. From here on in this chapter, isotopes in the context of mass spectrometry are understood to mean any isotope, stable or radioactive, with a natural abundance of at least 10^{-6} and hence possibly contributing to observed mass spectra (*see* **Note 1**).

Nearly all mass spectrometers have at least "unit resolution" for some mass-to-charge ratio range (m/z), meaning they have at least the ability to resolve two species differing in mass by 1 Da in that m/z range. As the isotopes of an element differ in mass by ~1 Da (or a small integral multiple of ~1 Da), any combination of isotopes will differ from any other combination of isotopes by either a very small fraction of 1 Da or nearly an integral multiple of 1 Da. Different isotopic compositions of the same compound are called *isotopologs*. For small molecules, these isotopologs differing in mass by 1 Da or more are easily resolved by any modern mass spectrometer. But the mass spectrometer also records information on the relative abundance of these isotopologs in the isotopic distribution. How do we calculate the masses and relative abundance of isotopologs in an isotopic distribution given an elemental composition? How can we make use of the information from measured isotopic envelopes for identifying unknown compounds? Can we use isotopes and isotopic distributions for quantifying or characterizing the structure of molecules? These are the topics that are discussed in this chapter.

2   Theory

The atomic isotopic distribution of an arbitrary element is defined as the abundances of the isotopes of that element. An isotopic distribution could be thought of as a list of identifiers with associated abundances or alternatively as a probability function: isotopic abundance as a function of mass. The isotopic distribution of atomic carbon could be characterized by an identifier for one of the isotopes, ^{12}C, and the abundance associated with that isotope, 0.989, and a second identifier, ^{13}C, and the abundance associated with that isotope, 0.011. Note that mass is not explicitly part of the description, although the integer (nominal) mass is part of the identifier. In the second, or probability function view, one could say that the isotopic distribution of atomic carbon is a mass-dependent abundance function that is everywhere zero except for a peak at centered at (exactly) 12 Da with an abundance or height of 0.9889 together with a second peak at 13.0033554 Da with an abundance of 0.0111. The symbolic peak identifier is not an explicit part of this picture, although there is of course some peak identity associated with each peak. The latter corresponds with experimental measurements since the only thing the mass spectrometer tells us directly is mass (strictly speaking, mass-to-charge ratio m/z) and abundance. However, for theoretical and certain conceptual or algebraic purposes the first way of thinking about isotopic distributions can sometimes be very useful.

One can also think of isotopic abundance as the probability of any one atom, when selected at random, being a particular isotope. The relative abundance of a molecular isotopolog with, say, two heavy isotopes is the probability of exactly two atoms in the same molecule being of that heavy isotope. Calculating the relative abundance of isotopologs is a *combinatorial* problem, analogous to the familiar textbook problems typically involving drawing various numbers of balls of different colors from a bag (Fig. 1). One can think of the balls representing atoms drawn from different bags to construct molecules, with one bag per element containing balls of different colors in relative abundance depending on the isotopic composition of the element. (The balls should also be returned to the bag after drawing each ball, unless the bags contain an infinite number of balls.) This immediately suggests a possible method for estimating isotopic distributions—namely, by simulating such a stochastic process. We will return to this method, known as a Monte Carlo method, later in the chapter.

Before turning our attention to specific methods for calculating isotopic distributions of molecules, let us briefly consider the underlying mathematical problem of isotope distribution calculations. The problem of calculating molecular isotopic distributions can be discussed in terms of the mathematical operation of *convolution* which is normally defined by the following equation:

$$h(t) = f(t) * g(t) = \int_{-\infty}^{\infty} f(t)g(t - \tau)d\tau \tag{1}$$

where $f(t)$ $f(t)$ and $g(t)$ $g(t)$ are the two functions being convoluted and $*$ represents the convolution operator. When discussing isotopic distributions, these functions would naturally be functions of mass, not time or any other physical quantity.

Convolutions are common in mathematics, science, engineering, and statistics. For example, the peak shape of an ideal magnetic sector mass spectrometer is the convolution of the entrance slit

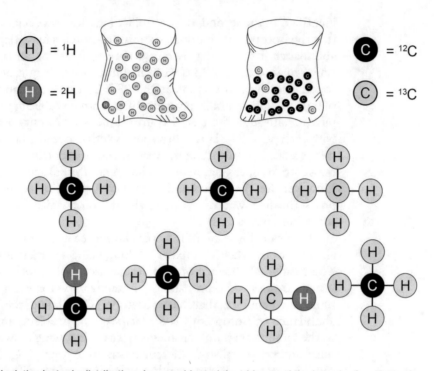

Fig. 1 Calculating isotopic distributions is a combinatorial problem and the branch of mathematics concerned with these types of problems is *combinatorics*. One can think of the task being analogous to building molecules by drawing differently colored balls representing the atoms from different bags, for instance methane (CH_4) by drawing four hydrogens and one carbon (returning each ball after it was drawn). Hydrogen and carbon have two isotopes each: $^1H/^2H$ and $^{12}C/^{13}C$. This gives many possible combinations of isotopes, or isotopologs, three of which are illustrated above. The problem of calculating or estimating isotopic distributions is generally about predicting the frequencies by which these isotopologs occur

(typically a boxcar function) and the exit slit (also a boxcar function). As another example, if two types of objects, A and B, are repeatedly weighed, the result for the weighing of each object can be characterized by a probability distribution function. If one were to weigh object A, then separately (independently) weigh object B, and finally add the two results, then repeat this operation multiple times the sum of the two weighings would be characterized by a probability distribution. That probability distribution would be given by the convolution of the probability distribution for A with the probability distribution for B.

Extending the weighing analogy, suppose there are several subclasses of objects of type A (A_1, A_2, etc.), each having a different mass. Further, suppose that there are also several subclasses of objects of type B (B_1, B_2, etc.), each having a different mass. Taking objects A as an aggregated class, there is a probability distribution that characterizes the mass distribution for A, and likewise there is a probability distribution that characterizes the mass distribution for B. If we randomly select one item of A and one item of B and weigh them together there are a number of outcomes, the probabilities of which are determined by taking all combinations from A and B ($A_1 + B_1$, $A_1 + B_2$, $A_2 + B_1$, $A_2 + B_2$, etc.). The mass distribution that characterizes the combined mass measurement is given by the convolution of the probability distributions of A and B. This is an exact analogy to the problem of calculating the isotopic distribution of a molecule such as CO from the isotopic distributions of the elements of which the molecule is composed.

Applying this to isotope distribution calculations, one would replace the variable t in the above integral with mass. The functions being convoluted are isotopic profiles expressed as probability function, that is, abundance as a function of mass. The isotopic distribution of a molecule of composition E_2 (where E represents an arbitrary generic element) is the convolution of the elemental isotopic distribution of E with itself. The convolution process can be repeated for more complex molecules: The isotopic distribution of E_3 is the convolution of the isotopic distribution of E with the isotopic distribution of E_2, and so forth. If one were to add an atom of element, F, to compound E_3 to form a new compound, E_3F, the resulting isotopic distribution would be the isotopic distribution of element F convoluted with the isotopic distribution of E_3.

This process can be extended to molecules of any complexity, and the convolutions can be done in any order as the convolution operator is both *commutative* and *associative*. For example, the isotopic distribution of hypothetical compound E_eF_f is the isotopic distribution of E_e convoluted with the isotopic distribution of F_f, and the isotopic distribution of $E_eF_fG_g$ is the convolution of the isotopic distribution of E_eF_f with the isotopic distribution of G_g. Since the convolutions can be done in any order the isotopic distribution of $E_eF_fG_g$ is also given by the convolution of the isotopic distribution of $E_eF_fG_{g-1}$ with the isotopic distribution of G.

There are two broad classes of algorithms for calculating isotopic distributions: exact methods and approximate methods. In the context of this chapter an "exact" method is one that would produce an error-free result on a perfect computer, that is, in the absence of round-off errors. As will shortly become evident, exact methods exist that can be performed in a finite number of operations. However, it will also become equally clear that exact calculations quickly become intractable, and we therefore must resort to approximate methods, some of which could be said to use a particular exact algorithm as a conceptual starting point.

Turning first to principally exact methods of calculation, we first consider a simple case, that of a single-element compound where the element contains only two isotopes. The isotopic distribution of a molecule made from single element with two isotopes would be *binomial*, and the convolution calculations can be performed using the *binomial theorem*. The isotopic distribution of a hypothetical molecule consisting of n atoms of element E (having two isotopes), E_n, could thus be written as:

$$\binom{n}{k} p^k (1-p)^{n-k} \qquad (2)$$

where k ($0 \leq k \leq n$) is the number of atoms in the molecule belonging to the isotope with natural abundance p and $\binom{n}{k}$ the *binomial coefficient*.

For example, the isotopic distribution of Cl_2 can be written as

$$\binom{2}{k} 0.2423^k (1-0.2423)^{2-k} \qquad (3)$$

if $k = 0$, 1 or 2 is the number of ^{37}Cl atoms (natural abundance 24.23%) in the Cl_2 molecule. For C_{60}, the isotopic distribution can be written

$$\binom{60}{k} 0.9889^k (1-0.9889)^{n-k} \qquad (4)$$

where 0.9889 is the natural abundance of ^{12}C and k the number of ^{12}C atoms. Note the symmetry of these expressions and remember $\binom{n}{k} = \binom{n}{n-k}$. It does not matter if we let k denote the number of light/common or heavy/rare isotope, as long as the correct isotopic abundance is used.

A word regarding the interpretation of the binomial expansion formula presented: once the expansion is performed one does not collect the terms into a single number, but rather the individual terms represent the probabilities for each isotopic peak. Also, the isotopic composition is not explicitly included in the formula. One must keep track of this implicitly by a strict ordering of the terms in the expansion. Similarly, mass must be accounted for by some bookkeeping scheme as an adjunct to the binomial expansion.

One can generalize this to compounds composed of a single element with more than two isotopes. It is useful at this point to introduce a more generalized notation in order to systematically present more complex expressions than presented above. Let E again be a generic symbol for an element, and let $E(i, j)$ a specific isotope of a specific element, with i representing the atomic number and j representing the nucleon (or mass) number. These are strictly symbolic representations and the various $E(i, j)$ are not assigned numerical values. To illustrate, the isotope represented by ^{13}C in conventional notation is represented here by $E(6,13)$, where 6 is the atomic number of carbon and 13 is the nucleon number.

We use the notation $P(i, j)$ to represent elemental isotopic abundances. Thus, $P(6,12)$ is the abundance of ^{12}C with a value of 0.9889, and $P(6,13)$ is the abundance of ^{13}C with a value of

0.0111. Elemental isotopic abundances sum to unity for a given element:

$$\sum_j P(i, j) = 1 \tag{5}$$

We define another quantity, N_i, to be a composition parameter in a chemical formula, where i refers to the atomic number and N_i is the number of atoms of that element in the compound. For example, oxygen has an atomic number of $i = 8$, and there are three oxygen atoms in the compound O_3, so the composition parameter for O_3 would be $N_i = N_8 = 3$.

The isotopic distribution for a polyatomic single-element compound is implicitly given in compact form by the unexpanded polynomial (or multinomial)

$$\left(\sum_j P(i, j) E(i, j) \right)^{N_i} \tag{6}$$

This expression may be expanded to a sum of individual terms using the "multinomial theorem":

$$\left(\sum_{j=m}^{n} P(i, j) E(i, j) \right)^{N_i} = \sum_{(\sum_{i=m}^{n} k_i) = N_i} \left(\binom{N_i}{k_m, k_{m+1}, \ldots, k_n} \prod_{1 \leq t \leq n} P(i, t)^{k_t} E(i, t)^{k_t} \right) \tag{7}$$

where $\binom{N_i}{k_m, k_{m+1}, \ldots, k_n} = \frac{N_i!}{k_m! k_{m+1}! \cdots k_n!}$. Expanding this polynomial and calculating the coefficients for each term (isotopolog) will yield the isotopic distribution.

The rather formidable-looking polynomial expansion contains a complex structure of sums and products controlled by a set of multiple interlocking subscripts. Its meaning can be made more intuitive by illustrated by examining a single term from a specific example and discussing how each part found in the term relates to the chemical formula and the isotopes of the element in the formula. For this example, consider the ozone isotopolog $^{16}O^{17}O_2$, which is one of several isobaric components of the nominal mass 50 of O_3. The term in the polynomial expansion corresponding to $^{16}O^{17}O_2$ (which, for consistency with the combinatorial expression we can express as $^{16}O^{17}O_2\ ^{18}O_0$) is given by

$$\binom{3}{1,2,0} 0.99759^1 0.000374^2 0.002036^0 E(8, 16)^1 E(8, 17)^2 E(8, 18)^0 \tag{8}$$

where $\binom{3}{1,2,0} = \frac{3!}{1!2!0!} = 3$. This expression simplifies to $4.19 \times 10^{-7} E(8, 16)^1 E(8, 17)^2$ which means that the fractional composition of $^{16}O^{17}O_2$ is 4.19×10^{-7}.

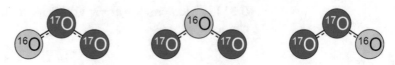

Fig. 2 Three isotopomers of $^{16}O^{17}O_2$. Chemically speaking, two of these are indistinguishable, but all three need to be considered in this exercise

The products of the abundances of ^{16}O and two ^{17}O is the probability of observing a one ^{16}O, two ^{17}O, and zero ^{18}O atoms *in a particular order*. However, there are three different *isotopomers* of $^{16}O^{17}O_2$ (*see* Fig. 2) which cannot be resolved by mass spectrometry (MS[1]):

The multinomial coefficient $\begin{pmatrix} 3 \\ 1,2,0 \end{pmatrix} = 3$ accounts for the number of ways of arranging the particular order of one ^{16}O, two ^{17}O, and zero ^{18}O atoms in the molecule. The relative abundance or probability of the $^{16}O^{17}O_2$ isotopolog is therefore the probability of observing a one ^{16}O, two ^{17}O, and zero ^{18}O atoms in a specific order multiplied with the number of isotopomers with one ^{16}O, two ^{17}O atoms and zero ^{18}O atoms. This is the number of permutations of three atoms ($3! = 6$) divided by the product of the number of permutations of 1 ^{16}O atom ($1! = 1$), the number or permutations of two ^{17}O atoms ($2! = 2$), and the number of permutations of zero ^{18}O atoms ($0! = 1$). This is the origin of the coefficient 3 in the equations above. Remember that this isotope notation, here $E(8,16)$ and $E(8,17)$ for ^{16}O and ^{17}O, only serve as (unique) symbols of the two isotopes in the polynomial expansion and do not take on any numerical value.

Mass can be included in this overall scheme by introducing variables of the form $M(i,j)$, representing the mass of an element specified by i, with the isotope specified by j. Thus, $M(8,16)$ and $M(8,17)$ represent the masses of ^{16}O and ^{17}O, respectively.

With infinite mass spectrometer resolving power, the different isotopologs of the same nominal mass would always be resolved. However, with many real instruments, this *isotopic fine structure* is difficult or impossible to resolve in practice, and $^{16}O^{17}O_2$ may not be separated from $^{16}O_2{}^{18}O$ at nominal mass 50. The relative abundance of the nominal isotopic peak at 50 Da is therefore the sum of the abundances of the two isotopologs with a nominal mass of 50, namely, $^{16}O^{17}O_2$ and $^{16}O_2{}^{18}O$:

$$\begin{pmatrix} 3 \\ 1,2,0 \end{pmatrix} 0.99759^1 0.000374^2 0.002036^0 E(8,16)^1 E(8,17)^2 E(8,18)^0 +$$

$$\begin{pmatrix} 3 \\ 2,0,1 \end{pmatrix} 0.99759^2 0.000374^0 0.002036^1 E(8,16)^2 E(8,17)^0 E(8,18)^1$$

$$(9)$$

This approach can be generalized to compounds containing more than one element. For elements with more than two contributing isotopes, the distribution can be written in a compact form as

$$\prod_i \left(\left(\sum_j P(i,j)E(i,j) \right)^{N_i} \right) \tag{10}$$

We may refer to this expression as a "compound polynomial" or, equivalently, "compound multinomial" to emphasize that it is a product of single-element polynomial expressions. The product is taken over all elements in the molecular formula, where the index i refers to the chemical element.

Expanding this expression to a sum of individual terms would result in a very complex expression, which is left as an exercise for the ambitious reader.

Each term in the multielement polynomial expansion contains factors taken from the single-element expansions. Each single-element expansion contributes one and only one factor to a given term in the multielement expansion. For example, if we are considering a three-element expansion then one of its terms will be the first term in the single-element expansion of element 1 multiplied by the first term in the single element expansion of element 2 multiplied by the first term in the single element expansion of element 3. Another term in the three-element expansion will be the first term in the single-element expansion of element 1 multiplied by the first term in the single element expansion of element 2 multiplied by the second term in the single element expansion of element 3. These combinations are taken over all possibilities with each element represented in the multielement expansion exactly as many times as there are atoms of that element in the molecular formula.

The coefficients in the multinomial expansion provide the abundance for the corresponding fine structure peak. The mass, or position, of the peak is simply the sum of the masses of the isotopes which compose the peak. For peaks corresponding to several unresolved species, the mass is the composition-weighted average of the isotopologs contributing to that peak. For example, CO_2 at nominal mass 46 consists of $E(6,12)^1E(8,17)^2$, $E(6,13)^1E(8,16)^1E(8,17)^1$, and $E(6,12)^1E(8,16)^1E(8,18)^1$ with exact masses 45.99826, 45.99740, and 45.99407 Da, and relative abundances 1.4×10^{-7}, 8.42×10^{-6}, and 3.95×10^{-3}, respectively. The weighted average of the masses of CO_2 at nominal mass 46 is therefore 45.99408 (*see* **Note 2**).

At least two reflections can be made from the CO_2 example: (1) one isotopolog may dominate the contributions to the intensity of an isotopic peak and (2) the averaged mass of the peak may therefore be very close to the exact mass of the dominant isotopo-log (here $^{12}C^{16}O^{18}O$). This means the relative abundance and positions of isotopic peaks can sometimes be approximated by considering only the major contributing isotopologs, and that small terms may be ignored. This will become a necessity if this polynomial expansion method will be used to calculate isotopic distributions of larger molecules with many atoms, as we shall see later.

In principle it is possible to use all terms in the multinomial expansion to perform an exact calculation of an isotopic distribu-tion. This is known as the *polynomial expansion method* for the calculation of isotopic distributions, and it provides a mathemati-cally exact and rigorous method of calculating isotopic distributions in a finite number of operations. The result of such a theoretical calculation is an infinite-resolution list of all isotopic fine structure peaks, with information on relative abundance as well as exact mass of every theoretical peak in a mass spectrum.

The full peak list thus contains all the information about an isotopic distribution and we only need to keep track of all the terms in the expansion or convolution, which is simply a problem of counting, and calculating the corresponding coefficients. So, is the problem solved? Not quite. Although it is easy to write down a mathematical expression for an isotopic distribution in the com-pact form (as above), experimental scientists are usually more inter-ested in calculating the distributions numerically. Calculating isotopic distributions using any polynomial expansion method is trivial for the relatively simple compounds used in the examples above, but staggeringly difficult for even relatively small proteins like bovine insulin ($H_{378}C_{234}N_{65}O_{75}S_6$, molecular weight 5494.3), using the naïve approach of expanding the polynomial expression and adding up all coefficients for all isotopologs of the same nominal mass.

The reason for this is the enormous number of terms that then would be included, a so-called *combinatorial explosion*. To see why this is the case, consider the number of terms for a single-element compound given by the number of terms in the multinomial expansion

$$\#\text{terms} = \binom{n+m-1}{n} = \frac{(m+n-1)!}{n!(m-1)!} \tag{11}$$

where m is the number of isotopes of the element and n the number of atoms in the molecule or cluster. Applying (10) to a fictitious molecule O_{75} we have $m = 3$ and $n = 72$. The expression for O_{75} therefore has $\binom{75+3-1}{75} = \frac{(3+75-1)!}{75!(3-1)!} = 2926$ terms.

While the number of terms in this expression may still be manageable, the number of terms in a multielement expansion is the product of the number of terms in the single-element expansions for the individual elements. For example, bovine insulin has 378 hydrogen, 234 carbon, 65 nitrogen, 75 oxygen, and 6 sulfur atoms. The single-element polynomial expansions for these single elements have 379, 235, 66, 2926, and 84 terms, respectively. Multiplying these numbers together to calculate the number of terms for the multielement polynomial expansion we get 1.44×10^{12} unique terms. It is not practical to calculate or store such large number of terms on computers that are easily accessible to most researchers. If we were to store each term as a mass and abundance pair, each represented by 64 bits, the memory required to store all terms for bovine insulin would be 1.85×10^{14} bits or 2.31×10^{13} 8-bit bytes. This is 23 terabytes! If one would like to retain the isotopic composition as well as the mass/abundance information for each fine structure peak the memory requirement, would be even greater.

To put these numbers into perspective, the computing resources available to most laboratory scientists are limited to desktop computers with far less memory than the 2.31×10^{13} bytes required to hold all mass/abundance information for bovine insulin. Most desktop computers do not even have enough hard disk space to hold the complete results of a single calculation, let alone a series of calculations. Furthermore, the problem gets rapidly worse as molecular weight increases. For example, a full calculation of human holomyoglobin ($C_{803}H_{1247}N_{213}O_{225}S_4Fe$) would require calculation and storage of 7.71×10^{14} terms, which would require 1.24×10^{16} bytes. It would not be possible to store this amount of data in memory of even the most powerful supercomputers currently available, having ~1 petabyte (10^{15} bytes) of total memory. Furthermore, the CPU time required for such a large calculation would be prodigious. Clearly, the naïve approach of brute force multinomial expansion keeping all terms is not practical for large organic or biological molecules. Also, for inorganic clusters of elements with multiple isotopes, this calculation quickly becomes prohibitively expensive. For example, tin has ten isotopes, mercury seven isotopes, and lead four isotopes. A hypothetical $(SnHgPb)_{11}$ cluster with a molecular weight of ~5792 Da, similar to that of insulin, would have a multinomial expansion with 7.57×10^{11} terms. In the following section we will discuss solutions to this problem.

3   Practical Calculation of Isotopic Distributions

The *practical* problem of calculating isotopic distributions has been addressed in a number of ways. Since the detection of isotopic distributions in a mass spectrometer is a stochastic process, one might consider mimicking the natural detection processes by simulating an isotopic distribution using a Monte Carlo method. The Monte Carlo method is fairly straightforward to implement on a computer using pseudo-random number generator functions to simulate the random drawing of isotopes from a pool.

The Monte Carlo method has a number of attractive features. Most importantly, it breaks the tyranny of the "combinatorial explosion" discussed above. Briefly, the combinatorial explosion refers to the fact that in many cases the computational effort to calculate an exact result rapidly becomes intractable as the isotopic complexity of the problem increases. The Monte Carlo method typically requires only a modest computational effort to calculate a useful, though not exact result. It is also relatively straightforward to include isotopic fine structure into the calculation. Furthermore, it is possible to correlate the number of samples in the Monte Carlo method with the number of ions detected in an experiment. For example, if 10^6 ions were detected in an experimental measurement of an isotopic distribution, and if the number of sampling points in the Monte Carlo method were several times the experimental number of ions, then the disagreement between the simulated distribution and the true distribution will be smaller than the error in the experimental distribution relative to the true distribution. Another useful feature of Monte Carlo methods is that if well-implemented they can produce an unbiased estimate of the true isotopic distribution. This contrasts with other approaches which tend to suppress certain regions of the isotopic profile.

However, although the computational effort to calculate a useful level of accuracy using the Monte Carlo method is relatively modest, it only converges to the true solution as a function of the square root of the number of trials. For example, to decrease the average error by a factor of two requires an increase in the number of samples by a factor of four. Nevertheless, Monte Carlo methods may be generally useful and are probably underutilized given that they implicitly model the sampling, or counting, errors inherent in measuring real isotopic distributions through detection of a finite number of ions. A Monte Carlo method for calculating isotopic distributions was included in early versions of Senko's IsoPro computer program (Michael Senko, personal communication), though later versions of the program (http://sites.google.com/site/isoproms/home) have omitted that feature.

Turning to more deterministic methods, for relatively small molecules, nothing prohibits straightforward implementations of the polynomial method as described above. This is also simple to program. However, as the polynomial expansion method becomes impractical for large molecules, we need another method for practical calculation of isotopic distributions. Consequently, the problem of calculating isotopic distributions reduces to finding a method that produces an approximate solution with an acceptable degree of accuracy with an acceptable computational effort. It should be mentioned at the outset that what is sufficiently accurate depends on the research question, instrumentation used, whether we care more about the relative intensities or the positions (accurate masses) of the isotopic peaks, and to what degree we can resolve any isotopic fine structure. A number of strategies have been developed over the last decades to achieve these somewhat conflicting goals.

One class of solutions is constituted by modified versions of the polynomial expansion method that avoid the problems discussed in the previous section. These methods generally seek computational efficiency by "pruning" the calculation, that is, including only terms that will have a major contribution to the isotopic abundances of nominal (with unresolved fine structure) isotopic peaks. The methods differ primarily in their approach to selecting which terms to include in the calculation and which terms to omit [6, 7]. Most of these methods have the advantage that they are capable of including isotopic fine structure in the final result, and the isotopic fine structure is essentially of infinite resolution.

The disadvantages of these pruning methods are that not all the isotopic abundance is included in the result and that minor fine structure peaks simply will not appear in the predicted isotopic distribution, though the major, observable, fine structure peaks should be included. This distortion results from the fact that pruning methods generally produce a biased estimate of the isotopic distribution, that is, a bias in favor of isotopic fine structure peaks of high abundance and against those of low abundance. This bias is not necessarily evenly distributed, either within a nominal isotopic peak or between nominal isotopic peaks. Generally speaking the higher-mass nominal isotopic peaks tend to contain more isotopic fine structure peaks, and most of the fine structure peaks are of lower absolute abundance compared to the fine structure peaks of the lower-mass nominal isotopic peaks. Consequently, pruning tends to affect the higher-mass nominal isotopic peaks more than the lower-mass nominal isotopic peaks, thus biasing the simulation against the higher-mass nominal isotopic peaks. The effect can even result in the loss of some of the nominal isotopic peaks, particularly those at higher mass, and in some cases the distortions can be quite noticeable [8].

The isotopic distributions calculated by the polynomial-based methods are not typically directly comparable to experimental data, so additional processing must be applied if one is to compare the results to experimentally measured mass spectra. The type of processing required will depend on the type of data produced by the mass spectrometer. If the mass spectrum is of the form of centroided, or "stick," spectra then the fine structure peaks must be grouped together or binned so that peaks of the same nucleon number can be added to a single "stick." If the mass spectrometer produces profile-mode data then the peak list must be convoluted with a realistic peak profile matching the peak shape and resolution of the experimentally acquired data.

In summary, as the number of terms in the polynomial method quickly become too numerous to be stored in memory, all terms below a certain threshold are pruned. This means we can only approximate the frequencies (intensities) in the distribution. On the other hand, we are considering each combination of isotopes separately, so the masses of the peaks in the distribution are calculated with arbitrary precision, which may have some utility for small molecules and very high resolving power and mass measurement accuracy.

Another class of methods surrenders the calculation of isotopic fine structure, focusing instead on treating the isotopic distribution as if there were no isotopic fine structure. Most of these methods produce very accurate peak abundances. However, they differ in their mass accuracies and in their computational speeds. One of the early methods was introduced by Kubinyi [9], who broke down the task into four steps: (1) in the first phase of the calculation, treat the elemental isotopic masses as if the isotopes are all integer masses, (2) perform the sequence of direct convolutions, (3) select the sequence of direct convolutions in a computationally efficient order, and (4) apply a mass correction at the end of the calculation.

The primary goal in Kubinyi's method was computational efficiency, and the method is sufficiently fast to make it practically useful. In the literature, this class of algorithms is sometimes referred to as "ultrahigh speed" methods. In the context of this chapter, we understand "ultrahigh speed" to mean that the algorithm is among the fastest known and (perhaps more importantly) that the computational effort for the methods has favorable scaling properties as the mass of the molecule increases. Kubinyi achieved computational efficiency primarily by optimizing step (3) in the above list, taking advantage of the fact that the convolutions for a single element can be ordered as a binary sequence. To illustrate a binary sequence using a generic element E, the peak abundances of E_2 can be calculated by direct convolution of E with itself. E_4 can be calculated by a direct convolution of E_2 with itself. E_8 can be calculated by direct convolution of E_4 with itself, and so forth. E_n can be calculated by direct convolution of selected distributions from the binary sequence, using the binary representation of the composition parameter, n, to select which distributions from the binary sequence are convoluted to produce the result for E_n. Building an isotopic distribution of a multielement compound is slightly more complex, and there are several efficient schemes for doing this that are compatible with the binary sequence concept introduced by Kubinyi, the details of which need not concern us here.

Because the first phase of the Kubinyi method assumes that isotope masses are integers, keeping track of the masses in the calculation is a trivial task. Three consequences of the assumption of integer masses are that (1) the spectra are stick spectra, (2) there is no isotopic fine structure, and (3) the nominal isotopic peaks are centered at integer masses during the first phase of the calculation rather than the correct noninteger accurate masses. In the last phase of the calculation the masses are corrected to provide a better approximation of the accurate masses.

Kubinyi's strategy for the mass correction was to add a constant offset to all masses in the distribution, selected to give the mono-isotopic mass the correct value. The resulting distribution has peaks at noninteger masses, though mass differences between the peaks are integer valued. The integer mass spacing inherent in this scheme introduces mass errors into the accurate masses of most isotope peaks. For example, using this scheme the first four peaks of C_{60} have mass errors of 0, −4.6, −9.3, and −13.8 ppm, respectively, and the first five peaks of bovine insulin have errors of 0, −0.5, −0.9, −1.3, and 1.6 ppm error, respectively. In summary, Kubinyi's algorithm is extremely fast, generates stick spectra with very accurate abundance ratios, but sometimes predicts noninteger masses with errors that are incompatible with requirements for accurate-mass work.

In 1996, Rockwood and Van Orden [10] introduced a different ultrahigh speed method. This method also assumes integer masses during the first part of the calculation. However, it differs in the method of performing the convolutions and in correcting integer masses to noninteger masses. The convolutions were performed using fast Fourier transform techniques and took advantage of the integer-mass assumption to use small arrays for the calculation. Consequently, the method, like the Kubinyi method, can be classed as an ultrahigh speed method of generating stick spectra.

The correction to noninteger mass used by Rockwood's method is generally much more accurate than Kubinyi's method. Rockwood's method transforms the integer masses to noninteger masses using two parameters, one being a mass offset and the other being a mass scaling factor. These two parameters are chosen to give the isotope distribution the correct mean mass and standard deviation. As in the Kubinyi method the peaks are still equally spaced, but the spacing between the peaks is, on average, more accurate than those from Kubinyi's method, and the peak placement is also, generally speaking, more accurate. For example, rather than errors of 0, −4.6, −9.3, and −13.8 ppm for the first four isotope peaks of C_{60} as in Kubinyi's method, this method produces 0.00 ppm error for the same peaks, and instead of 0, −0.5, −0.9, −1.3, and 1.6 ppm for the first five peaks of bovine insulin as in Kubinyi's method this method produces error of 0.35, 0.14, 0.03, −0.01, and −0.02 ppm. Like Kubinyi's method, this method is an ultrahigh speed method that produces stick spectra with no isotopic fine structure. Though its methods of calculation differ, the main functional difference compared to Kubinyi's method is a more accurate placement of masses. However, Kubinyi's method could easily be modified to use the same mass correction strategy, and in that case it both methods would have the same mass errors.

There are at least three methods for calculating semiaccurate stick spectra with one stick per mass unit. (This is distinct from the problem of calculating stick spectra that includes isotopic fine structure.) In one method one calculates the isotopic composition of each nominal isotopic peak and then uses that information, together with the accurate masses of the elemental isotopes to calculate the positions of the molecular nominal isotopic peaks [11]. A second method by Roussis and Proulx [12] uses direct convolutions that omit the isotopic fine structure but retains the accurate masses of the nominal isotopic peaks throughout the calculation. In their method, Roussis and Proulx defined and used "superatoms" during intermediate calculations. Superatoms can be thought of as isotopic patterns that resemble elemental isotopic patterns in the sense that there is an accurate mass associated with each nominal isotopic peak. Superatoms differ from atoms in the sense that they represent multiatom clusters rather than single atoms. A third method [13] combines concepts from Roussis and Proulx' method (such as maintaining accurate masses in a series of convolutions) with concepts from Kubinyi [9] (such as arranging the calculation into a binary series) to develop a very fast algorithm that produces stick spectra whose nominal isotopic peaks have both accurate masses and accurate abundances. In cases where one is interested only in abundances and accurate masses of nominal isotopic peaks the algorithm described in reference [13] and implemented in the program emass probably represents the current state of the art for accuracy and speed. However, if one needs to include information on peak profiles and/or isotopic fine structure in the calculation then other algorithms need to be considered.

It is also important to keep in mind that a line spectrum with one "stick" per nominal mass is not a good representation of high-resolution spectra showing isotopic fine structure. The position of the major peak, corresponding to the major contributing isotopolog at that nominal mass, will depend on the resolution if the latter is sufficient to partially but not completely resolve isotopic fine structure.

One frequently implemented class of methods for calculating isotopic distribution relies on the convolution theorem [14] from Fourier analysis. This allows us to use the fast Fourier Transform (FFT) algorithm [15] and its relatives to convolute the isotopic distributions of the elements. Conceptually, this is done by transforming the isotopic distribution of the elements, multiplying these in the Fourier domain (i.e., the inverse mass domain or "μ-domain" as we call it here) and inverse transforming the product to give the convolution of the isotopic distribution of the elements, that is, the isotopic distribution of the molecule [10, 16, 17].

The *convolution theorem* in Fourier analysis relates convolution in the normal, or mass domain to multiplication in the Fourier or mass frequency domain or μ-domain. The convolution $f * g$ of two functions f and g can be calculated by using the pair of formulas:

$$F[f * g] = F[f]F[g] \tag{12}$$

and

$$F^{-1}[F[f * g]] = F^{-1}[F[f]F[g]] \tag{13}$$

where F is the Fourier transform and F^{-1} the inverse Fourier transform.

In descriptive terms this theorem tells us that we can convolute two functions if we perform a Fourier transform on each of the input functions, then multiply the resulting functions, and finally transform the result back to the original domain. This theorem is generalizable to any number of convolutions, each convolution in the original domain corresponding to a multiplication in the transformed domain. Since the calculation of isotopic distributions of molecules is done by convolution of the atomic isotopic distributions, the convolution theorem has direct application to the calculation of isotopic distributions. In particular, it is possible to formulate the problem so as to use the computationally efficient fast Fourier transform.

There are several variations of FFT-based methods. Let us first consider a simple case in which the input functions, f and g, are vectors representing the isotopic distributions of the elements comprising the molecule with the vector indices representing the nucleon number of the isotopes and the vector components their relative abundance. For example, the isotopic distribution of carbon with atomic number 6, can be written

$$\mathbf{p}_6 = (0\ 0\ 0\ 0\ 0\ 0\ 0\ 0\ 0\ 0\ 0\ 0.98889\ 0.01111\ 0\ \dots\ 0\ ) \tag{14}$$

The isotopic distribution, $\mathbf{p}_{\text{molecule}}$, of a molecule with N_i atoms of element i, is calculated as an N_i-fold convolution in the mass domain. This corresponds to an N_i-fold multiplication in the transformed or mass frequency domain. (We henceforth refer to the mass frequency domain as the inverse mass domain or the μ-domain.) In other words, an N_i-fold convolution in the mass domain corresponds to raising the μ-domain representation to the power of N_i. The overall process is therefore given by the following expression:

$$\mathbf{p}_{\text{molecule}} = F^{-1}\left(\prod_i F[\mathbf{p}_i]^{N_i} \right) \tag{15}$$

where F^{-1} is the inverse Fourier transformation.

The generalization to multielement compounds is straightforward. Each atom (of whatever element) added to the chemical composition results in a multiplication in the μ-domain.

Using the convolution theorem reduces the algorithmic complexity of numerically calculating convolutions. The algorithmic complexity of direct implementations of a single convolution step is of order n^2 (i.e., $O(n^2)$), where n is the number of elements in a single vector. If multiple convolutions are to be performed, then the complexity is even higher. However, in methods that use the convolution theorem the algorithmic complexity is determined primarily by the Fourier transform step, and if one uses FFT-based methods the algorithmic complexity of a single convolution is $O(n\log n)$. Furthermore, the complexity is only a weak function of the number of convolutions that need to be performed, so the advantage of FFT-based methods over direct convolutions is even greater for multiple convolutions. Implementation of FFT-based methods is facilitated by the availability of commonly available and highly efficient libraries for the FFT, such as the FFTW [18] or the FFT included in the free GNU Scientific Library [19] (*see* **Note 3**). Figure 3 shows an example of the FFT method in MATLAB and Fig. 4 a workflow incorporating the same algorithm in an Rshell script for use with the Taverna scientific workflow manager. This workflow is available on myExperiment.org.

The method just described can be classed as one of the ultra-high speed methods of generating stick spectra. It also serves as an introduction to FFT-based methods. This method generates a μ-domain function by first building mass-domain arrays and then transforming them to the frequency domain. This method places masses at integer values.

The FFT-based methods can be modified to for higher resolution, even to the level of calculating isotopic fine structure, using finer sampling of the mass domain. These methods, described more fully in **Note 3**, calculate more accurate and higher resolution isotopic profiles. Through convolution, a peak profile function describing the peak shape and width is also easily embedded in FFT-based methods. This enables an investigator to match the peak profile and resolution to that of experimentally obtained profile-mode spectra, which aids comparison between theoretical and experimental isotopic profiles.

More recently a series of novel algorithms have been published. A number of these have been reviewed in a review article [20] and still more recently another algorithm has been described [21], the latter also including a set of accuracy benchmarks.

Until now we have only considered natural isotopic abundances. Given their popularity and utility in the field, a word is in order about the calculation of isotopic distributions for labeled compounds. One method is to define fictitious elements to represent the labeled atoms. For example, suppose the compound

```
%
% calculate isotopic distributions of molecules using the FFT
%
% (c) Magnus Palmblad, 1999
%

MAX_ELEMENTS=5; MAX_MASS=2^13;          % fast radix-2 fast-Fourier transform

M=[378 234 65 75 6];                    % empirical formula #H #C #N #O #S

A=zeros(MAX_ELEMENTS,MAX_MASS);                      % isotopic abundances
A(1,2:3)=[0.9998443 0.0001557];                      % H
A(2,13:14)=[0.98889 0.01111];                        % C
A(3,15:16)=[0.99634 0.00366];                        % N
A(4,17:19)=[0.997628 0.000372 0.002000];             % O
A(5,33:37)=[0.95018 0.00750 0.04215 0 0.00017];      % S

tA=fft(A,[],2);                         % FFT along each element's distribution

ptA=ones(1,MAX_MASS);
for i=1:MAX_ELEMENTS,
  ptA=ptA.*(tA(i,:).^M(i));             % multiply transforms (componentwise)
end

riptA=real(ifft(ptA));                  % inverse FFT to get convolutions

id=zeros(1,MAX_MASS);
id(1:MAX_MASS-1)=riptA(2:MAX_MASS);     % shift to real mass

bar(id);                                % bar plot of isotopic distribution
```

Fig. 3 MATLAB program for calculating and plotting the isotopic distribution of bovine insulin. The program is easily modified to calculate the isotopic distribution of any peptide or protein, or indeed any molecule, by changing the definitions of M and A, respectively

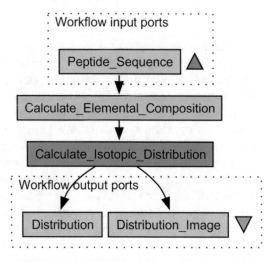

Fig. 4 Taverna workflow (available on myExperiment.org) for calculating and plotting the isotopic distribution of a peptide using an R version of the algorithm in Fig. 3

$C_6H_{12}O_6$ were labeled with ^{13}C at three sites. One would therefore calculate the isotopic distribution of $C_3X_3H_{12}O_6$, where X represents the positions labeled with ^{13}C. The label will of course never be isotopically pure, so it will have an isotopic distribution of its own. For example, X above may be composed of 98% ^{13}C and 2% ^{12}C. Therefore, this elemental isotopic profile for X must be included in the calculation of $C_3X_3H_{12}O_6$, treating X just as one would treat any other element and including its isotopic composition as part of the full calculation. If the isotopic distribution of X is not precisely known, Palmblad and coworkers have described methods to derive this from the measured data [22, 23].

It is also possible to apply isotope-related calculations to tandem mass spectrometry. Consider the problem of calculating the isotopic distribution of a product ion that would result from the dissociation of a single isotope-selected precursor ion. If the precursor ion is selected as the lightest isotopic peak then the product ion is composed entirely of all-light isotopes and the isotopic profile of the product ion (if we limit ourselves to consideration of a single dissociation channel) is composed of a single isotopic peak. However, if other isotopic peaks are chosen for dissociation then the product ion peaks may have complicated isotopic patterns. An early method for calculating the expected distributions was published by Tou and coworkers [24]. However, this method was restricted to the consideration of certain special cases. Other groups have used information from isotopic content or distributions in tandem mass spectra and more recently a general method for calculating the isotopic distributions of MS-MS spectra has been published [25].

4 Applications

The classical application for measurements of isotopic patterns has been to identify or confirm the elemental composition of organic compounds [26]. In this technique one compares experimental measurements of relative isotopic abundances to calculated isotopic abundances. A nonmatching pattern may rule out a candidate composition, whereas a matching pattern provides confirmatory evidence for a candidate composition. Although isotopic patterns can be very useful for confirmation of elemental composition, it is fair to say that accurate mass measurements have provided an alternative approach that for many years largely supplanted the use of isotopic patterns for confirming elemental composition. Indeed, a commonly accepted standard for elemental composition identification is that the measured monoisotopic mass should match the calculated monoisotopic mass candidate of a candidate composition. However, composition determination by accurate mass measurements rapidly becomes ambiguous for molecular weights greater than ~500 Da.

This issue has become the topic of intense scrutiny. For example, based on their evaluation of a database of 1.6 million compounds Kind and Fiehn have shown that even at a 1 ppm mass measurement error (i.e., roughly the current state of the art for near-routine accurate mass measurement) the ambiguity of chemical formula assignment increases sharply above ~400 Da, and by 500 Da even 0.1-ppm uncertainty is typically insufficient to uniquely assign a chemical composition. At 900 Da a 1-ppm uncertainty would leave 345 candidate compositions, and a 0.1 ppm mass uncertainty (which is currently beyond the state of the art for near-routine measurements) would still leave 32 candidates.

However, if one includes isotopic abundances in the evaluation the ability to assign chemical composition improves dramatically. The usefulness of combining mass information and abundance information was recognized at least as early as 1950 [27]. Recently there has been a renewed interest in this topic and systematic studies have been performed on the topic, most notably by Kind and Fiehn. They found, for example, that for compounds at ~900 Da measured with 3 ppm mass uncertainty the inclusion of isotopic abundance at a 2% relative abundance accuracy shortens the list of possible candidates to just 18 compared to 1045 possible candidates if isotopic abundances are not taken into account. In fact, Kind and Fiehn found that an analysis based on a combination of 3 ppm mass uncertainty and 2% isotopic abundance accuracy would equal or outperform an analysis based solely on 1 ppm uncertainty over the full molecular weight range they evaluated (150–900 Da), with the advantage being greater as molecular weight increased, and in most cases an analysis based on the combination of accurate mass and isotopic abundance also equaled or outperformed an analysis based on 0.1 ppm mass accuracy alone.

There are a number of approaches to use isotopic abundance information in combination with accurate mass information for assigning molecular formula. These range from heuristic approaches [28] to approaches that rely primarily on mathematical approaches [29]. In the latter, Wang and Gu compare an experimental spectrum and simulated spectra using a scoring parameter referred to as" spectral accuracy" and a related quantity," spectral error." The scoring scheme is modeled closely on methods used in numerical analysis and vector analysis that measure the similarity between vectors.

The spectrum is treated as a multidimensional vector, with mass being mapped to vector index and abundance being mapped to vector elements. (In this case the mapping of masses would associate noninteger masses with integer vector indices.) After taking into account some mathematical details, such as vector normalization, an algorithm is performed to calculate the spectral error, As described in the literature the algorithm amounts to an optimization calculation to minimize the distance between the tip of one vector (normalized to unit length) and the tip of the second vector whose length is scaled to minimize the distance between the tips of the two vectors. The spectral error (SE) is the distance between the two points after the optimization has been performed. The SE can then be used as a measure to compare any two isotopic distributions, such as an experimental result and a calculated result based on a trial chemical formula. If the SE is small, then the trial formula becomes a good candidate to be the correct formula. Additional details about SE are given in **Note 4**. Wang's treatment of the comparison between spectra using SE provides an additional layer of sophistication because it goes somewhat beyond the simple concept of spectral error as discussed above and also implements methods of adjusting peak shape and peak width as part of the comparison process. One must keep in mind that if relative abundance patterns are to be used for chemical formula determination then it is important that the experimental method not distort isotopic abundance patterns. Surprisingly, one cannot necessarily assume that this condition will hold for all types of instruments operated under all possible conditions. For example, a 2009 publication showed that in some cases the isotopic patterns obtained in the Orbitrap™ is distorted [30].

The use of isotopic information has been relatively neglected in tandem mass spectrometry. Indeed, the more usual approach is to deliberately eliminate isotopic information in product ion spectra by selecting a precursor ion consisting of all-light isotopes (commonly though not always correctly referred to as the "monoisotopic" ion). Thus, product ion mass spectra are composed of all-light isotopes, and such spectra contain no isotopic information. Among the relatively few applications of isotopic information in tandem mass spectrometry, Todd et al. [31] measured the isotopic patterns from the dissociation of individual isotopic peaks of multiply halogenated hydrocarbons.

One factor that may have prevented a wider acceptance of isotopic techniques in tandem mass spectrometry was the absence of a fully general theory to predict isotopic patterns in tandem mass spectra of isotope-selected precursor ions, although a fully general algorithm for performing such calculations was published over a decade ago [25]. This method constructs a two dimensional array (a "dissociation matrix") that contains the prediction of isotopic patterns for all common tandem mass spectrometry scan functions (product ion, precursor ion, and neutral loss scans). The work also showed that isotope-selected tandem mass spectra are very information-rich and potentially useful for characterization of chemical structure and elucidation of dissociation pathways. For example, Fig. 5 compares the predicted isotopic pattern for dissociation of the M + 2 peak of $C_{12}H_{23}O_4^+$ (protonated dibutyl ester of methylmalonic acid) for two possible dissociation pathways.

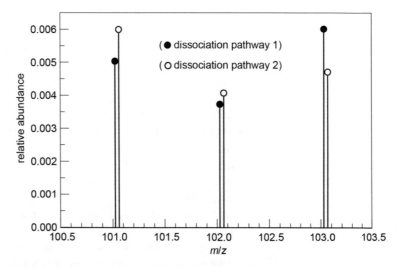

Fig. 5 Theoretical isotopic distribution resulting from dissociation of a protonated dibutyl ester of methylmalonic acid via two pathways: $C_{12}H_{23}O_4^+ \rightarrow C_4H_5O_3^+ + C_8H_{18}O$ (pathway 1) and $C_{12}H_{23}O_4^+ \rightarrow C_5H_9O_2^+ + C_7H_{14}O_2$ (pathway 2). These two spectra are distinctly different. Based just on the abundance patterns (and ignoring accurate mass differences), the spectral contrast angle between these two spectra is $10.8°$. By comparison with an experimental measurement one could judge which of the hypothetical dissociation pathways would be supported by experimental data

However, despite this the technique has been little used. One possible explanation for the dearth of such applications is that no computer program has been available that implemented the computational algorithm presented by Rockwood et al. [25]. More recently a computer program, IsoPatrn [32], has become available that contains options for calculating isotopic patterns in tandem mass spectrometry. This or similar software may facilitate a more widespread use of isotopic information in tandem mass spectrometry.

Isotopes are inextricably linked with mass spectrometry from their very discovery. The applications of stable isotopes in mass spectrometry are probably so numerous it would be challenging to cover all of them, even if the rest of this book was dedicated to the topic. In many applications, enriched (usually) heavy stable isotopes are used to label compounds to trace their metabolism or to provide standards for accurate quantitation in an analytical technique which typically does not give information about absolute concentrations of a compound. In most of these quantitative applications, the shapes of the natural and artificially enriched distributions are not taken into account, only the relative intensity of these, usually two, usually separable, distributions. We will therefore here concentrate on applications where the isotopic *distributions* are used to give information in an experiment and where calculation of these distributions are essential.

One of the most obvious examples uses of isotopic distributions as a probe comes from H/D-exchange [33]. Briefly, the hydrogen–deuterium exchange, particularly of amide hydrogens in proteins in deuterated water, takes place on an experimentally tractable time scale (seconds to hours). The rate of exchange (or back-exchange to hydrogen) is dependent on the three-dimensional structure and dynamics of the molecule and binding of substrates or cofactors. Measuring the exchange rates therefore gives information about these properties, which is above and beyond what is normally revealed in proteomics experiments and important in understanding the function and roles of proteins in biological systems.

In the era of low-resolution instruments, only the shift in average molecular mass was used to calculate the deuterium incorporation. In 2000, Palmblad, Buijs, and Håkansson adopted the rapid Fourier transform method by Rockwood et al. for calculating isotopic distributions to fit, using a least-squares method, measured distributions with unknown isotopic composition, specifically, the isotopic composition of hydrogen for particular hydrogens in a peptide or protein, thereby deriving the incorporation of deuterium as well as identifying the species by accurate mass [22]. This was the first demonstration of combining mass and spectral accuracy for identification of unknowns, albeit the candidates were limited to enzymatic fragments of the protein under study. The method was incorporated in the AUTOHD software [34]. The same algorithm was later used in other programs for H/D exchange data analysis, such as DEX [35] and HD Desktop [36]. Figure 6 shows the identification and theoretical fit to a measured peptide isotopic distribution using the AUTOHD software.

Another application of isotopic distributions, combined with measurement of deuterium incorporation in AUTOHD, is the identification of unknown species from accurate mass measurement and the entire measured isotopic distribution. These methods typically extracts the monoisotopic mass, which is well-defined and independent on instrument resolution and isotopic abundances, along with the measured relative intensities of all peaks considered part of the isotopic envelope. In practice, this is more difficult than it would first appear, as there is often some chemical background in the spectra that could be mistaken for isotopic peaks. The (accurately) measured monoisotopic mass and isotopic distribution are then compared against calculated monoisotopic masses and against calculated isotopic distributions, the latter typically using a least-squares method. This has been extended to include isotopic fine structure as well, in more recent versions of DataAnalysis (Bruker Daltonics)—a powerful illustration that resolving power can be

A

#	m/z	z	M (exp.)	ΣI	# peaks	fragment	sequence	M (calc.)	wSSE	p(D)	MME
22	615.31764	2	1228.62073	5.28	4	461-473	IQKIAEGTAGEGG	1229.62512	0.3127	0.140	2.21
						572-584	EASVTIASPAQPG	1226.61426	0.5540	0.471	-6.62
						574-586	SVTIASPAQPGEA	1226.61426	0.5540	0.471	-6.62
						566-577	QAAAYVEASVTI	1221.62415	0.2548	0.877	1.41
						292-305	**TAVAAAELGFAAGA**	**1218.62451**	**0.0942**	**0.961**	**1.71**
23	617.33864	4	2465.32544	116.08	12	**16-38**	**AHTIVKPAGPPRVGQPSWNPQRA**	**2463.32471**	**0.0048**	**0.586**	**-1.91**
24	633.80049	2	1265.58643	186.45	8	311-322	CLFGNGERTGNV	1265.58227	0.0547	0.357	-4.76
						313-324	FGNGERTGNVCL	1265.58227	0.0547	0.357	-4.76
						375-386	FSGSHQDAINKG	1259.58948	0.0219	0.886	0.93

B <u>autohd2</u> **spectrum view**

autohd2 fit of isotopic cluster at m/z=617.3386 (12 peaks) by peptide AHTIVKPAGPPRVGQPSWNPQRA [p(amide D)=0.586]

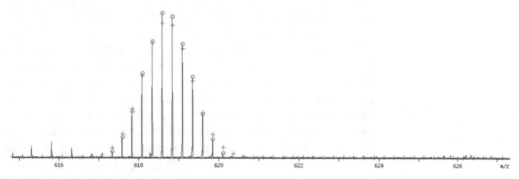

C sequence view :

MTTSESPDAY**TESFGAHTIVKPAGPPRVGQPSWNPQRA**SSMPVNRYRPF**AEEVEPIRL**RNRTWPDRVIDR
APLW**CAVDLRDGNQALIDPMS**PARRRRMFDLLVRMGYKEIEVGFPSASQTDFDFVREIIEQGAIPDDVTI
QVLTQCRPELIERTFQACSGA**PRAIVHFYNSTSILQRRVVFRANRAEVQAIATDGARKCVBQAAKY**PGTQ
WRFEYSPESYTCTELEYAKQVCDA**VGEVIAPTPERPIIFNLPATV**EMTTPNVYADSIEWMSRNLANRESV
ILSLHPHNDRG**TAVAAAELGFAAGA**DRIEGCLFGNGERTGNVCLVTLGLNLF**SRGVDPQIDFSNI**DEIRR
TVEYCNQLPVHERHPYGGDLVYTA**FSGSHQDAINKG**LDAMKLDADAADCDVDDMLWQVPYLPIDPRDVGR
TYEAVIRVNSQSGKGGVAYIMKTDHGLSLPRRLQIEFSQVIQKIAEGTAGEGGEVSPKEMWDAFAEEYLA
PVRPLERIRQHVDAADDDGG**TTSITATVKINGVETEI**SGSGNGPLAAFVHALADVGFDVAVLDYEHAMS
AGDDAQAAAYVEASVTIASPAQPGEAGRHASDPVTIASPAQPGEAGRHASDPVTSKTVWG**VGIAPSITTA**
SLRAVVSAVNRAAR

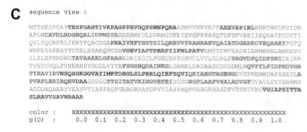

color :   XXXXXXXXXXXXXXXXXXXXXXXXXXXXXXXXXXXXXXXXXXXXXXXXXXXXXXXXXXXXXX
p(D) :   0.0 0.1 0.2 0.3 0.4 0.5 0.6 0.7 0.8 0.9 1.0

Fig. 6 Representative output from autohd2 and similar programs for calculating deuterium incorporation in peptides from mass spectrometry data. Measured data are compared, time point by time point, with predicted peptides using both accurate mass and agreement between measured and calculated isotopic distributions to identify the peptides. In H/D exchange experiments, the fraction of amide deuterium incorporated into the peptide is unknown and therefore allowed to vary to find the best fit to the measured isotopic distribution, thereby also extracting the information on deuterium incorporation (or back-exchange) in the peptide. In (**a**) we see part of the peptide overview in autohd2, with the peptide candidate with the smallest mass measurement error in blue, and the best fit to the isotopic in bold. Bold and blue peptides are the therefore the best fit both to both the masses and intensities of the measured isotopic distribution (**b**). The deuterium incorporation can be summarized in a heat map, either projected onto the sequence or an existing structure (**c**). (H/D exchange data courtesy of Dr. Patrick Frantom)

more important than mass accuracy for identifying unknowns (*see* **Note 5**).

A third application of calculation and fitting of isotopic distribution is the use of overlapping isotopic envelopes from differential labeling in quantitative proteomics. One of the first demonstrations of this was the subtle modification of isotope ratio proteomics (SMIRP) method by Whitelegge and coworkers in 2004 [37]. This was generalized to multiplexed quantitative proteomics in 2007 by Palmblad et al. [23], explicitly using rapid calculations of isotopic distributions to first calculate the degree of ^{15}N incorporation in different samples and then use this information to dissect relative abundances between samples from overlapping isotopic distributions. The method allows for high degrees of multiplexing, in principal limited only by the number of nitrogen atoms in the peptide. This algorithm was implemented in the muxQuant software [38] and reintroduced in 2011 as "next-generation quantification" (NGQ) by Kirchner et al. [39]. Note that these methods are different from classic isotope dilution or stable isotope labeling by amino acids in cell culture (SILAC) [40] methods, which typically assume or require that the isotopic envelopes are resolved and therefore need not be modeled or calculated. In methods such as iTRAQ [41] there is an overlap between reporter ions, but because the reporter ions are always the same this can easily be compensated for by constant correction factors, at least in data from low- and medium-resolution mass spectrometers such as ion traps and time-of-flight (TOF) instruments. Although requiring additional computational effort, methods such as SMIRP and muxQuant are experimentally inexpensive and make use of the "bandwidth" of the mass spectrometer. Methods like muxQuant and NGQ allow higher degrees of multiplexing than SILAC and are therefore particularly well-suited experiments involving more than three or four conditions. Figure 7 illustrates 6-plex quantitation using muxQuant. One should be aware that data is always noisy and the risk of making major errors in the quantitation increase with the degree of multiplexing. Attempting to reach the *theoretically possible* 30- or 40-plexing is discouraged.

In high-resolving power instruments, such as high-field FTICR, isotopic fine structure can routinely be resolved for small species. Isotopic fine structure provides a lot of information on the chemical composition of the species and is often sufficient to uniquely identify the empirical formula of an unknown compound, even with limited mass measurement accuracy. The isotopic fine structure can be calculated using the high-resolution FFT technique [8] or any of the alternative methods proposed [12, 13]. Figure 8 shows the isotopic fine structure of two isobaric peptides MIM and HTH (the H_8S_2 vs N_4O split) at m/z 394–396 resolved in a 12 T FTICR instrument and calculated using the FFT method.

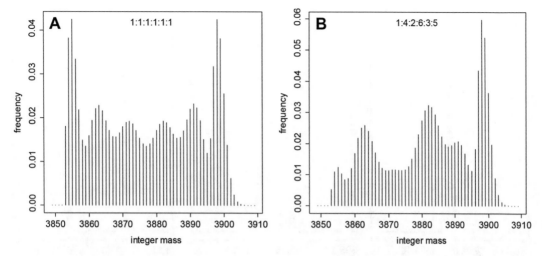

Fig. 7 Calculated distributions of a predicted tryptic peptide VTPQPGVPPEEAGAA-VAAESSTGTWTTVWTDGLTSLDR ($H_{261}C_{168}N_{44}O_{60}$) from the *C. sativa* ribulose-1,5-bisphosphate carboxylase large subunit in a hypothetical 6-plex muxQuant experiment with differential N-15 labeling. For these calculations it was assumed that the N-15 incorporation was 0.366% (natural abundance) 18.0%, 38.5%, 61.5%, 82.0%, and 99.0% respectively, in the six samples, not far from which can be practically achieved. If the protein is present in the same concentration in all samples and the samples are mixed in equal proportions, a nearly symmetric distribution results (**a**). If, however, the protein is differentially abundant, for example in ratios 1:4:2:6:3:5, an asymmetric distribution results (**b**). In either case, the relative contributions of the peptides from the six samples can be dissected by muxQuant or similar algorithm. Mathematically this is a well-conditioned problem as long as the number of samples is smaller than the number of nitrogen atoms in the molecule, or, strictly speaking, smaller than or equal to the number of isotopic peaks in the combined distributions

5   Concluding Remarks

Isotope patterns are the basis for a long-standing method of determining or confirming chemical compositions of compounds analyzed by mass spectrometry. Recent publications have shown that the use of isotopic patterns in conjunction with accurate mass measurements is far more effective than using accurate mass alone, particularly as molecular weight increases. Often only the monoisotopic mass and the relative intensity of the isotopic peaks are considered when comparing measured spectra with those predicted from different chemical species. However, more sophisticated methods such as those described by Wang make full use of abundance and accurate mass information over the full isotopic distributions.

The applications of isotopes and isotopic distributions are so numerous that they cannot be described in detail in a single book chapter. One can distinguish between passive use of natural (or given) isotopic abundances and use these to identify unknowns or partitioning the contributions from two or more overlapping

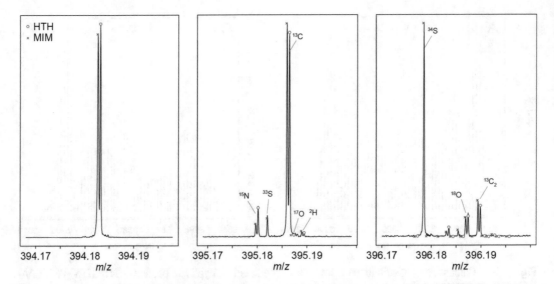

Fig. 8 Isotopic fine structure of two isobaric ($\Delta m \approx 0.00047$ Da) peptides MIM ($H_{32}C_{16}N_3O_4S_2$) and HTH ($H_{24}C_{16}N_7O_5$) resolved by a 12 T Bruker Daltonics solariX FTICR mass spectrometer in broadband mode with some of the major isotopologs are indicated by the heavy isotope or isotopes they contain. Note that only one of the peptide contains sulfur and the singlet peaks in the overlapping isotopic fine structure of the two peptides therefore must contain one or more heavy sulfur isotopes. The isotopic fine structure was calculated using the high-resolution FFT method described above, with a bin size of 0.00001 Da and vector length 16 Da. The source code of the program is available in MATLAB and Python on www.ms-utils.org. The Bruker DataAnalysis software also allows simulation of isotopic fine structure and use of this for unambiguous determination of elemental formulae. Given resolved isotopic fine structure, mass measurement accuracy becomes far less important. On the other hand, instruments capable of resolving isotopic fine structure are typically also able to measure masses with low-ppm or sub-ppm mass measurement uncertainty

species, for example, using the averagine model used in THRASH [42], and active, deliberate, manipulation of isotopic distributions by labeling standards with heavy isotopes for quantitation, to map biochemical pathways or in tracing studies, for example, in toxicology and pharmacology. We have given a few representative examples where measurement or calculation of isotopic distributions plays a nontrivial role. This includes determination of elemental composition, investigating fragmentation pathways, hydrogen–deuterium exchange, and multiplexed quantitation using partially overlapping isotopic distributions. We did not include the common uses of heavy isotopes as labels and in internal standards, as these do not require precise measurement or calculation of isotopic distributions.

Virtually all applications of isotopic information require the theoretical calculation of isotope patterns. In general, Fourier transform methods for calculating isotopic distributions are simple to implement using the FFT and efficient both in CPU time and memory. Depending on the type of FFT-based method used, the calculated masses can be semiaccurate, or highly accurate.

The Fourier methods are accurate in predicting the relative intensities of isotopic peaks, as there is no pruning, save for machine epsilon. The Fourier methods are easily adopted for isotopic fine structure and accurate mass calculation of all fine structure peaks and do not use more memory than is available on modern desktop computers. With the method published in 2010 by Fernandez-de-Cossio [43], this was also made more efficient in memory. Finally, the full profile spectrum can be simulated by convoluting the isotopic distribution with the instrument peak shape. The peak shape is included in the calculation as one atom of an "element" with a very narrow but very smooth distribution.

Monte Carlo methods can be seen as simulating the actual mass spectrometry measurement of a finite number of ions. The number of detected ions in the mass spectrometer can sometimes be controlled [44] and often inferred from the signal intensity. If repeatedly performed using the same number of samples as the number of measured ions, a Monte Carlo method would also give the expected variation (or error) in the measurement caused by counting statistics or measuring a finite rather than an infinite number of ions.

Methods based on polynomial expansions are very practical for molecules of moderate molecular weight. However, as molecular weight increases the polynomial-based methods often become computationally intractable due to a combinatorial explosion of the number of terms that need to be considered. Pruning of small terms can extend the useful range of polynomial methods at the cost of accuracy, but as molecular weight increases accuracy may be compromised to an unacceptable degree.

If one is only interested in abundance and centroided masses of peaks, neglecting any information arising from isotopic fine structure or nominal peak profile then the algorithm implemented in emass (https://github.com/princelab/emass) is probably an optimal combination of speed and accuracy.

Other methods may have a few advantages. Some claim increased speed and/or accuracy, or other advantages, such as providing a direct route to the exact mass and relative intensity of major (and completely resolved) isotopic peaks and being based on principles simpler to explain, not requiring understanding of Fourier transforms or convolutions. The accurate mass and intensity of individual major isotopic peaks may be obtained using a simple pocket calculator using one of these methods, but this is of course not the same as deriving the entire distribution.

Isotopic information can also be used in the interpretation of MS/MS experiments. This potentially very powerful approach to structural and elemental composition determination has been relatively little used, possibly because computational methods for predicting isotopic distributions of product ions of isotopically selected precursor ions have not been well developed until relatively recently. Early computational methods were limited to special cases. However, within the last decade more general methods have become available for performing these calculations, and this may lead to more widespread adoption of this approach to structural and elemental composition determination.

6 Notes

1. A note on stable and unstable isotopes. Of the 114 elements [45] already included in the periodic Table (1–112, 114 and 116), 55 have at least two stable isotopes. These include the elements that make up most of the organic compounds: hydrogen, carbon, nitrogen, oxygen and sulfur. Current data suggest 25 elements have one single stable isotope: beryllium, fluorine, sodium, aluminum, phosphorous, scandium, vanadium, manganese, cobalt, rubidium, arsenic, yttrium, niobium, rhodium, iodine, cesium, lanthanum, praseodymium, terbium, holmium, thulium, lutetium, tantalum, rhenium, and gold. The other 34 elements (technetium, promethium, bismuth, and all heavier elements) lack stable isotopes, although several have half-lives long enough to be common in nature.

   Some observationally radioactive isotopes have very long half-lives, such as bismuth-209, $t_{1/2} \approx 1.9 \times 10^{19}$ years—more than a billion times the age of the universe [46], or vanadium-50 which has a half-life of 1.4×10^{17} years—ten million times the age of the universe. Although vanadium-50 and bismuth-209 are observationally radioactive, they can be considered stable for nearly all intents and purposes. Other important long-lived isotopes are thorium-232 with a half-life of 14.1×10^9 years, very nearly the same as the estimated age of the universe, or 13.8×10^9 years. Uranium-238 (isotopic abundance 99.3%) has a half-life equal to the age of the Earth, 4.5×10^9 years. Because of their long half-lives, isotopes such as Va-50, Bi-209, Th-232, and U-238 exist naturally on earth and are easily observable by mass spectrometry. Isotopes with half-lives $<10^8$ years will have decayed to undetectable levels in nature unless they are replenished by radioactive decay of long-lived isotopes or other processes such as cosmic ray bombardment. For example, all of the naturally occurring isotopes of radium have half-lives measured in thousands of years or less, but they are observable in nature because they are formed by radioactive decay of uranium and thorium.

   Of the 25 elements with a single stable isotope, vanadium and six other elements also have a long-lived radioactive isotope that would be measurable by a mass spectrometer: rubidium (Rb-87, $t_{1/2} \approx 4.75 \times 10^{10}$ years, 27.835% natural abundance), indium (In-115, $t_{1/2} \approx 4.41 \times 10^{14}$ years,

95.71% natural abundance), lanthanum (La-138, $t_{1/2} \approx 1.05 \times 10^{11}$ years, 0.0902% natural abundance), lutetium (Lu-176, $t_{1/2} \approx 4.00 \times 10^{10}$ years, 2.59% natural abundance), tantalum (Ta-180m∗, $t_{1/2} > 1.2 \times 10^{15}$ years, 0.012% natural abundance), and rhenium (Re-187, $t_{1/2} \approx 4.35 \times 10^{10}$ years, 62.60% natural abundance).

*The "m" in Ta-180m stands for *metastable*. Ta-180m is an excited nuclear state of Ta-180. Theoretically, Ta-180m could decay to one of three products: Ta-180 by emission of a gamma ray, W-180 by emission of a beta particle, and Hf-180 by electron capture. As there is no change in nucleon number for any of these processes the decay products all have the same nominal mass. However, to refer to any of these three elements as "decay products" of Ta-180m is a bit of exaggeration because the stated half-life $t_{1/2} > 1.2 \times 10^{15}$ years is only a lower limit to the half-life and no radioactive decay of Ta-180m has ever been observed. Interestingly, that the ground state, Ta-180, is itself unstable and decays with a half-life of just 8 h. Thus, the metastable excited state, Ta-180m, has a half-life at least 18 orders of magnitude longer than that of the ground state, Ta-180. This remarkable stability of Ta-180m arises from a nuclear spin mismatch in the decay paths. Curiously, the ground state Ta-180 has a half-life of just 8 h.

2. As mass spectrometry measures mass-to-charge ratio of ions, we would also have to subtract the mass of an electron to calculate the mass of CO_2^+. As a word of warning, some computer programs do not necessarily follow a uniform standard regarding inclusion or exclusion of the electron mass. The electron mass must be taken into consideration for the most accurate results.

3. The main body of the chapter described an FFT-based method that starts with mass-domain elemental isotopic profiles wherein the isotopes are placed at integer values rather than accurate masses. These arrays are then transformed to the μ-domain using the FFT and then processed further, with the data ultimately being transformed back to the mass domain using an FFT. (Aside from a few minor considerations, the FFT is its own inverse, so for the purposes of this chapter we often refer to both the forward and reverse transforms as FFTs.) The final result is an array of mass–abundance pairs with masses placed on integer values.

However, using this approach the granularity of the original array limits the mass accuracy of subsequent calculations. It is also possible to use a broadly similar approach with the exception that the mass-domain point spacing is a small fraction of a Dalton rather than 1 Da. The finer granularity of the mass axis allows higher mass accuracy to be achieved.

However, using this approach the mass accuracy is still dependent on the granularity of the original array. For example, if the array indexes were to have a granularity of 0.01 Da then it is possible for an atomic mass to be off by as much as 0.005 Da. For example, oxygen would be assigned a mass of 15.99 Da, which would underestimate its value by 0.0049 Da. The mass error would be magnified in a final result for a compound containing many oxygen atoms. For example, the mass of the monoisotopic peak of bovine insulin, which contains 75 oxygen atoms, would be underestimated by $75 \times 0.0049 = 0.37$ Da, which represents a sizable error.

One can avoid this issue by generating the μ-domain function by direct calculation in the μ-domain using the elemental isotopic information. For example, the μ-domain representation of CCl_4 would be $(0.99e^{i2\pi \times 12\mu} + 0.01e^{i2\pi \times 13.0033554\mu}) \times (0.76e^{i2\pi \times 34.9688531\mu} + 0.24e^{i2\pi \times 36.9659034\mu})^4$ if we round off the abundances to two decimal places and i refers to the imaginary unit.

The μ-domain representation can be sampled at appropriately chosen equally spaced points in the μ-domain, and the complex exponential functions can be calculated at the sampling points using Euler's relations and trigonometric functions. Prior to raising the resulting complex-valued numbers to integer powers and performing the final multiplications it is convenient to convert the complex numbers to (r,θ) format. Once these calculations are completed the resulting complex-valued vector can be apodized and transformed to the mass domain using the FFT. This scheme has the advantage that the masses can be entered into the calculation at any desired level of accuracy. We have used seven decimal places in this example. The precision of the masses entered into the μ-domain expression is not required to match any particular point spacing in either the μ-domain or mass domain. Additional details about the method can be found in a paper by Wan et al. [47].

Some FFT-based methods include a peak shape function in the calculation. This is done by multiplying the μ-domain function by an apodization function. The mass-domain peak profile is related to the μ-domain apodization function by a Fourier transform/inverse Fourier transform pair. Thus, the apodization function can be tailored to match any desired mass-domain peak width and shape. An apodization function is almost mandatory in the calculation anyway in order to avoid the risk of Gibbs oscillations in the mass domain result, so the additional computational cost of matching a desired mass-domain peak profile is insignificant. It should be noted however that this method for generating a desired mass-domain peak profile gives the same shape and width over the full mass range of the calculation. Thus, if an experimental peak profile were significantly mass-dependent over the mass range of the isotopic distribution then the simulated peak profile would not match the experimental profile over the full isotopic distribution. This is not likely to be a frequently encountered problem, but the possibility should be evaluated on a case by case basis.

Fourier transform-based methods can also be extended to ultrahigh resolution, that is, a resolution regime appropriate to resolve isotopic fine structure. There are at least three methods for doing this. The most straightforward is to simply increase the array size and adjust the sampling interval appropriately. To illustrate, the program mercury uses an array size of 2048 complex points, and for a typical calculation the mass range would self-adjust to several tens of Daltons (e.g., 32 Da for a bovine insulin calculation), and the default resolution parameter is eight points for near-baseline to near-baseline peak width (resolution ~82,000 for bovine insulin). If the number of points were simply scaled up by a factor of 128 and the point density increased by the same factor, thus keeping the overall mass range the same, then the resolution would increase to 10.5 million and isotopic fine structure is resolved. A second way is to zoom in on a single nominal isotopic peak, performing the calculation over a limited mass range and using digital filtering in the μ-domain to prevent aliasing [17]. This scheme has been implemented as an option in the mercury program. Fernandez-de-Cossio has addressed the problem another way by more efficiently packing the vectors in memory [48].

4. It can be shown that spectral error (SE) is the sine of the angle between the two vectors. This angle has been referred to in the literature as the" spectral contrast angle" [47]. One can calculate the cosine of the spectral contrast angle by first normalizing the two vectors (i.e., the two spectra) to unit length and then calculating the dot product between the two normalized vectors. The spectral contrast angle can then be calculated using an arccosine operation. These relationships provide a direct path to the calculation of SE without treating it as an optimization problem. The dot product, also known as the inner product, is a classical method of calculating the similarity between two vectors. Consequently, the approach using spectral error is closely related to classical methods used in numerical analysis and vector analysis.

Of the three methods of comparing spectra just discussed, the cosine method is the easiest to compute. However, it has the characteristic of being a nonlinear quantity, by which we mean that if a small perturbation is added to a reference vector the response function (the cosine of the angle between the reference vector and the perturbed vector) is not a linear function of the scaling factor for the perturbation. In the small angle approximation the cosine is a quadratic function of the scaling factor. The other two comparison parameters (the spectral contrast angle and the SE) are linear functions of the perturbations in the small angle approximation. Because of the linearity issue there may be slight conceptual advantage in favor of SE and/or spectral contrast angle for comparing mass spectra. However, because there is a well-defined mathematical relationship between the three quantities, the information content is the same, regardless of which of the three methods are chosen to compare spectra. It is also worth pointing out that because mass spectra contain no negative values the dot product between two normalized vectors representing mass spectra is always nonnegative. Furthermore, the spectral contrast angle between two spectra always lies between $0°$ and $90°$, and the SE is therefore always between 0 and 1, with 0 occurring of two vectors are identical and 1 occurring if two vectors are orthogonal.

5. Finally, we provide a note on mass measurement accuracy and resolving power. Mass accuracy in absence of high resolving power is not always as high as claimed by instrument manufacturers (or their marketers), something that is easily realized by considering a spectrum of two compounds with masses 1000.00 and 1000.01 on an instrument with a resolving power of 20,000. Any talk about low-ppm (or indeed sub-ppm) mass measurement errors in such cases should be taken with a generous helping of salt, as the two species differ by 10 ppm and would not be resolved by this instrument.

References

1. Reimer PJ et al (2009) IntCal09 and Marine09 radiocarbon age calibration curves, 0–50,000 years cal BP. Radiocarbon 51(4):1111–1150

2. de Laeter JR (1998) Mass spectrometry and geochronology. Mass Spectrom Rev 17 (2):97–125

3. Kreuzer-Martin HW, Jarman KH (2007) Stable isotope ratios and forensic analysis of microorganisms. Appl Environ Microbiol 73 (12):3896–3908

4. McDermott F (2004) Palaeo-climate reconstruction from stable isotope variations in speleothems: a review. Q Sci Rev 23 (7–8):901–918

5. Harris SB et al (2012) List of observationally-stable isotopes. Available from: http://en. wikipedia.org/wiki/Stable_isotope#List_of_ observationally-stable_isotopes

6. Yergey JA (1983) A general approach to calculating isotopic distributions for mass Spectrometry. Int J Mass Spectrom Ion Phys 52:337–349

7. Hsu CS (1984) Diophantine approach to isotopic abundance calculations. Anal Chem 56 (8):1356–1361

8. Rockwood AL, VanOrden SL, Smith RD (1996) Ultrahigh resolution isotope distribution calculations. Rapid Commun Mass Spectrom 10(1):54–59

9. Kubinyi H (1991) Calculation of isotope distributions in mass spectrometry. A trivial solution for a non-trivial problem. Anal Chim Acta 247:107–119

10. Rockwood AL, Van Orden SL (1996) Ultrahigh-speed calculation of isotope distributions. Anal Chem 68:2027–2030

11. Rockwood AL, Van Orman JR, Dearden DV (2004) Isotopic compositions and accurate masses of single isotopic peaks. J Am Soc Mass Spectrom 15(1):12–21

12. Roussis SG, Proulx R (2003) Reduction of chemical formulas from the isotopic peak distributions of high-resolution mass spectra. Anal Chem 75(6):1470–1482

13. Rockwood AL, Haimi P (2006) Efficient calculation of accurate masses of isotopic peaks. J Am Soc Mass Spectrom 17(3):415–419

14. Arfken G (1985) Convolution theorem §15.5, in mathematical methods for physicists. Academic Press, Orlando, FL

15. Cooley JW, Tukey JW (1965) An algorithm for the machine calculation of complex Fourier series. Math Comput 19:297–301

16. Rockwood AL (1995) Relationship of Fourier transforms to isotope distribution calculations. Rapid Commun Mass Spectrom 9:103–105

17. Rockwood AL, Van Orden SL, Smith RD (1995) Rapid calculation of isotope distributions. Anal Chem 67:2699–2704

18. Frigo M, Johnson SG (1998) FFTW: an adaptive software architecture for the FFT. In: The IEEE international conference on acoustics, speech and signal processing, Seattle

19. GSL—GNU scientific library (2007)

20. Valkenborg D et al (2012) The isotopic distribution conundrum. Mass Spectrom Rev 31 (1):96–109

21. Claesen J et al (2012) An efficient method to calculate the aggregated isotopic distribution and exact center-masses. J Am Soc Mass Spectrom 23(4):753–763

22. Palmblad M, Buijs J, Håkansson P (2001) Automatic analysis of hydrogen/deuterium exchange mass spectra of peptides and proteins using calculations of isotopic distributions. J Am Soc Mass Spectrom 12(11):1153–1162

23. Palmblad M, Mills DJ, Bindschedler LV (2008) Heat-shock response in Arabidopsis thaliana explored by multiplexed quantitative proteomics using differential metabolic labeling. J Proteome Res 7(2):780–785

24. Tou JC, Zakett D, Caldecourt VJ (1983) Tandem mass spectrometry of industrial chemicals. In: McLafferty FW (ed) Tandem mass spectrometry. John Wiley and Sons, New York, pp 441–446

25. Rockwood AL, Kushnir MM, Nelson GJ (2003) Dissociation of individual isotopic peaks: predicting isotopic distributions of product ions in MSn. J Am Soc Mass Spectrom 14(4):311–322

26. Beynon J (ed) (1960) Mass spectrometry and its applications to organic chemistry. Elsevier Science, Amsterdam, pp 294–302

27. Zemany PD (1950) Punched card catalog of mass spectra useful in qualitative analysis. Anal Chem 22(7):920–922

28. Kind T, Fiehn O (2007) Seven golden rules for heuristic filtering of molecular formulas obtained by accurate mass spectrometry. BMC Bioinformatics 8:105

29. Wang Y, Gu M (2010) The concept of spectral accuracy for MS. Anal Chem 82 (17):7055–7062

30. Erve JCL et al (2009) Spectral accuracy of molecular ions in an LTQ/Orbitrap mass spectrometer and implications for elemental composition determination. J Am Soc Mass Spectrom 20(11):2058–2069

31. Todd PJ, Barbalas MP, Mclafferty FW (1982) Collisional activation mass-spectra of ions containing polyisotopic elements. Org Mass Spectrom 17(2):79–80

32. Software for Mass Spectrometry (2012) [cited 2012 20120509]. Available from: http://tarc. chemistry.dal.ca/soft_down.htm

33. Tsutsui Y, Wintrode PL (2007) Hydrogen/deuterium, exchange-mass spectrometry: a powerful tool for probing protein structure, dynamics and interactions. Curr Med Chem 14(22):2344–2358

34. Palmblad M (2000) AUTOHD [cited 2012]. Available from: http://www.ms-utils.org/autohd.html

35. Hotchko M et al (2006) Automated extraction of backbone deuteration levels from amide H/(2) H mass spectrometry experiments. Protein Sci 15(3):583–601

36. Pascal BD et al (2009) HD desktop: an integrated platform for the analysis and visualization of H/D exchange data. J Am Soc Mass Spectrom 20(4):601–610

37. Whitelegge JP et al (2004) Subtle modification of isotope ratio proteomics; an integrated strategy for expression proteomics. Phytochemistry 65(11):1507–1515

38. Palmblad M (2007) muxQuant [cited 2012]. Available from: http://ms-utils.org/muxQuant

39. Kirchner M et al (2011) Changing the rules of the game: next generation quantification (NGQ) enables complete isotopic multiplexing for quantitative functional and dynamic proteomics. In: 59th ASMS conference on mass spectrometry and allied topics, Denver, CO.

40. Ong SE et al (2002) Stable isotope labeling by amino acids in cell culture, SILAC, as a simple and accurate approach to expression proteomics. Mol Cell Proteomics 1(5):376–386

41. Ross PL et al (2004) Multiplexed protein quantitation in Saccharomyces cerevisiae using amine-reactive isobaric tagging reagents. Mol Cell Proteomics 3(12):1154–1169

42. Horn DM, Zubarev RA, McLafferty FW (2000) Automated reduction and interpretation of high resolution electrospray mass spectra of large molecules. J Am Soc Mass Spectrom 11(4):320–332

43. Fernandez-de-Cossio Diaz J, Fernandez-de-Cossio J (2012) Computation of isotopic peak center-mass distribution by Fourier transform. Anal Chem 84(16):7052–7056

44. Schwartz JC, Zhou X-G, Bier ME (1996) Method and apparatus of increasing dynamic range and sensitivity of a mass spectrometer. US

45. Element 114 is named flerovium and element 116 is named Livermorium (2012). Available from: http://www.iupac.org/news/news-detail/article/element-114-is-named-flerovium-and-element-116-is-named-livermorium.html

46. de Marcillac P et al (2003) Experimental detection of alpha-particles from the radioactive decay of natural bismuth. Nature 422 (6934):876–878

47. Wan KX, Vidavsky I, Gross ML (2002) Comparing similar spectra: from similarity index to spectral contrast angle. J Am Soc Mass Spectrom 13(1):85–88

48. Fernandez-de-Cossio J (2010) Efficient packing Fourier-transform approach for ultrahigh resolution isotopic distribution calculations. Anal Chem 82(5):1759–1765

Chapter 4

Retention Time Prediction and Protein Identification

Alex Henneman and Magnus Palmblad

Abstract

In bottom-up proteomics, proteins are typically identified by enzymatic digestion into peptides, tandem mass spectrometry and comparison of the tandem mass spectra with those predicted from a sequence database for peptides within measurement uncertainty from the experimentally obtained mass. Although now decreasingly common, isolated proteins or simple protein mixtures can also be identified by measuring only the masses of the peptides resulting from the enzymatic digest, without any further fragmentation. Separation methods such as liquid chromatography and electrophoresis are often used to fractionate complex protein or peptide mixtures prior to analysis by mass spectrometry. Although the primary reason for this is to avoid ion suppression and improve data quality, these separations are based on physical and chemical properties of the peptides or proteins and therefore also provide information about them. Depending on the separation method, this could be protein molecular weight (SDS-PAGE), isoelectric point (IEF), charge at a known pH (ion exchange chromatography), or hydrophobicity (reversed phase chromatography). These separations produce approximate measurements on properties that to some extent can be predicted from amino acid sequences. In the case of molecular weight of proteins without posttranslational modifications this is straightforward: simply add the molecular weights of the amino acid residues in the protein. For IEF, charge and hydrophobicity, the order of the amino acids, and folding state of the peptide or protein also matter, but it is nevertheless possible to predict the behavior of peptides and proteins in these separation methods to a degree which renders such predictions useful. This chapter reviews the topic of using data from separation methods for identification and validation in proteomics, with special emphasis on predicting retention times of tryptic peptides in reversed-phase chromatography under acidic conditions, as this is one of the most commonly used separation methods in bottom-up proteomics.

Key words Liquid chromatography, Mass spectrometry, Prediction, Retention time, Peptide, Protein identification

1 Introduction

In proteomics, peptides and proteins are often identified by comparing specific properties predicted from translated genome sequences with the same properties measured by analytical techniques such as mass spectrometry. Ideally, individual proteins are isolated, sequenced, and characterized in a "top-down" fashion, but due to intrinsic limitations of separation and mass spectrometric technology, proteins are more commonly digested by one or

Rune Matthiesen (ed.), *Mass Spectrometry Data Analysis in Proteomics*, Methods in Molecular Biology, vol. 2051,
https://doi.org/10.1007/978-1-4939-9744-2_4, © Springer Science+Business Media, LLC, part of Springer Nature 2020

more specific enzymes into shorter peptides. The masses of many such peptides or mass and tandem mass spectra generated from one or more such peptides are then compared with those predicted for putative peptides derived from genome sequences. These peptide comparisons are then used to infer the protein and possibly reconstruct the protein sequence and any posttranslational modifications from the "bottom-up." When introducing a complex sample such as a total protein enzymatic digest to a mass spectrometer, ionization suppression and detection dynamic range limit the ability to detect low-abundant peptides. To reduce the complexity and separate abundant species from less abundant ones, methods such as one- or two-dimensional gel electrophoresis of proteins [1] and liquid chromatography (LC) or capillary electrophoresis of peptides [2–4] have been used with great success. The primary function of using these separations method prior to mass spectrometry is to increase the analytical dynamic range and coverage of the sample. In the process, the separations themselves provide information on the analytes. In sodium dodecyl sulfate polyacrylamide gel electrophoresis (SDS-PAGE), this is the molecular weight of the protein, in isoelectric focusing the isoelectric point (the pH at which the protein or peptide has no net charge) and in reversed-phase chromatography, the hydrophobic nature of the peptides or proteins. Importantly, these are all properties that in principle can be predicted from amino acid sequences, just like molecular masses but less accurately. This information can be used to constrain database searches and for protein identification, adjust initial probabilities for peptide-spectrum matches [5, 6] or experimental feedback and quality control. In the case of SDS-PAGE of proteins, it is simple to calculate the molecular weight of the protein from the sequence in the database used in the identification of the tandem mass spectra or for peptide mass fingerprinting. Figure 1 shows a comparison between position on the SDS-PAGE gel and calculated molecular weight. In general, this kind of analysis may be useful to identify protein complexes, find large posttranslational modifications (if the measured molecular weight significantly exceeds the predicted weight), splicing events or posttranslational processing of the protein (if the molecular weight indicated by SDS-PAGE is smaller than that predicted from the database entry). The main point here is that any property on the peptide or protein level that can be both predicted and measured, however approximately or inaccurately, provides useful information.

A recurring problem in peptide identification is the size of the search space, which can lead to a considerable number of false positives. Increasing the number of features that we can accurately predict for each candidate peptide allows us to embed our peptide candidate distribution in a feature space with additional dimensions and possibly dispersing false positives from the correct identifications. In addition to precursor and product ion masses, relative

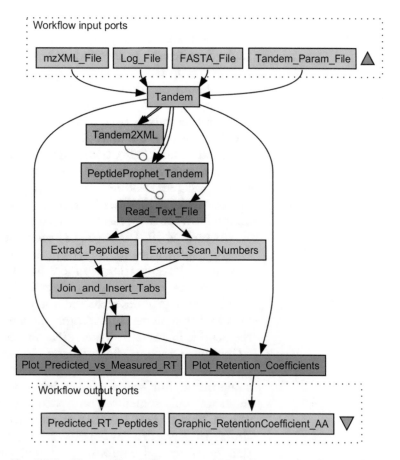

Fig. 1 Scientific workflow in Taverna workbench for retention time prediction, building on Workflow 1 in de Bruin et al. [7]. The workflow starts from LC-MS/MS data in mzXML, a FASTA sequence file and input parameters. The data is first searched with X!Tandem against the sequences in the FASTA file using the supplied X!Tandem parameters. The results from X!Tandem are then converted to pepXML using the Tandem2pepXML converter in the Trans-Proteomic Pipeline [8]. The peptide-spectrum matches are then validated using PeptideProphet [9]. The user can here supply a minimum probability for the peptide-spectrum matches to be included in the PeptideProphet output. The PeptideProphet output, still in pepXML, are read by the Read_Text_File local service in Taverna, and two XPath services are used to extract peptide sequences and retention times from the pepXML file. These are then joined in a tab-delimited file used as input to the existing rt. software [10], here called via the Taverna tool invocation. The predictions from rt. are tested with an embedded Rshell script and the retention coefficients in the linear rt. model plotted by another Rshell script. For executing the Rshell scripts, Taverna requires access to an R server (which could also be running on the local computer). The Rshell scripts were written by W. Plugge

intensities, and perhaps isotopic distributions measured in a mass spectrometer, a number of additional information can also be used when available, such as protein molecular weight, peptide or protein isoelectric point, or chromatographic retention times.

Proteins can be identified by comparing likelihood ratios of the observed masses and retention times against those predicted from the proteome under study and against a large number of unrelated or random proteins of the same size and amino acid distributions. The random or "decoy" protein database can be generated by filtering sequences with little homology from other organisms [11] or constructed by reversing or scrambling the sequences in the database or using Markov chains [12]. A conservative measure of statistical significance is thus given by the frequency of random (false) matches to experimental data by random protein sequences. Unlike accurate mass and time tags, complete knowledge of all posttranslational modifications, nontryptic peptides, and contamination from other species is not assumed or necessary.

As early as 1951 Pardee [13] succeeded in predicting peptide retention properties for paper chromatography, as measured by Knight [14], using a linear model based on peptide length only. These studies, referring to measurement and prediction of retention factors of peptides separated using paper chromatography, also indicated that taking into account the amino acid composition of the peptide in a model, in particular for short peptides, could lead to improved accuracy. Soon Sanger [15] showed that even such a model would not suffice, as peptides with identical amino acid composition but where the amino acids occur in a different sequence could sometimes be separated using paper chromatography. In this chapter we are not so much concerned with paper chromatography of a few peptides but primarily with peptide retention times in high-performance liquid chromatography, as this separation technique is more commonly used in proteomics. The retention time of a peptide is the time after the start of a chromatographic separation the peptide is detected by the mass spectrometer. Often the injection of the sample on the column is used to start the clock, but other events could be used as well, such as the delayed start of data acquisition in the mass spectrometer or the start of a chromatographic gradient. Retention times are specific to the instrumental setup under conditions, such as the column packing material, mobile phases, flow rate, pH, temperature, and dead volumes in valves and connecting tubing. In reversed-phase liquid chromatography, the retention time is often considered a measure of the hydrophobicity of a peptide—in fact, several hydrophobicity scales are derived from elution order or retention times in reversed-phase chromatography [16]. The absolute peptide retention times lose much of their meaning when any changes are made to either liquid chromatography system or conditions, and differ somewhat even between consecutive runs. This means it is important to normalize retention time from different chromatographic runs when comparing measured and predicted retention times or when combining measurements for training of predictors of peptide retention times. The chromatographic gradient, or volume fraction of (in reversed-phase chromatography) organic to aqueous solvent, is often changed linearly in time. This allows for, if there are enough common peptides in both runs, application of linear regression or piecewise linear alignment [17] to map the retention times between separate chromatographic runs to a common time scale. This common time can be translated into a parameter of a more universal nature, such as normalized elution time (NET) [18].

Almost three decades after the works of Pardee, Knight, and Sanger, Meek [19] demonstrated that the retention times of 25 peptide in reversed liquid chromatography, measured within a single run, could be accurately fitted using a linear model with one retention coefficient for each amino acid type describing the effect on retention time for each instance of an amino acid in the peptide. The values of the retention coefficients were determined by consecutive minimization of the squared error for each coefficient. The promising results encountered with the linear model led to extensive efforts with hundreds to even several thousands of peptides and different chromatographic conditions [20–25] to determine the values of these retention coefficients. In addition to confirming the validity of the model to predict peptide retention times, individual amino acid retention coefficients were found to be highly dependent on chromatographic conditions, something which have recently been employed in quality control [26]. Although not the most accurate, the simple linear retention model is capable of roughly predicting the retention time of peptides under a wide range of conditions using a training set of modest size. After over a decade of almost no attention, peptide retention time prediction was brought into the proteomics era using least-squares linear regression for the first time train a model from peptides identified by mass spectrometry [10] and use retention time predictions to improve protein identification accuracy.

Success of the linear model predicting protein structural properties [27], such as transmembrane protein domains, confirms the partial correctness of the assumption of linear dependence of hydrophobicity on the contributions from amino acids separately. On the other side, the fact that peptides with the same amino acids in different sequence can be separated using HPLC [28–31] exposes the incompleteness of the model. This together with the accuracy loss in predicting peptides longer than 15–20 amino acids hints at that not all amino acids in the peptide contribute equally to the hydrophobicity of the peptide and also of mutual interaction effects between amino acids. Although the linear retention coefficient model can roughly predict peptide retention times, considerable improvement can be gained by including nonlinear elements into the prediction of retention times.

One way of including nonlinearity for improving retention time prediction, is to start with a linear retention coefficient model and add nonlinear corrections to describe the departure of real data from linear behavior. Starting from a linear retention coefficient model [32] corrects an increasing overestimation in the predicted retention times for peptides with lengths beyond 15–20 amino acids, by adding a length-dependent correction term and achieving better prediction performance. Krokhin et al. [33] extended this approach further by starting from the linear model determined experimentally by Guo et al. [22] and adding specific corrections to eliminate the worst discrepancies between predicted and experimental hydrophobicities. The model takes into account enhanced contributions of the first, second, and third N-terminal amino acids, and correction factor for peptides shorter than ten amino acids and a different correction factor for peptides longer than 20 amino acids. Furthermore, peptides resulting in a predicted hydrophobicity that is larger than a threshold value of 38 are multiplied by an attenuation factor of 0.7. The values of the correction constants and factors are found by maximization by hand of the Pearson correlation coefficient between measured and predicted values of a fixed set of peptides. This approach is taken further by Krokhin in 2006 [34, 35] in which also the first three C-terminal amino acids are boosted in their contribution. Also isoelectric point, nearest neighbor effect of charged side chains (K, R, and H amino acids) and propensity to form helical structures are used to construct terms that improve the prediction accuracy. The current model bears the name SSRC and can be used via a Web interface (https://hs2.proteome.ca/SSRCalc/SSRCalcX.html). This approach may seem ad hoc, but the model generalizes surprisingly well to different data sets.

Instead of starting off from linearity and adding in nonlinearity, it is possible to start from a model capable of predicting retention time nonlinearly from the peptide features directly. Artificial neural networks (ANNs) are capable of this. The first application of these models was very simple [18] using 20 input nodes, each encoding the number of amino acids of each type, two hidden layer neurons and a single output neuron. The model was trained using 7000 peptides and the results were of comparable accuracy to those of linear retention coefficient-based models. This still relatively simple model was used 1 year later by Strittmatter et al. to enhance peptide identification accuracy [36]. Two years later, an ANN with a more complex network [37] produced what is still one of the most accurate predictors of retention time. The authors investigated a large number of (derived) peptide properties as possible candidates to use as features for predicting retention time, but decide to encode peptides only using the first 25 N-terminal and last 25 C-terminal amino acids, the peptide length and hydrophobic moment as introduced by Eisenberg [38–40], as these give the best results. This led to a network consisting of 1052 input nodes, 24 hidden nodes and 1 output node. The vast number of degrees of freedom of the model requires an extensive training set and was trained using approximately 345,000 peptides and tested using 1303 peptides achieving a prediction accuracy of approximately 1.5%. The large number of necessary training samples for a network of this complexity might be prohibitive for many applications and required a careful normalization strategy for unifying the retention data oft 12,059 separate HPLC runs.

Support vector machine [41] (SVM) methods can also incorporate nonlinear effects and have been applied to retention time prediction. In these methods an efficient optimization procedure determines the size of weights to all training samples, retaining a sub-set of the training samples to support a hyperplane. The training samples in this subset are known as the support vectors. In a linear SVM the vectors correspond to the representation of the training peptides. A suitable kernel function can be used to map the training samples in which case the support vectors support a plane in the space in which the training samples are mapped. Klammer et al. [42] constructed a support vector regressor (SVR) using linear and Gaussian kernels and were able to train their predictor using peptides from a single run. In their model, peptide sequences are represented by a 63-component vector. The first 20 components encode the amino acid composition of the peptide. The following 40 components are of binary nature and encode the identity of the N-terminal and the identity of the penultimate C-terminal amino acid. The next two components are also of binary nature and encode whether the C-terminal amino acid is an R or a K. The last component is the length of the peptide.

Pfeifer et al. [43] used a SVR [44] with a more appropriately tailored oligo-border kernel [45] to construct their SVR, to make their programs RTModel and RTPredict, which are included into TOPP/OpenMS [46] suite of programs. This model represents peptides by their 30 most N-terminal and 30 most C-terminal amino acids of the peptide sequence. In this representation, k-mers, subsequences of length k, in these so-called borders are mapped by an oligo-kernel into the space in which the support vector regression takes place.

A program called ELUDE by Moruz et al. [47] also uses SVR to predict retention time. The program constructs a regression model when there is enough data available to train a new model. According to its authors, ELUDE achieves the best accuracy when at least 1000 peptides are used for training. If there is not enough data available, a pretrained model is selected from a built-in library, calibrated, and used for retention time prediction. In this case, the model giving the best results for the data at hand is chosen and calibrated using a FAST-least trimmed squares [48] method. The ELUDE model uses a 60-component representation for the peptide sequence. In a first step, a 20-dimensional representation of all training samples is used to derive 20 linear retention coefficients using linear support vector regression. Then 17 values that are constructed from retention coefficients are evaluated. These values are the sequence sum of retention coefficients, the sequence average retention coefficient, the N-terminal and C-terminal retention coefficient, the sequence sum of the squared difference of the retention coefficients of adjacent amino acids, the sum of the retention coefficients of amino acids adjacent to K and R, and the same for amino acids D and E, and the largest and smallest sum of retention coefficients belonging to adjacent amino acids. Also calculated are the largest and smallest hydrophobic moment [38–40] for α-helices and β-sheets over 11 amino acid windows found in the sequence, and the largest and smallest values of their own normalized α-helical retention coefficient that is evaluated over windows of nine amino acids. These values calculated for two retention coefficient sets, both the retention coefficients found by SVR, as well as using the retention coefficients published by Kyte and Doolittle [27]. These 34 values are supplemented by peptide length, sum of bulkiness coefficients [49] number of polar residues (R, K, D, E, N, Q), the number of consecutive occurrences of the same polar residues and analogous values for hydrophilic amino acids (A, I, L, M, F, W, Y, V, C) and the amino acid composition.

In 2010, the Trans-Proteomic Pipeline [8] (TPP) program called RTCalc replaced SSRCalc for retention time prediction and includes both a simple linear retention coefficient approach and an ANN-based predictor. The Pyteomics [50] and Skyline [51] frameworks now also include retention time predictors. More recently, Dorfer and colleagues demonstrated the use of retention time prediction to simplify peptide identification from chimeric spectra [52]. Peptide retention time prediction is not limited to reversed phase. Krokhin and coworkers have also recently developed successful predictors for hydrophilic interaction liquid chromatography [53, 54] as well as strong cation exchange [55]. Hydrophilic strong anion exchange (hSAX) was studied by Giese, Ishihama, and Rappsilber using retention time modeling and inspection of model parameters to conclude that the retention in hSAX is driven by charged and aromatic residues [56].

Given a measured mass of a peptide from an enzymatic digest and mass measurement accuracy, there is a limited number of possible matching peptides from proteins in any given sequence database that are within the measurement error of the observed mass. If mass accuracy is very high, better than 1 ppm, close to which can be achieved in Fourier transform ion cyclotron resonance (FTICR) mass spectrometry [57], there may exist a peptide of each protein that is unique in the organism's proteome within the measurement uncertainty. These peptides can then be used as "accurate mass tags" for protein identification [58]. In general, however, mass accuracy is insufficient to identify proteins based on a single tryptic peptide mass, requiring either a pattern of several peptides or additional information on the peptides, such as short sequence tags or otherwise informative tandem mass spectra, for unambiguous protein identification.

The accurate mass measurements, together with information on protein size, peptide or protein isoelectric point, or peptide retention times, form a multidimensional protein-dependent *pattern*, and as stated above, these patterns can also be predicted from protein and peptide sequences that in turn are predicted from genome sequence databases. It is this fact that enables protein identification using pattern recognition. Pattern recognition, or the detection of patterns using knowledge of the rules generating those patterns, includes all existing methods for protein identification by mass spectrometry. When good-quality tandem mass spectra are available, these are in general sufficient to identify the protein (or at least the expressed gene) based on short sequence tags. In absence of such tandem mass spectra, proteins may still be identified based on accurately measured mass and retention times for one or more peptides. Accuracy in retention times refers to the closeness to predicted retention times.

The retention time of peptides in reversed phase chromatography using linear gradients is often observed to have a linear mass dependence. Nonlinearities, such as from using nonlinear gradients, require at least a nonlinear scaling function of either measured or predicted retention times. An intuitive and simple model with a small number of parameters for establishing a quantitative structure–activity relationship between amino acid sequences and retention times in liquid chromatography are thus linear functions of amino acid composition, sometimes with special consideration of terminal residues or other modifications [10]. Using a model similar to that described by Hodges et al. [30, 59, 60], we showed how retention time prediction of peptides can be combined with accurate mass from FTICR mass spectrometry to improve the confidence in identifications of human proteins [10]. A similar accurate mass and time tag (AMT) approach was subsequently demonstrated on prokaryotic proteomes by LC-FTICR mass spectrometry by Petritis et al. [18] and LC-time-of-flight mass spectrometry by Strittmatter et al. [36], using normalized retention times and the ANN taking the amino acid composition as input and with sigmoid transfer and output functions. The improvement from adding nonlinear nodes in a hidden layer in the ANN was found to be small [18], and the linear model [10] and the ANN with no hidden layers [18] are very similar, since observed retention times were mapped to the most linear part of the sigmoid output of the ANN. However, optimizing the output function or the part of the output function to which measured data is mapped, as well as using a larger number of input neurons, taking the amino acid *sequence* into account, should improve the accuracy of the predictor. Chromatographic reproducibility of relative (internally calibrated) retention times sets a lower limit on achievable prediction imprecision.

Retention time prediction can also be used as a discriminant function in SEQUEST MS/MS database searches, although the benefit is less significant than when using accurate mass alone.

A number of papers [34, 37, 43, 47] have compared different methods for retention time prediction, mostly in reversed-phase chromatography. There is no apparent consensus in what metric should be used to describe the accuracy of retention time prediction and authors have used the coefficient of determination after linear regression and even standard error of the mean difference between measured and predicted retention. In general, we are not interested if the model is linear or not, as a nonlinear regression or correction would be simple to apply if motivated. Neither do we care about standard errors of the mean, as these say nothing about the uncertainty of predictions, only that there is no systematic error. As we should be primarily interested in assigning an uncertainly or confidence interval to a predicted retention time, we should use a metric such as standard error to avoid confusion. A graph comparing measured and predicted retention times for a training set is also helpful, as it illustrates global trends, such as nonlinearity, outliers and time or peptide size dependence of prediction accuracy.

2   Linear Retention Time Predictor

To illustrate how retention time prediction is implemented in practice, we will use what is probably the simplest useful retention time predictor, namely a linear combination of unique retention coefficients for each amino acid according to

$$t_{calc} = \sum_{i=1}^{20} n_i c_i + t_0 \qquad (1)$$

where c_i is the retention coefficient of amino acid i, of which there are n_i in the peptide, and t_0 compensates for void volumes and any delay between sample injection and acquisition of mass spectra. The coefficients c_i are similar to the weights in the 20-0-1 neural network used by Petritis et al. [18]. The order of the amino acids in the sequence is ignored in this minimal model, including which amino acids are terminal.

To predict retention times of peptides with some of the most frequently studied posttranslational modifications, four modified amino acids (methionine-S-oxide, phosphoserine, phosphothreonine, and phosphotyrosine) were added to the right-hand side of Eq. 1:

$$t_{calc} = \sum_{i=1}^{24} n_i c_i + t_0 \qquad (2)$$

The 20 or 24 c_i and t_0 were determined by least-squares fitting t_{calc} to measured retention times of 100–500 peptides from standard proteins, abundant proteins in the samples or as identified by MS/MS. In the latter case, retention time prediction is subsequently used to increase the confidence in identification of proteins for which no MS/MS data of sufficient quality is available. All software was written in C and run on standard single-processor platforms under Cygwin [61] or Linux.

In contrast to the sequential method used by Meek and Rossetti [20], the least squares multidimensional linear regression problem can be elegantly solved by elementary linear algebra methods. The retention coefficients, regarded as components of a vector, and the value of the dead time t_0 can be written in closed form in terms of the peptide data. The retention coefficients are obtained from the multiplication of a matrix with a vector in a space with the dimensionality of the number of amino acid types taken into account.

The score or probability assigned to a peptide-spectrum match can be adjusted by the likelihood of observing the putatively identified peptide at the measured retention time using the linear model in Eq. 1, or any other model, to predict the retention time from the amino acid composition. The error of retention time prediction for the correctly identified spectra can be estimated from highly confident (probable) peptide-spectrum matches or standard proteins. The error in predictions for false identifications can be estimated using random or decoy proteins. A conservative measure of statistical significance is thus given by the frequency of random (false) matches to experimental data by random protein sequences. Unlike accurate mass and time tags, complete knowledge of all posttranslational modifications, nontryptic peptides and contamination from other species is not assumed or necessary.

Powerful and efficient linear algebra packages such as LAPACK make possible the construction of predictors that can be trained in a very short amount of time. The amount of time necessary to train a linear retention coefficient model even considering several PTMs as additional amino acid types, is even for several thousands of training peptides a fraction of a second on modern desktop machines and increases linearly with training set size. The necessary training time for ANN or SVR based predictors increases is super linear and can be prohibitive for many applications involving many training samples.

The source code for the retention time predictor, or "rt" for short, is available under the GNU General Public License at https://www.ms-utils.org/rt.html, along with instructions for compilation and usage. In its simplest form, rt. takes one argument on the command-line,

```
rt <training set>
```

where *training set* is a space delimited list of retention times in arbitrary units and peptide sequences in the one-letter code with lower case letters for the modified amino acids and one measured retention time and peptide sequence pair per row in an ASCII text file, e.g.

```
28.536 AHGHSmsDPAISY
32.14 SHLtWFCTMKLD
34.763 AGASyTDVAYK
```

etc.

The output from rt. is a list of amino acids in the one-letter code and the corresponding retention coefficients c_i. The chromatographic retention time `peptide[i].t` of a candidate peptide `peptide[i].sequence` (as a text string in the one-letter code) is then predicted by applying Eq. 2:

```
peptide[i].t=c[24];
for(j=0;j<strlen(peptide[i].sequence);j++)
    {
      peptide[i].t+=c[(24-
        strlen(strchr("ARNDCEQGHILKMFPSTWYVmsty",peptide[i].se-
quence[j])))];
    }
```

where `c[]` is a vector of the retention coefficients c_i, "ARNDCEQ-GHILKMFPSTWYVmsty" is a string defining the order of amino acids (in one-letter code) in `c[]`, and `c[24]` is the constant term t_0 in Eqs. 1 and 2). The Taverna workflow shown in Fig. 1 is available on myExperiment.org.

3   Results and Discussion

To illustrate how retention time prediction can be used in a typical proteomics data analysis, we implemented a simple workflow in Taverna [62] (Fig. 1). In this workflow rt. is running on the local machine, but to facilitate integration into scientific workflows, we also made it available as a Web service (http://www.proteoserver.org:8080). The resulting rt. model or retention coefficients can be used to predict retention times of peptides. Figure 2 shows the retention coefficients and Fig. 3 the validation after training rt. on a dataset containing 9630 peptide-spectrum matches of 1049 unique peptides. Allowing redundancy in the training set will generally only have a small effect on the retention coefficients in the model.

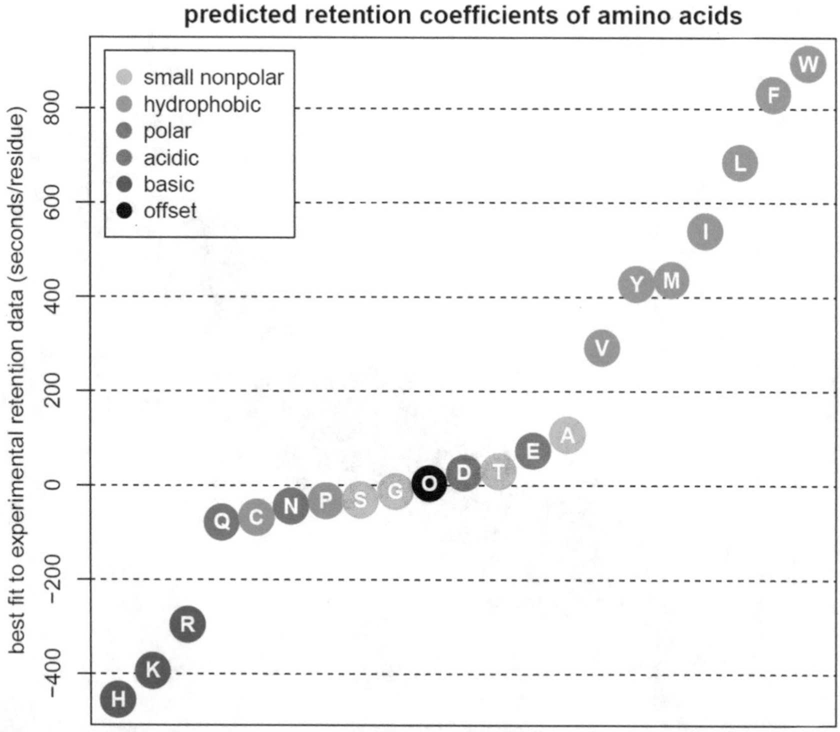

predicted retention coefficients of amino acids

Fig. 2 Retention coefficients for the 20 common amino acids and offset (O) after training of the predictor using the workflow in Fig. 1 and 9630 peptide-spectrum matches of 1049 unique peptides in the training set. The amino acids are colored according to the scheme of Lesk [63] (Rshell script courtesy of W. Plugge). It could be noted that the basic residues are positively charged and hence the least hydrophobic. The acidic residues aspartic and glutamic acid, with side chain pK_a around 4.0 and 4.1 respectively are largely neutral at the pH (~3.0) of the mobile phase. Hence, they behave similarly to the other polar neutral residues. The cysteines were carbamidomethylated, making the residues considerably more hydrophilic. Proline constrains the conformation of the peptide and has an atypical influence on retention time. The offset typically takes on a small positive value close to the void time t_0

In our experience, a good rule-of-thumb is that the training set for simple predictors with 20–30 free parameters, each amino acid should be present in at least 10–15 different peptides in the training set for the parameters in the model to be robustly determined. The c_i values in the model scale with hydrophobicity as expected from literature, that is, the more hydrophobic, the higher the retention coefficient c_i. One should be aware that this is a somewhat circular argument since hydrophobicity and hydrophobicity scales are often defined experimentally by methods not dissimilar to reversed-phase chromatography [16]. The retention coefficients are strongly dependent on chromatographic conditions such as stationary phase, mobile phase composition, and pH. The pH directly influences the charge of amino acid side chains and termini, and hence the hydrophobicity (charged residues being more hydrophilic) and retention coefficients. The t_0 or offset value is often near the observed void time. The overall positive c_i also implies a size

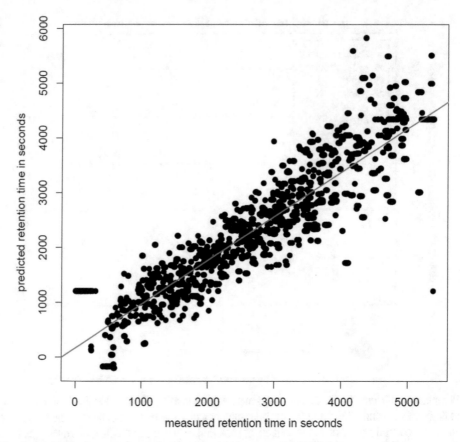

Fig. 3 Predicted versus measured retention time for 1049 unique *E. coli* peptides separated by reversed-phase chromatography, identified by tandem mass spectrometry and analyzed by the workflow in Fig. 1 (Rshell script courtesy of W. Plugge). More refined predictors can produce more accurate predictions, given a sufficiently large training set

dependency of the retention of tryptic peptides, that is, the longer the peptide, the longer the retention time.

To train more sophisticated retention time predictors, many more measured peptide retention times become necessary. For example, Petritis and coworkers used more than 600,000 peptides to train their ANN predictor [37]. The actual software (executable) used for training the retention time predictor can be easily replaced in the workflow in Fig. 1 for more advanced predictors, such as those that take the actual peptide sequence into account or predicted global properties of the peptide [64].

Finally, reversed-phase is not the only separation method used prior to mass spectrometry. In proteomics, it is not uncommon to first separate the proteins using SDS-PAGE. The separation in SDS-PAGE is primarily dependent on the chain lengths, that is, molecular weight, of denatured proteins. A small script can be used to match the position of the gel of protein X (the slice or dataset with the largest number of identified spectra for protein X) with the calculated molecular weight of the sequence of protein X in the

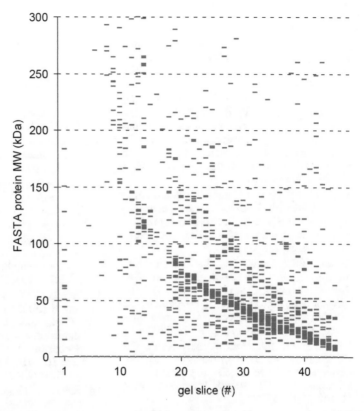

Fig. 4 Reversed-phase retention times of peptides are not the only additional information that can be used to improve or qualify peptide or protein identifications. When fractionating proteins according to size using SDS-PAGE, it is easy to compare the calculated molecular weight with the position on the gel where the identified protein is most abundant, as determined by the number of peptide-spectrum matches to that protein. Deviations from the main distribution could indicate a protein fragment, large posttranslational modification or an erroneous identification (data showing 5609 protein identifications from 48 SDS-PAGE slices—data provided by E. Mostovenko)

FASTA file used in the peptide identification. Figure 4 illustrates how this relationship typically looks. The majority of proteins fall along a curve. The hits lying outside this curve are not necessarily completely false, but may indicate that the protein we have identified is only a fragment of the full protein as present in the FASTA file, or that the FASTA file sequence only covers part of the actual protein sequence, or that the protein is heavily modified.

Information on physicochemical properties of peptides and proteins, such as hydrophobicity, pI, or molecular weight, which to some degree can be predicted from the amino acid sequence, can also assist or validate protein identification by peptide mass finger-printing. This information is already available in LC-MS and LC-MS/MS datasets with no additional experimental cost.

References

1. Laemmli UK (1970) Cleavage of structural proteins during assembly of head of bacteriophage-T4. Nature 227 (5259):680–685

2. Stacey CC, Kruppa GH, Watson CH, Wronka J, Laukien FH, Banks JF, Whitehouse CM (1994) Reverse-phase liquid chromatography/electrospray-ionization Fourier-transform mass spectrometry in the analysis of peptides. Rapid Commun Mass Spectrom 8:513–516

3. Voyksner RD (1997) Combining liquid chromatography with electrospray mass spectrometry. In: Cole RB (ed) Electrospray ionization mass spectrometry. John Wiley & Sons, New York, pp 323–341

4. Jensen PK, Pasa-Tolic L, Peden KK, Martinovic S, Lipton MS, Anderson GA, Tolic N, Wong KK, Smith RD (2000) Mass spectrometric detection for capillary isoelectric focusing separations of complex protein mixtures. Electrophoresis 21(7):1372–1380

5. Käll L, Storey JD, MacCoss MJ, Noble WS (2008) Posterior error probabilities and false discovery rates: two sides of the same coin. J Proteome Res 7(1):40–44. https://doi.org/10.1021/Pr700739d

6. Käll L, Storey JD, MacCoss MJ, Noble WS (2008) Assigning significance to peptides identified by tandem mass spectrometry using decoy databases. J Proteome Res 7(1):29–34. https://doi.org/10.1021/Pr700600n

7. de Bruin JS, Deelder AM, Palmblad M (2012) Scientific workflow management in proteomics. Mol Cell Proteomics. https://doi.org/10.1074/mcp.M111.010595. M111.010595 [pii]

8. Keller A, Eng J, Zhang N, Li XJ, Aebersold R (2005) A uniform proteomics MS/MS analysis platform utilizing open XML file formats. Mol Syst Biol 1:2005.0017

9. Nesvizhskii AI, Keller A, Kolker E, Aebersold R (2003) A statistical model for identifying proteins by tandem mass spectrometry. Anal Chem 75(17):4646–4658

10. Palmblad M, Ramström M, Markides KE, Håkansson P, Bergquist J (2002) Prediction of chromatographic retention and protein identification in liquid chromatography/mass spectrometry. Anal Chem 74(22):5826–5830

11. Eriksson J, Chait BT, Fenyö D (2000) A statistical basis for testing the significance of mass spectrometric protein identification results. Anal Chem 72(5):999–1005

12. Victor B, Gabriel S, Kanobana K, Mostovenko E, Polman K, Dorny P, Deelder AM, Palmblad M (2012) Partially sequenced organisms, decoy searches and false discovery rates. J Proteome Res 11(3):1991–1995. https://doi.org/10.1021/pr201035r

13. Pardee AB (1951) Calculations on paper chromatography of peptides. J Biol Chem 190(2):757–762

14. Knight CA (1951) Paper chromatography of some lower peptides. J Biol Chem 190(2):753–756

15. Sanger F, Thompson EOP (1953) The amino-acid sequence in the glycyl chain of insulin. Biochem J 53:353–374

16. Cornette JL, Cease KB, Margalit H, Spouge JL, Berzofsky JA, DeLisi C (1987) Hydrophobicity scales and computational techniques for detecting amphipathic structures in proteins. J Mol Biol 195(3):659–685

17. Palmblad M, Mills DJ, Bindschedler LV, Cramer R (2007) Chromatographic alignment of LC-MS and LC-MS/MS datasets by genetic algorithm feature extraction. J Am Soc Mass Spectrom 18(10):1835–1843. https://doi.org/10.1016/j.jasms.2007.07.018. S1044-0305(07)00624-1 [pii]

18. Petritis K, Kangas LJ, Ferguson PL, Anderson GA, Pasa-Tolic L, Lipton MS, Auberry KJ, Strittmatter EF, Shen Y, Zhao R, Smith RD (2003) Use of artificial neural networks for the accurate prediction of peptide liquid chromatography elution times in proteome analyses. Anal Chem 75(5):1039–1048

19. Meek JL (1980) Prediction of peptide retention times in high-pressure liquid chromatography on the basis of amino acid composition. Proc Natl Acad Sci U S A 77(3):1632–1636

20. Meek JL, Rossetti ZL (1981) Factors affecting retention and resolution of peptides in high-performance liquid-chromatography. J Chromatogr 211(1):15–28

21. Browne CA, Bennett HP, Solomon S (1982) The isolation of peptides by high-performance liquid chromatography using predicted elution positions. Anal Biochem 124(1):201–208. 0003-2697(82)90238-X [pii]

22. Guo DC, Mant CT, Taneja AK, Parker JMR, Hodges RS (1986) Prediction of peptide retention times in reversed-phase high-performance liquid-chromatography .1. Determination of retention coefficients of amino-acid-residues of model synthetic peptides. J Chromatogr 359:499–517

23. Guo DC, Mant CT, Taneja AK, Hodges RS (1986) Prediction of peptide retention times in reversed-phase high-performance liquid-chromatography .2. Correlation of observed

and predicted peptide retention times and factors influencing the retention times of peptides. J Chromatogr 359:519–532

24. Wilce MCJ, Aguilar MI, Hearn MTW (1991) High-performance liquid-chromatography of amino-acids, peptides and proteins .107. Analysis of group retention contributions for peptides separated with a range of Mobile and stationary phases by reversed-phase high-performance liquid-chromatography. J Chromatogr 536(1–2):165–183

25. Wilce MCJ, Aguilar MI, Hearn MTW (1993) High-performance liquid-chromatography of amino-acids, peptides and proteins .122. Application of experimentally derived retention coefficients to the prediction of peptide retention times – studies with Myohemerythrin. J Chromatogr 632(1–2):11–18

26. Mohammed Y, Palmblad M (2015) Method and software workflow for integrating paired CE-MS and LC-MS bottom-up proteomics data from SDS-PAGE pre-fractionated samples. Paper presented at the 21st international mass spectrometry conference, Toronto, Canada, 2016-08-22

27. Kyte J, Doolittle RF (1982) A simple method for displaying the hydropathic character of a protein. J Mol Biol 157(1):105–132

28. Terabe S, Konaka R, Inouye K (1979) Separation of some polypeptide hormones by high-performance liquid-chromatography. J Chromatogr 172:163–177

29. Hearn MTW, Aguilar MI (1987) High-performance liquid-chromatography of amino-acids, peptides and proteins. 69. Evaluation of retention and bandwidth relationships of myosin-related peptides separated by gradient elution reversed-phase high-performance liquid-chromatography. J Chromatogr 392:33–49

30. Hearn MT, Aguilar MI, Mant CT, Hodges RS (1988) High-performance liquid chromatography of amino acids, peptides and proteins. LXXXV. Evaluation of the use of hydrophobicity coefficients for the prediction of peptide elution profiles. J Chromatogr 438 (2):197–210

31. Mant CT, Hodges RS (2006) Context-dependent effects on the hydrophilicity/hydrophobicity of side-chains during reversed-phase high-performance liquid chromatography: implications for prediction of peptide retention behaviour. J Chromatogr A 1125(2):211–219. https://doi.org/10.1016/j.chroma.2006.05.063

32. Mant CT, Burke TWL, Black JA, Hodges RS (1988) Effect of peptide-chain length on peptide retention behavior in reversed-phase chromatography. J Chromatogr 458:193–205

33. Krokhin OV, Craig R, Spicer V, Ens W, Standing KG, Beavis RC, Wilkins JA (2004) An improved model for prediction of retention times of tryptic peptides in ion pair reversed-phase HPLC - its application to protein peptide mapping by off-line HPLC-MALDI MS. Mol Cell Proteomics 3(9):908–919. https://doi.org/10.1074/mcp.M400031-MCP200

34. Krokhin OV, Ying S, Cortens JP, Ghosh D, Spicer V, Ens W, Standing KG, Beavis RC, Wilkins JA (2006) Use of peptide retention time prediction for protein identification by off-line reversed-phase HPLC-MALDI MS/MS. Anal Chem 78(17):6265–6269. https://doi.org/10.1021/Ac060251b

35. Krokhin OV (2006) Sequence-specific retention calculator. Algorithm for peptide retention prediction in ion-pair RP-HPLC: application to 300-and 100-angstrom pore size C18 sorbents. Anal Chem 78(22):7785–7795. https://doi.org/10.1021/Ac060777w

36. Strittmatter EF, Ferguson PL, Tang K, Smith RD (2003) Proteome analyses using accurate mass and elution time peptide tags with capillary LC time-of-flight mass spectrometry. J Am Soc Mass Spectrom 14(9):980–991

37. Petritis K, Kangas LJ, Yan B, Monroe ME, Strittmatter EF, Qian WJ, Adkins JN, Moore RJ, Xu Y, Lipton MS, Camp DG 2nd, Smith RD (2006) Improved peptide elution time prediction for reversed-phase liquid chromatography-MS by incorporating peptide sequence information. Anal Chem 78 (14):5026–5039. https://doi.org/10.1021/ac060143p

38. Eisenberg D, Weiss RM, Terwilliger TC (1982) The helical hydrophobic moment – a measure of the amphiphilicity of a helix. Nature 299(5881):371–374

39. Eisenberg D, Weiss RM, Terwilliger TC (1984) The hydrophobic moment detects periodicity in protein hydrophobicity. Proc Natl Acad Sci U S A 81(1):140–144

40. Eisenberg D (1984) 3-Dimensional structure of membrane and surface-proteins. Annu Rev Biochem 53:595–623

41. Cortes C, Vapnik V (1995) Support-vector networks. Mach Learn 20(3):273–297

42. Klammer AA, Yi XH, MacCoss MJ, Noble WS (2007) Improving tandem mass spectrum identification using peptide retention time prediction across diverse chromatography conditions. Anal Chem 79(16):6111–6118. https://doi.org/10.1021/Ac070262k

43. Pfeifer N, Leinenbach A, Huber CG, Kohlbacher O (2007) Statistical learning of peptide retention behavior in chromatographic separations: a new kernel-based approach for computational proteomics. BMC Bioinformatics 8.

https://doi.org/10.1186/1471-2105-8-468. Artn 468

44. Scholkopf B, Smola AJ, Williamson RC, Bartlett PL (2000) New support vector algorithms. Neural Comput 12(5):1207–1245

45. Meinicke P, Tech M, Morgenstern B, Merkl R (2004) Oligo kernels for datamining on biological sequences: a case study on prokaryotic translation initiation sites. BMC Bioinformatics 5. https://doi.org/10.1186/1471-2105-5-169. Artn 169

46. Kohlbacher O, Reinert K, Gropl C, Lange E, Pfeifer N, Schulz-Trieglaff O, Sturm M (2007) TOPP – the OpenMS proteomics pipeline. Bioinformatics 23(2):E191–E197. https://doi.org/10.1093/bioinformatics/btl299

47. Moruz L, Tomazela D, Kall L (2010) Training, selection, and robust calibration of retention time models for targeted proteomics. J Proteome Res 9(10):5209–5216. https://doi.org/10.1021/Pr1005058

48. Rousseeuw PJ, Van Driessen K (2006) Computing LTS regression for large data sets. Data Min Knowl Disc 12(1):29–45. https://doi.org/10.1007/s10618-005-0024-4

49. Zimmerman JM, Eliezer N, Simha R (1968) Characterization of amino acid sequences in proteins by statistical methods. J Theor Biol 21(2):170–201

50. Goloborodko AA, Levitsky LI, Ivanov MV, Gorshkov MV (2013) Pyteomics—a Python framework for exploratory data analysis and rapid software prototyping in proteomics. J Am Soc Mass Spectrom 24(2):301–304. https://doi.org/10.1007/s13361-012-0516-6

51. MacLean B, Tomazela DM, Shulman N, Chambers M, Finney GL, Frewen B, Kern R, Tabb DL, Liebler DC, MacCoss MJ (2010) Skyline: an open source document editor for creating and analyzing targeted proteomics experiments. Bioinformatics 26(7):966–968. https://doi.org/10.1093/bioinformatics/btq054

52. Dorfer V, Maltsev S, Winkler S, Mechtler K (2018) CharmeRT: boosting peptide identifications by chimeric spectra identification and retention time prediction. J Proteome Res. https://doi.org/10.1021/acs.jproteome.7b00836

53. Krokhin OV, Ezzati P, Spicer V (2017) Peptide retention time prediction in hydrophilic interaction liquid chromatography: data collection methods and features of additive and sequence-specific models. Anal Chem 89(10):5526–5533. https://doi.org/10.1021/acs.analchem.7b00537

54. Spicer V, Krokhin OV (2018) Peptide retention time prediction in hydrophilic interaction liquid chromatography. Comparison of separation selectivity between bare silica and bonded stationary phases. J Chromatogr A 1534:75–84. https://doi.org/10.1016/j.chroma.2017.12.046

55. Gussakovsky D, Neustaeter H, Spicer V, Krokhin OV (2017) Sequence-specific model for peptide retention time prediction in strong cation exchange chromatography. Anal Chem 89(21):11795–11802. https://doi.org/10.1021/acs.analchem.7b03436

56. Giese SH, Ishihama Y, Rappsilber J (2018) Peptide retention in hydrophilic strong anion exchange chromatography is driven by charged and aromatic residues. Anal Chem 90(7):4635–4640. https://doi.org/10.1021/acs.analchem.7b05157

57. Bruce JE, Anderson GA, Wen J, Harkewicz R, Smith RD (1999) High-mass-measurement accuracy and 100% sequence coverage of enzymatically digested bovine serum albumin from an ESI-FTICR mass spectrum. Anal Chem 71(14):2595–2599

58. Conrads TP, Anderson GA, Veenstra TD, Pasa-Tolic L, Smith RD (2000) Utility of accurate mass tags for proteome-wide protein identification. Anal Chem 72(14):3349–3354

59. Hodges RS, Parker JM, Mant CT, Sharma RR (1988) Computer simulation of high-performance liquid chromatographic separations of peptide and protein digests for development of size- exclusion, ion-exchange and reversed-phase chromatographic methods. J Chromatogr 458:147–167

60. Mant CT, Burke TW, Zhou NE, Parker JM, Hodges RS (1989) Reversed-phase chromatographic method development for peptide separations using the computer simulation program ProDigest-LC. J Chromatogr 485:365–382

61. The Cygwin homepage. http://www.cywin.com/

62. Oinn T, Addis M, Ferris J, Marvin D, Senger M, Greenwood M, Carver T, Glover K, Pocock MR, Wipat A, Li P (2004) Taverna: a tool for the composition and enactment of bioinformatics workflows. Bioinformatics 20(17):3045–3054

63. Lesk AM (2008) Introduction to bioinformatics, 3rd edn. Oxford University Press, New York

64. Rost B (2001) Review: Protein secondary structure prediction continues to rise. J Struct Biol 134(2–3):204–218

Chapter 5

Comparing Peptide Spectra Matches Across Search Engines

Rune Matthiesen, Gorka Prieto, and Hans Christian Beck

Abstract

Mass spectrometry is extremely efficient for sequencing small peptides generated by, for example, a trypsin digestion of a complex mixture. Current instruments have the capacity to generate 50–100 K MSMS spectra from a single run. Of these ~30–50% is typically assigned to peptide matches on a 1% FDR threshold. The remaining spectra need more research to explain. We address here whether the 30–50% matched spectra provide consensus matches when using different database-dependent search pipelines. Although the majority of the spectra peptide assignments concur across search engines, our conclusion is that database-dependent search engines still require improvements.

Key words Database dependent search, Peptide assignments

1 Introduction

A large number of freely available and open souce database-dependent search engines is currently available. We adress here if these search engines produce consensus solutions or if they produce largely different outputs. Relatively few reports have addressed this question [1–5]. Furthermore, both software and MS instruments are rapidly evolving which justifies revisiting this question.

Shteynberg and coworkers concluded—based on past research—that overall divergence of spectra matches is quite high among the different search engines. Despite high overall disagrement, high agreement among high confidence matches were reported. Consequently, they evaluated different methods to combine search results from different search engines in proteomics in order to improve the overall search result [2]. This lead to several efforts to combine search results from different search engines [1, 6–12], and as a result a number of commercial solutions are now available (e.g., Scaffold). The question is if the problem has scaled with increasing complexity of the proteomics data sets, or if the search data have improved due to for example improved mass accuracy and sensitivity of the MS instruments.

Rune Matthiesen (ed.), *Mass Spectrometry Data Analysis in Proteomics*, Methods in Molecular Biology, vol. 2051,
https://doi.org/10.1007/978-1-4939-9744-2_5, © Springer Science+Business Media, LLC, part of Springer Nature 2020

Our aim is therefore to evaluate—at the peptide level—the degree of disagreement between different search pipelines to understand if search engines still need improvements. Comparing search engines is complicated due to several factors: (1) raw spectra processing is frequently build into the search engine, which means that a standard peak list cannot always be provided for all the search engines; (2) some false discovery rate (FDR) algorithms are more conservative than others and therefore not comparable [13]; (3) search engines have intrinsic parameters such as min and max peptide length considered for matching with MSMS data, and additional modifications automatically included in the search; (4) protein inference is still a complicated issue on which the different search engines varies greatly. This issue is the main reason for only comparing on the peptide level in the present study.

Concerning issue one, listed above, we restrain the peak processing to Proteome Discoverer and RawConverter [14]. For FDR calculation Percolator [15], PeptideShaker [16] and Proteome Discoverer were used. Spectra assignment from a search engine that match modifications not directly specified for all search engines were filtered out and not included in the comparison. Protein identification was not compared do to the lack of consensus methods for protein inference.

A key issue when comparing search engine output is a proper reference set of peptide spectra matches. Synthetic peptides serve as a excellent benchmark for peptide identification. However, such data set with high mass accuracy in both MS and MSMS were not available to us [17]. Shteynberg et al. [2] used a manual validated data set. However, this approach is cumbersome and prone to individual bias. We decided to use identifications from five search engines at a 1% FDR threshold that were identified by the highest number of search engines as a reference. The search engines used were Mascot [18], SEQUEST [19], X!Tandem [20], OMSSA [21], and MyriMatch [22]. Our main objective was to evaluate the agreement between the different search engines on a complex data set obtained with high mass accuracy in both MS and MSMS. To conclude on which search engines is the most accurate, synthetic peptides are preferable [17], but our approach should enable us to conclude on the current consensus performance of the search engines.

One of the major problems when comparing peptides identified from different search engines is that only one match is reported for spectra matching peptides containing leucine (L) or isoleucine (I), and the different search engines frequently report I/L-containing peptides from different proteins. We therefore transformed all I and L to X before comparing the peptides, and location of modifications were ignored. After transforming the peptides, a consensus list of peptides was made, which also included the number of search engines supporting the spectra match. Figure 1

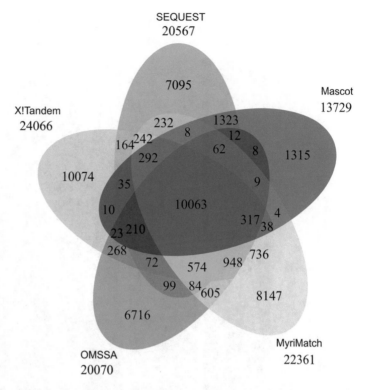

Fig. 1 Venn diagram comparing spectra matches from the five tested search engines

compares the five search engines in a Venn diagram. We observe that Mascot has a significantly lower number of unique peptide spectrum matches than the other search engines. Mascot and SEQUEST have the largest number of shared spectrum assignments between two search engines which means that the identifications from Mascot and SEQUEST are more correlated. We next counted the number of unmatched peptides, the number of peptide spectra assignments in agreement with the consensus, and the number of spectra assignments differing from consensus (*see* Fig. 2). We observe that Mascot differ substantially from the other search engines by assigning very few spectra that are in disagreement with the consensus and, on the other hand, a large number of unassigned spectra. The high number of spectrum matches in disagreement with the consensus for SEQUEST, X! Tandem, OMSSA, and MyriMatch strongly suggests that the FDR calculation underestimates the errors made by the pipelines for these search engines. For Mascot the FDR estimation looks more reasonable considering the low number of spectra matches in disagreement with the consensus. The FDR estimation was made using PeptideShaker [16] for X!Tandem, OMSSA, and MyriMatch, whereas SEQUEST and Mascot employs the Percolator node

Agreement with consensus

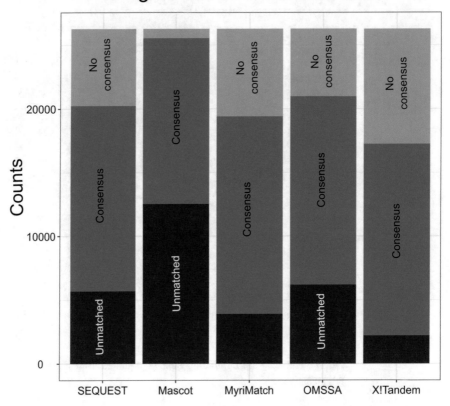

Fig. 2 Stacked bar plot of depicting the number of unmatched spectra (black), consensus-matched spectra assignments (dark blue), and spectra matches in disagreement with the consensus (light blue)

integrated into ProteomeDiscoverer for FDR determination. We are unable to conclude on which search engine is optimal because two or more search engines that aligne/correlate well might bias the consensus list. However, it is evident that the true error rate is higher than 1% for at least some of the four search engines because of the high number of spectrum assignments which are in disagreement with the consensus. To obtain a better insight concerning whether current search engines require improvement over current standard, we plotted X!Tandem score distributions for consensus spectra assignments supported by one and up to five search engines (*see* Fig. 3). As expected, the score distribution move toward higher scores depending on the number of search engines supporting the spectrum assignments. The score distributions for spectrum assignments supported by two to four search engines presents a considerable overlap of the score distribution from spectra assignments supported by all search engines. This makes us speculate that at least some of the peptides in the category supported by two to four search engines ideally should be identified by all search engines.

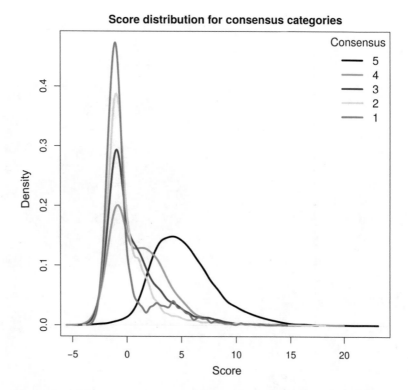

Fig. 3 Score distributions for consensus categories. The score distribution for spectra assignments supported by one (red), two (yellow), three (blue), four (green), and five (black) search engines are depicted as density versus score

Finally, consensus peptide spectrum assignments were made for all combinations of two search engines, and these combinations were evaluated against the consensus for all the five search engines (*see* Fig. 4). We see that specific combination of search engines maintain a similar level of peptide spectrum assignments in disagreements with the consensus as individual search engines but increase the number of spectrum assignments in agreement with the consensus (e.g., Mascot–OMSSA, Mascot–MyriMatch, and SEQUEST–OMSSA). The individual search engine pipelines cluster in three different areas in the plot: (1) Proteome Discoverer–Mascot (red); (2) Proteome Discoverer–SEQUEST, RawConverter–MyriMatch and –OMSSA (blue); (3) X!Tandem (green). X! Tandem had the most disagreement from the consensus list supported by all five search engines. X!Tandem reports some spectrum matches for which the annotated parent ion mass is not in agreement with the observed parent ion mass, which partly explains the larger disagreement for X!Tandem. However, we observed very few cases where X!Tandem reports incorrectly assigned parent ion masses when manually validated.

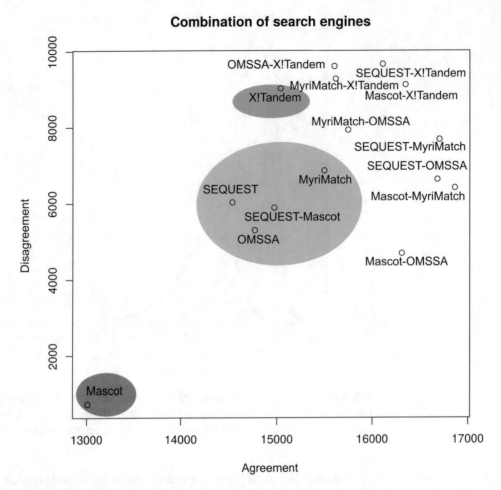

Fig. 4 Consensus peptide spectra assignments for all combination of one to two search engines were evaluated against the consensus for all five search engines

2   Materials and Software

2.1   Noncommercial

1. Data set: Raw data file "b1929_293T_proteinI-D_09A_QE3_122212.raw" from ProteomeXchange project "PXD001468."

    http://proteomecentral.proteomexchange.org/cgi/GetDataset?ID=PXD001468.

2. UniProt sequence database.

3. RawConvert [14]: http://fields.scripps.edu/rawconv/.

4. X!tandem: https://www.thegpm.org/TANDEM/index.html version Sledgehammer (2013.09.01.1).

5. Percolator: https://github.com/percolator/percolator/releases.

2.2 Commercial	1. Mascot version 2.4.
	2. SEQUEST.
	3. Proteome Discoverer v1.4.1.14.

3   Methods

3.1 Download the Raw Data File b1929_293T_ proteinID_09A_ QE3_122212.raw	1. Go to the web page Pride archive [23]: https://www.ebi.ac.uk/pride/archive/projects/PXD001468.
	2. Click on "Download Project Files" in the above right corner.
	3. Scroll down to the raw file "b1929_293T_proteinID_09A_QE3_122212.raw" and select for download.
3.2 Download UniProt Human Sequence Database	1. http://www.uniprot.org/uniprot/?query=*&fil=organism%3A%22Homo+sapiens+%28Human%29+%5B9606%5D%22+AND+reviewed%3Ayes.
	2. *See* Chapter 6 on how to format the sequence database containing reverse sequences and so the database is compatible with Percolator (a program to quality filter identifications).
3.3 Process Raw Data File "b1929_293T_ proteinID_09A_QE3_ 122212.raw" to mgf Format	1. After installing RawConverter the program is found in the start menu in Microsoft Windows.
	2. Press "+" button to select the raw file. "b1929_293T_proteinID_09A_QE3_122212.raw".
	3. Set experiment type "DDA."
	4. Set destination output directory by clicking the button "Browse."
	5. Choose MGF as output format.
	6. Set the charge states "Options → DDA → Charge States" (*see* **Note 1**).
	7. Press the button "Go!".
3.4 Database Dependent Search Using X!Tandem Sledgehammer for Reference Scores (See Note 2)	1. Download and compile X!Tandem according to the instructions in the "INSTALL" file.
	2. Open the file "input.xml" and change "protein, taxon" from yeast to human, "spectrum, path" to the file path of the mgf file created above in Subheading 3.3 and set the output path "output, path".
	3. Open the file "default_input.xml" and modify "spectrum, fragment monoisotopic mass error" from 0.4 to 0.01, "spectrum, parent monoisotopic mass error plus" to 5, "spectrum, parent monoisotopic mass error minus" to 5.

4. Scroll down to "residue, potential modification mass" and set the value to "15.99491462@[M],0.984016@[QN]" without the quotes.

5. Scroll down to "refine, potential N-terminus modifications" and set the value to "42.01056469@[".

6. Open taxonomy.xml file and change "<taxon label = "yeast">" to "<taxon label = "human">". Modify the path in the line "<file format = "peptide" URL = "../fasta/scd.fasta.pro" />" to contain the path for the UniProt sequence database and delete the next two lines.

7. Execute "tandem.exe input.xml" on the command line (*see* **Note 3**).

3.5 MyriMatch, X! Tandem and OMSSA Searches Interfaced from SearchGUI

1. Start PeptideShaker.

2. Click button "Start search."

3. Click the "Add" button to select the mgf files produced in Subheading 3.3.

4. Click the next "Add" button to configure the search settings.

5. Click the button "Spectrum matching."

6. Click the "edit" button for database FASTA files.

7. Click the "Browse" button and select the file from Subheading 3.2, **step 1**.

8. Click "Decoy" to create a FASTA file with decoy sequence and close by pressing "OK" (*see* **Note 4**).

9. Click edit to select the FASTA file with decoy sequences.

10. Select Carbamidomethylation of C as fixed modifications.

11. Select Deamidation of N, Deamidation of Q, Acetylation of protein N-term, and Oxidation of M as variable modifications.

12. Leave the remaining settings as default.

13. Click the PeptideShaker checkbox and fill in the project information and choose an output file.

14. Repeat **steps 1–13** for MyriMatch, X!Tandem and OMSSA and press" Start the search!"

15. For each of the searches click "Export → Identification Features" in PeptideShaker.

3.6 Mascot and SEQUEST Searches from Proteome Discoverer

1. Start Proteome Discoverer.

2. Click button "Workflow Editor."

3. Click the "New workflow" to create a Workspace pane.

4. In the annotation area select the "Spectrum Files," "Mascot," "Sequest," and "Percolator" panes and drag them to the Workspace pane one by one.

5. Connect the "Spectrum Files" pane with the "Spectrum Selector" pane by clicking on the "Spectrum Files" pane and drag the cursor from the small square that appears to the "Spectrum Selector" pane.

6. Connect the "Spectrum Selector" pane with the "Sequest" and the "Mascot" pane, and connect the "Sequest" and the "Mascot" pane with the "Percolator" pane as described in **step 5**.

7. Click the "Spectrum File" pane and browse for and select the raw file: "b1929_293T_proteinID_09A_QE3_122212.raw".

8. Click "Mascot" pane in the Workspace pane and then click "Protein Database" in right panel and select FASTA database, click "Enzyme name" and select "Trypsin, click "Maximum missed cleavage" and select "1", click Precursor Mass Tolerance and select "5 ppm", click Fragment Mass Tolerance and select "0.05 Da", click "Dynamic modifications" and select "methionine oxidation" "NQ deamidation", and "N-terminal acetylation", click "Static Modification" and click "Carbamidomethyl-C".

9. Click "Mascot" pane in the Workspace pane and then repeat as in **step 8**.

10. Leave the remaining settings as default.

11. Place cursor in the text box to the right of "Name" and type "Search name."

12. Click the "Start Search button."

13. Click "Administration" pane and select "Show Job Queue" pane.

14. Click "Search Name" and click "Open report."

15. For each of the searches click "Export →" then select appropriate file format.

4   Notes

1. Charge states from +1 to +6 were specified in RawConverter "Options → DDA → Charge States".

2. There exists a number of graphical interfaces to X!Tandem (e.g., *see* Chapter 6). However, it is relatively easy to modify the xml files directly or make your own script to change the parameter files as indicated in Subheading 3.4.

3. The compiled program outputted by the program make on Linux is also named "tandem.exe" which is confusing, but it also works in Linux-based systems.

4. The way the decoy sequences are created heavily affects the FDR estimations. For example, some programs eliminate

reverse or randomized sequences that are corresponding to a true peptide sequence. Furthermore, various FDR filtering algorithms exists [24]. It has been suggested that the decoy database can either be either concatenated and search together with the target database or search separately. If searched together the decoy sequences must compete with the target sequences, which for most search engines means that the score distribution for decoy sequence is biased toward lower scores than if searched separately. FDR estimation is also not valid or comparable if several rounds of optimizations are performed (methods applied by many search engines). Finally, postprocessing of peptide scores can also bias the FDR distributions (e.g., increasing peptide score if matching a protein with many other peptide matches).

Acknowledgments

R.M. is supported by Fundação para a Ciência e a Tecnologia (FCT investigator program 2012), iNOVA4Health—UID/Multi/04462/2013, a program financially supported by Fundação para a Ciência e Tecnologia/Ministério da Educação e Ciência, through national funds and is cofunded by FEDER under the PT2020 Partnership Agreement. This work is also funded by FEDER funds through the COMPETE 2020 Programme and National Funds through FCT – Portuguese Foundation for Science and Technology under the projects number PTDC/BTM-TEC/30087/2017 and PTDC/BTM-TEC/30088/2017.

References

1. Keller A, Eng J, Zhang N, Li XJ, Aebersold R (2005) A uniform proteomics MS/MS analysis platform utilizing open XML file formats. Mol Syst Biol 1:2005.0017. https://doi.org/10.1038/msb4100024

2. Shteynberg D, Nesvizhskii AI, Moritz RL, Deutsch EW (2013) Combining results of multiple search engines in proteomics. Mol Cell Proteomics 12(9):2383–2393. https://doi.org/10.1074/mcp.R113.027797

3. Paulo JA (2013) Practical and efficient searching in proteomics: a cross engine comparison. WebmedCentral 4(10). https://doi.org/10.9754/journal.wplus.2013.0052

4. Kapp EA, Schutz F, Connolly LM, Chakel JA, Meza JE, Miller CA, Fenyo D, Eng JK, Adkins JN, Omenn GS, Simpson RJ (2005) An evaluation, comparison, and accurate benchmarking of several publicly available MS/MS search algorithms: sensitivity and specificity analysis.

Proteomics 5(13):3475–3490. https://doi.org/10.1002/pmic.200500126

5. Balgley BM, Laudeman T, Yang L, Song T, Lee CS (2007) Comparative evaluation of tandem MS search algorithms using a target-decoy search strategy. Mol Cell Proteomics 6(9):1599–1608. https://doi.org/10.1074/mcp.M600469-MCP200

6. Alves G, Wu WW, Wang G, Shen RF, Yu YK (2008) Enhancing peptide identification confidence by combining search methods. J Proteome Res 7(8):3102–3113. https://doi.org/10.1021/pr700798h

7. Kwon T, Choi H, Vogel C, Nesvizhskii AI, Marcotte EM (2011) MSblender: a probabilistic approach for integrating peptide identifications from multiple database search engines. J Proteome Res 10(7):2949–2958. https://doi.org/10.1021/pr2002116

8. Searle BC, Turner M, Nesvizhskii AI (2008) Improving sensitivity by probabilistically combining results from multiple MS/MS search methodologies. J Proteome Res 7(1):245–253. https://doi.org/10.1021/pr070540w

9. Shteynberg D, Deutsch EW, Lam H, Eng JK, Sun Z, Tasman N, Mendoza L, Moritz RL, Aebersold R, Nesvizhskii AI (2011) iProphet: multi-level integrative analysis of shotgun proteomic data improves peptide and protein identification rates and error estimates. Mol Cell Proteomics 10(12):M111.007690. https://doi.org/10.1074/mcp.M111.007690

10. Sultana T, Jordan R, Lyons-Weiler J (2009) Optimization of the use of consensus methods for the detection and putative identification of peptides via mass spectrometry using protein standard mixtures. J Proteomics Bioinform 2(6):262–273. https://doi.org/10.4172/jpb.1000085

11. Dagda RK, Sultana T, Lyons-Weiler J (2010) Evaluation of the consensus of four peptide identification algorithms for tandem mass spectrometry based proteomics. J Proteomics Bioinform 3:39–47. https://doi.org/10.4172/jpb.1000119

12. Nahnsen S, Bertsch A, Rahnenfuhrer J, Nordheim A, Kohlbacher O (2011) Probabilistic consensus scoring improves tandem mass spectrometry peptide identification. J Proteome Res 10(8):3332–3343. https://doi.org/10.1021/pr2002879

13. Serang O, Noble W (2012) A review of statistical methods for protein identification using tandem mass spectrometry. Stat Interface 5(1):3–20

14. He L, Diedrich J, Chu YY, Yates JR 3rd (2015) Extracting accurate precursor information for tandem mass spectra by RawConverter. Anal Chem 87(22):11361–11367. https://doi.org/10.1021/acs.analchem.5b02721

15. The M, MacCoss MJ, Noble WS, Kall L (2016) Fast and accurate protein false discovery rates on large-scale proteomics data sets with percolator 3.0. J Am Soc Mass Spectrom 27(11):1719–1727. https://doi.org/10.1007/s13361-016-1460-7

16. Vaudel M, Burkhart JM, Zahedi RP, Oveland E, Berven FS, Sickmann A, Martens L, Barsnes H (2015) PeptideShaker enables reanalysis of MS-derived proteomics data sets. Nat Biotechnol 33(1):22–24. https://doi.org/10.1038/nbt.3109

17. Quandta A, Espona L, Balasko A, Weissera H, Brusniak M, Kunsztb P, Aebersold R, Malmström L (2015) Using synthetic peptides to benchmark peptide identification software and search parameters for MS/MS data analysis. EuPA Open Proteom 5:21–31

18. Perkins DN, Pappin DJ, Creasy DM, Cottrell JS (1999) Probability-based protein identification by searching sequence databases using mass spectrometry data. Electrophoresis 20(18):3551–3567. https://doi.org/10.1002/(SICI)1522-2683(19991201)20:18<3551::AID-ELPS3551>3.0.CO;2-2

19. Eng JK, McCormack AL, Yates JR (1994) An approach to correlate tandem mass spectral data of peptides with amino acid sequences in a protein database. J Am Soc Mass Spectrom 5(11):976–989. https://doi.org/10.1016/1044-0305(94)80016-2

20. Craig R, Beavis RC (2004) TANDEM: matching proteins with tandem mass spectra. Bioinformatics 20(9):1466–1467. https://doi.org/10.1093/bioinformatics/bth092

21. Geer LY, Markey SP, Kowalak JA, Wagner L, Xu M, Maynard DM, Yang X, Shi W, Bryant SH (2004) Open mass spectrometry search algorithm. J Proteome Res 3(5):958–964. https://doi.org/10.1021/pr0499491

22. Tabb DL, Fernando CG, Chambers MC (2007) MyriMatch: highly accurate tandem mass spectral peptide identification by multivariate hypergeometric analysis. J Proteome Res 6(2):654–661. https://doi.org/10.1021/pr0604054

23. Ternent T, Csordas A, Qi D, Gomez-Baena G, Beynon RJ, Jones AR, Hermjakob H, Vizcaino JA (2014) How to submit MS proteomics data to ProteomeXchange via the PRIDE database. Proteomics 14(20):2233–2241. https://doi.org/10.1002/pmic.201400120

24. Aggarwal S, Yadav AK (2016) False discovery rate estimation in proteomics. Methods Mol Biol 1362:119–128. https://doi.org/10.1007/978-1-4939-3106-4_7

Chapter 6

Calculation of False Discovery Rate for Peptide and Protein Identification

Gorka Prieto and Jesús Vázquez

Abstract

Shotgun proteomics is the method of choice for large-scale protein identification. However, the use of a robust statistical workflow to validate such identification is mandatory to minimize false matches, ambiguities, and amplification of error rates from spectra to proteins. In this chapter we emphasize the key concepts to take into account when processing the output of a search engine to obtain reliable peptide or protein identifications. We assume that the reader is already familiar with tandem mass spectrometry so we can focus on the use of statistical confidence methods. After introducing the key concepts we present different software tools and how to use them with an example dataset.

Key words Shotgun proteomics, Bioinformatics, Peptide identification, Protein inference, False discovery rate, Target–decoy approach

1 Introduction

Shotgun proteomics (also called bottom-up or peptide-centric proteomics) is the method of choice for large-scale protein identification. An example of this is the recent publication of the first drafts of the human proteome [1, 2]. Although this approach can report an impressive number of identifications, the use of robust statistical validation methods is mandatory to keep down the number of false positives. As a matter of fact, the above mentioned human proteome drafts faced this problem as pointed out by further analyses [3, 4].

In this introduction we remark the key aspects to take into account when building up from peptide-spectrum matches (*see* **Note 1**) to protein-level identifications (Fig. 1). In Subheading 2 we present different software tools available, and in Subheading 3 we show how to use them with example datasets.

Rune Matthiesen (ed.), *Mass Spectrometry Data Analysis in Proteomics*, Methods in Molecular Biology, vol. 2051, https://doi.org/10.1007/978-1-4939-9744-2_6, © Springer Science+Business Media, LLC, part of Springer Nature 2020

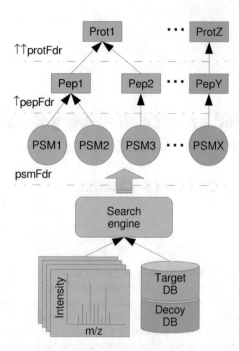

Fig. 1 Shotgun proteomics bioinformatic workflow from spectra to protein identifications. Different PSMs provided by the search engine can refer to the same peptide, and different peptide sequences can be mapped to a same protein. The target matches (green) tend to aggregate more than the decoy matches (red) since these last ones are random matches. FDR can be computed for each of the PSM, peptide, and protein levels

1.1 The Target–Decoy Database Search Strategy

The different database search engines return a match for the large majority of the input MS/MS spectra, but only a fraction of these PSMs are true positives. In order to assign a confidence score to these PSMs and keep low the number of false positives, the most accepted method is the calculation of the false discovery rate (*see* **Note 2**).

For FDR calculation the most popular strategy in shotgun proteomics is the target–decoy approach (TDA). In this approach we use a target database containing every possible protein sequence in the sample and a decoy database that can be used separately or concatenated to the first one. The decoy database contains fake proteins that should not be detected in the sample, and should have the same size and the same statistical properties than the target database, so that the chance of a random match is the same for the two databases.

There are several strategies for building the decoy database [5]:

- Stochastic methods relying on amino acid frequencies or training Markov chain models with the target database to build the decoy database.

- Reverse the target protein sequences to build the decoy sequences. This method is simple and preserves amino acid frequencies and protein lengths.

- Pseudoreverse the target protein sequences by performing an in silico digestion of proteins into peptides. These peptide sequences are reversed but their position in the protein and the amino acids marking the cleavage site are maintained. This strategy not only preserves the amino acid distributions and protein lengths but also the peptide length distribution.

Apart from the statistical distribution of the decoy database, there are other aspects to take into account when building the databases that are to be used for the search. For instance, it is very common that, following any of the decoy database generation strategies already presented, peptides with the same sequence in the decoy than in the target database are generated. Filtering out these peptides is important for avoiding incorrect target–decoy assignments. In Subheading 2 we present the DecoyPyrat tool [6] for this purpose.

Another aspect than can be taken into account when building the databases is the case of amino acids with the same molecular mass, like leucine and isoleucine. These amino acids can lead to ambiguous matches with the same score for one spectrum to two peptides where the only difference between them is the substitution of a leucine by an isoleucine. Some software tools just take one of them, missing the other case that could be the real one, while other tools take into account the two possible PSMs reporting PSM-level ambiguities. One simple approach for managing this situation is to replace leucine (L) and isoleucine (I) in the sequences database with a common symbol (J for instance, or just L if J is not supported by the software). In this way PSM-level ambiguities are managed as peptide-level ambiguities as presented later.

1.2 PSM-Level FDR

Once the decoy database has been built, a search is carried out against each database separately or against the concatenated database. By comparing the number of PSMs matching the target or decoy protein entries we can estimate a FDR value.

Since the same spectrum yields multiple PSMs, they are ranked according to a specific search engine PSM-level score, and the one with the best score (rank = 1) is typically selected. PSM-level FDR is computed after this step.

Conceptually, the simplest method to apply the target–decoy strategy is to search separately against the target and decoy databases, and use the number of decoy PSMs above threshold (D) as an estimate of the number of FP target PSMs. If T is the number of target PSMs above threshold, the FDR can be simply estimated as:

$$\text{FDR} = \frac{D}{T} \tag{1}$$

This procedure has the main drawback that the mass spectra yielding high scores in the target database tend to produce relatively high scores also in the decoy database. Hence the FDR is over-estimated. To avoid this problem, the competition strategy was introduced [5]. The search is performed against the concatenated target + decoy database. Since only the best match is retrieved from each spectrum, the target and decoy sequences have to compete to match the mass spectra, so that the TP PSMs tend to match the target sequences, while the FP PSMs are equally distributed among target and decoy sequences. Since the number of decoy PSMs represents half the number of FP matches, the FDR in the competition strategy can be calculated as:

$$FDR = \frac{2D}{T + D} \tag{2}$$

The competition approach has the drawback that the population of identified peptides contains target and decoy sequences. A simple approach to exploit the competition strategy is to estimate FDR using (1). This approach misses some target PSMs that fall above threshold, but are not considered in the population of identified PSMs because a decoy PSM is competing with a higher score. This effect can be avoided by using the "refined FDR" as proposed in [7], where the search is performed separately against the target and decoy databases, and the competition strategy is applied a posteriori, in a way that all the target PSMs above threshold are considered. Although this method provides what is assumed to be a more accurate calculation of FDR, its implementation requires more computer resources. Since the results for small FDR values are similar than the ones obtained when using (1) with a concatenated database search, we will adhere to the traditional method for this chapter, which is also more conservative.

To control the PSM-level error we can apply a threshold to the PSM score and compute the resulting FDR. If we repeat this process using as a threshold the score of each PSM, we can compute a local FDR value for each PSM (*see* **Note 2**). However, this local FDR value is not a monotonically decreasing function[1] of the PSM score since increasing the PSM score threshold can locally result in a relatively larger increase of decoy PSMs than of target PSMs, producing a local FDR increase; this produces a "saw effect" when FDR is plotted against the score. To avoid this effect the use of a q-value was proposed [8]. The *q*-value for each PSM can be computed from the local FDR values of all the PSMs by computing the minimum FDR threshold at which a given PSM is accepted

---

[1] For this example we consider that a higher PSM score is better than a lower one. If the opposite case, the same consideration applies but comparing the local FDR to a monotonically increasing function, and the ≤ in (3) would be ≥.

$$q - \text{value}_i = \min_{\forall S_j \leq S_i} \text{FDR}(S_j), \tag{3}$$

where S_i is the score of PSM number i and FDR(S_j) is the resulting FDR when using S_j as the PSM score threshold. Using this q-value we can apply, for example, a 1% FDR data threshold to the reported PSMs by just accepting those ones with a q-value lower than 0.01.

1.3 Peptide Identification

In shotgun proteomics each peptide is almost always identified by more than one PSM (*see* **Note 3**). Furthermore, this is more common to happen within the target matches than within the decoy ones, leading to an amplification of the error rate [9–11]. Thus, at a given PSM-level score threshold, the peptide-level FDR is greater than the PSM-level FDR (Fig. 1).

In order to prevent the peptide error from growing up, a threshold can be applied to pepFDR (*see* **Note 4**) instead of (or apart from) using psmFDR. One simple way to achieve this goal is to use as the peptide score the score of its best PSM and compute the FDR applying (1) to the peptides.

There exist also other more elaborated approaches for controlling the peptide-level error, such as the ones followed by Percolator [12] or PeptideProphet [13]. Percolator uses semisupervised machine learning to discriminate between correct and decoy spectrum identifications by using a rich vector of 20 features for each PSM. PeptideProphet models the distribution of correctly and incorrectly identified spectra and calculates a Bayesian probability, which can be conceptually related to the FDR.

1.4 Protein Identification

Protein identification is usually the main purpose when using shotgun proteomics, being peptide identification an intermediate step. In this case special attention must be paid to two known issues: protein inference and amplification of error rate.

1.4.1 Protein Inference

In shotgun proteomics all the proteins in the sample are digested together into peptides before being analyzed by mass spectrometry. In this way peptides can be identified more efficiently but their connection to the originating protein is lost. The protein inference problem appears when going back from peptide identifications to a list of protein identifications. Since the same peptide sequence can result from the digestion of different proteins, there is an ambiguity when trying to infer the original protein from which the peptide is produced [14].

At the beginning, many studies obviated the protein inference problem and reported as identified all the proteins matched by each of the identified peptide sequences. This resulted in inflated protein counts, since not all the possible proteins were really present in the original biological sample.

A simple approach followed by newer workflows to cope with the protein inference problem was to assign to each peptide the most probable protein only. For instance, if a peptide sequence can be assigned to two proteins but one of them has a greater evidence (because of the presence of other peptides), then it is assumed that the peptide belongs to that protein, discarding the other one. Another approach is to use the Occam's razor (or law of parsimony) principle. This method only considers as true the minimal set of proteins that can explain the presence of the identified peptides. Both approaches produce more conservative protein counts, but do not solve the problem. For instance, there may be proteins that were actually present in the original sample but are discarded in favor of others, whose identified peptides could have been explained with a combination of the discarded proteins. In other cases, two proteins may be indistinguishable because they share the same peptides (e.g., isoforms) and several workflows prioritize just one of them following an arbitrary criterion.

Recent approaches manage this problem by making ambiguity groups of proteins that share peptides [15–19], thus avoiding the elimination of valid proteins but without inflating the protein count. The Proteomics Standards Initiative (PSI) of the Human Proteome Organization (HUPO) promotes the use of protein ambiguity groups in their standardized file format mzIdentML, used for reporting identifications [20].

1.4.2 Amplification of Error Rate

Another key issue that must be faced when reporting protein identifications is the amplification of the error rate [9, 10]. This is due to the fact that target peptides tend to match a smaller number of proteins than the same number of decoy peptides (Fig. 1). Therefore, for a given peptide-level FDR threshold, the resulting FDR at the protein level is always higher. As already mentioned, the two drafts of the human proteome [1, 2] published in 2014 suffered from this effect. A peptide-level FDR of 1% was used but no protein-level FDR was applied.

The solution used by PeptideAtlas [21] to obtain a 1% protein-level FDR for the human proteome [22] was to apply a stringent PSM-level FDR filter of 0.0002 [23]. Another option to keep under control the error rate is to compute a protein-level score and use (1) for applying a FDR threshold at the protein level. But computing this protein level score using the combined evidence of its peptides is not an immediate task and the simple "best" peptide approach seems to be actually more efficient [10, 24, 25], especially for large datasets. Furthermore, (1) overestimates the protFDR as explained in the following section.

1.4.3 Protein-Level FDR

When computing the protein-level FDR target proteins can contain true positive (TP) PSMs and false positive (FP) PSMs, while decoy proteins only contain false positive PSMs. Therefore, the number of decoy proteins identified above threshold cannot be used as an estimate of the number of FP target proteins. The original MAYU method was formulated to take into account this effect by using a complex hypergeometric distribution [9]. In the practice, however, the MAYU-corrected protFDR can be computed by replacing the numerator in (1) using [26]

$$FP = d \cdot \frac{T - t}{D - d},\qquad(4)$$

where d is the number of decoy proteins identified, t the number of target proteins identified, D the total number of proteins in the decoy database, and T the total number of proteins in the target database (note that D and T are usually the same number).

More recently a "picked protein FDR" approach has been presented [24]. In this approach target and decoy sequences of the same protein are treated as pairs rather than as independent entities, and it is chosen either the target or the decoy sequence depending on which receives the highest score. This competition approach has no built-in bias for the selection of either a decoy protein or a false positive target protein and increases the number of TP protein identifications.

2   Materials

In this section we present different software tools and example datasets to put into practice the aspects presented in the introduction for peptide and protein identification. All the software tools presented are free and multiplatform, so they can be used in GNU/Linux, macOS, and Windows.

2.1   Proteomics Dataset

The ProteomeXchange Consortium has been set up to provide a globally coordinated submission of mass spectrometry proteomics data to the main existing proteomics repositories [27, 28].

In order to execute the different software tools we will use the dataset submitted by the Spanish member of the Chromosome-based Human Proteome Project (C-HPP) [29] to ProteomeXchange as PXD000443 [30]. This dataset contains multiple files corresponding to experiments carried out by different laboratories, but for our purpose it is enough to use just one of them:

http://ftp.pride.ebi.ac.uk/pride/data/archive/2013/09/PXD000443/SPHPP_UPV_JURKAT_QE_CHAPS_RPB_R1_2_10.mgf.gz

2.2   Human Proteome Database

The human proteome database from UniProt [31] will be used as a target database from which a decoy database will be created. This database can be downloaded in fasta format by clicking the download button at the following link:

http://www.uniprot.org/uniprot/?query=*&fil=organism%
3A%22Homo+sapiens+%28Human%29+%5B9606%5D%22+AND
+reviewed%3Ayes

For compatibility with different software tools, the headers of
this fasta file need to be parsed with the following script offered by
the authors of this chapter:

https://raw.githubusercontent.com/akrogp/EhuBio/mas
ter/Projects/python/fasta/stripFasta.py

You will need to have Python 3.x installed to run this tool.

2.3 DecoyPyrat

The DecoyPyrat software tool [6] generates decoy sequences with
minimal overlap between target and decoy peptides. It achieves this
purpose by first switching proteolytic cleavage sites with preceding
amino acid, reversing the database and then shuffling any decoy
sequences that became identical to target sequences. This tool
consists on a Python script that can be downloaded from the
following link:

https://www.sanger.ac.uk/science/tools/decoypyrat

However, this original script generates decoys entries with a
name that has no relation with the originating target name. This
invalidates the use of several approaches like the picked FDR. To
solve this issue the authors of this chapter offer a modification to
the original script. This modified script can be downloaded from:

https://raw.githubusercontent.com/akrogp/EhuBio/mas
ter/Projects/python/fasta/decoyPYrat.py

You will need to have Python 3.x installed to run this tool.

2.4 SearchGUI

SearchGUI [32] is a user-friendly, open-source graphical user interface for configuring and running many of the freely available search
engines. It is a Java multiplatform application that can be run on
GNU/Linux, Windows and macOS. We will use it for obtaining
the search engine output that will be used as the starting point for
the different peptide or protein identification tools. SearchGUI can
be downloaded from:

http://compomics.github.io/projects/searchgui.html

2.5 Percolator

Percolator is a widely used software tool that increases yield in
shotgun proteomics experiments and assigns statistical confidence
measures, such as q-values and posterior error probabilities, to
peptide-spectrum matches (PSMs), peptides and proteins [25].

Percolator uses semisupervised machine learning to discriminate between correct and decoy spectrum identifications by using a
rich vector of 20 features for each PSM. The most recent version
3.0 also supports computing protein-level FDR using the picked
FDR approach.

Percolator is available for GNU/Linux, Windows and macOS.
Binary downloads are available at:

https://github.com/percolator/percolator/releases

2.6 PeptideShaker

PeptideShaker [33] is a user-friendly proteomics software tool that can be used for the analysis and interpretation of MS/MS data. It can use as input the output of SearchGUI or reanalyze a dataset already submitted to PRIDE [34].

Peptide-level scores are computed as the product of PSM PEPs (Posterior Error Probabilities) using (5).

$$Score_{peptide} = \Pi_{i=1}^{N} \widehat{PEP}_{PSM_i} \tag{5}$$

Protein ambiguity groups are created based on peptide unicity, such that peptides are unique to a group, and protein group probabilities are computed as the product of their peptide PEPs using (6).

$$Score_{protein} = \Pi_{i=1}^{N} \widehat{PEP}_{Peptide_i} \tag{6}$$

Finally FDRs are computed using the traditional approach (1).

Peptide shaker is distributed as a Java archive and can be downloaded from:

http://compomics.github.io/projects/peptide-shaker.html

3 Methods

In this section we describe how to use the software tools with the sample datasets already presented in the previous section. The objective is to show how to apply the different key aspects presented in the introduction for peptide and protein identification using these resources.

When running commands from a terminal we assume we are working inside a `workspace folder` unless stated otherwise.

3.1 Input Dataset

In this subsection we obtain the PSMs list that will be used as input for the different peptide and protein identification tools presented. To achieve this, we first create a target–decoy database and then carry out the MS/MS search.

3.1.1 Creation of a Target–Decoy Database

1. Create a `workspace` folder where the following resources will be downloaded into and from which the software will be executed.

2. Download the human proteome database into `workspace` and, after decompressing the file, rename it to `human.fasta`.

3. Download the `stripFasta.py` script into the same folder.

4. Open a terminal and run the following command in that folder (use the `cd` command to change to the folder):

```
python stripFasta.py -o target.fasta human.fasta
```

5. Download the `decoyPYrat.py` script into the same folder.

6. Open a terminal and run the following command in that folder:

```
python decoyPYrat.py -l 7 -d decoy -o decoy.fasta
target.fasta
```

where:

- `-l 7`: sets to seven the minimum length of peptides to compare between target and decoy.

- `-d decoy-`: sets the accession prefix for decoy proteins in output so that they start with `decoy-`.

7. Finally, to obtain a concatenated target–decoy database you can execute the following command in GNU/Linux or macOS:

```
cat target.fasta decoy.fasta > concatenated.fasta
```

or the following one in Windows:

```
type target.fasta decoy.fasta > concatenated.fasta
```

3.1.2 Generation of a PSMs List

1. Download the proteomics dataset `*.mgf.gz` file and uncompress the file to `peaks.mgf`.

2. Download SearchGUI as a zipped file and uncompress it. Open a terminal in the resulting folder and type:

```
java -jar SearchGUI-X.Y.Z.jar
```

where X.Y.Z corresponds to the number of the downloaded version.

3. Run the search from SearchGUI under the following configuration:

   (a) Select `peaks.mgf` as a `Spectrum File`.

   (b) Click the `Add` button in `Search Settings`, then `Spectrum Matching` and finally select: `concatenated.fasta` as the database (answer NO to generation of decoys), carbamidomethylation of C as fixed modification, oxidation of M as variable modification, 10 ppm as precursor tolerance and 0.05 Da as fragment tolerance. Leave the rest of parameters with their default values.

4. Select an `Output Folder`.

5. Select `X! Tandem` search engine.

6. Click the `Start the Search!` button and wait until it finishes.

7. When the search completes successfully a `searchgui_out.zip` will be generated in the output folder selected. Unzip this file and move the `peaks.t.xml` file into the `workspace` as `psms.t.xml`.

3.2 Identification with Percolator

1. After downloading the Percolator binaries install them in the case of GNU/Linux or macOS, or just copy the unzipped *.exe files into the workspace in the case of Windows,.

2. Convert the `psms.t.xml` file into the tab-delimited format used by percolator by running the following command:

```
tandem2pin -P decoy- -o psms.pin psms.t.xml
```

Note that it is an uppercase -P and not a lowercase -p.

3.2.1 Peptide Identification

1. Run Percolator to obtain peptide identifications:

```
percolator -r peptides.tsv psms.pin
```

2. The resulting `peptides.tsv` file can be opened using any spreadsheet program like LibreOffice or Microsoft Office. If it is not recognized automatically you can try renaming its extension from `tsv` to `csv`. This file contains one line per peptide, and the columns provide its score, q-value, posterior error probability, sequence with modifications, and a list of proteins in which the peptide is found.

3. Once in your favorite spreadsheet program, sort the peptides by their q-value and apply the desired pepFDR threshold to this q-value.

3.2.2 Protein Identification

1. Run Percolator to obtain protein identifications using the picked FDR approach and without discarding protein isoforms:

```
percolator -f concatenated.fasta -P decoy- -c -g -l
proteins.tsv -r peptides.tsv psms.pin
```

where:

– -P decoy-: defines the text pattern to identify decoy proteins in the database for the picked-protein algorithm.

– -c: makes Percolator report a comma-separated list of protein IDs, where a full-length protein B is first in the list and a fragment protein A is listed second.

– -g: tells Percolator to report the IDs of isoforms as a comma-separated list, instead of the default behavior of randomly discarding all but one of the proteins.

– -l proteins.tsv: outputs tab delimited results of proteins to proteins.tsv.

2. The resulting `proteins.tsv` file can be opened using any spreadsheet program like LibreOffice or Microsoft Office. If it is not recognized automatically you can try renaming its extension from `tsv` to `csv`.

3. Once in your favorite spreadsheet program, sort the protein groups by their q-value and apply the desired protFDR threshold to this q-value.

3.3 Identification with PeptideShaker

PeptideShaker is optimized for using the UniProt fasta file and building its own decoy entries using the reverse strategy. Thus, for this section we will use the original `human.fasta` and let PeptideShaker build the decoy database.

1. Repeat the steps of Subheading 3.1.2 but using `human.fasta` and let `PeptideShaker` build the decoys when asked. Also click the PeptideShaker checkbox in the Post Processing section of SearchGUI. This will ask for a project and a sample name (enter any) and launch automatically PeptideShaker once the search finishes.

2. For next executions of PeptideShaker it is not necessary to run SearchGUI again, just run PeptideShaker directly using the following command:

```
java -jar PeptideShaker-X.Y.Z.jar
```

where X.Y.Z corresponds to the number of the downloaded version, and then click `Open Project` for restoring the previous values.

3. You can browse and analyze the results directly within the GUI or export them to an Excel spreadsheet. To do this just click in `Identification Features` inside the `Export` menu and select the desired report: `Default Peptide Report` for peptides or `Default Protein Report` for proteins.

4 Notes

1. A peptide to spectrum match (PSM) refers to a possible peptide assignment for a given spectrum. Sometimes there are more than one possible peptide assignments for the same spectrum, so one spectrum may have more than one PSM. Each of these PSMs has an score computed by the search engine and it is based on the similarity between the observed spectrum and the theoretical peptide spectrum computed in silico. The PSMs corresponding to the same spectrum are ranked according to their score and assigned a *rank* number, being *rank* = 1 the best one.

2. The false discovery rate (FDR) can be interpreted as the expected proportion of false positives in the population of validated candidates. This is a statistic used to correct multiple-hypothesis testing, being less conservative than

familywise error rate (FWER) procedures, but with a greater power. A FDR value can be computed globally for a list of reported elements (e.g., PSMs), but it can also be computed locally for each element in the list. In this case the element score is used as the significance threshold for obtaining a virtual sublist used to compute the FDR. In case of ambiguities, we well refer to the first FDR value as the *global FDR* and to the second one as the *local FDR*.

3. Throughout this chapter when we use the term *peptide* we refer not only to a peptide sequence, but also to the different post-translational modifications the peptide could have. Thus two peptides with the same sequence but with different modifications are considered as two different peptides. When we are interested just in the sequence we refer to it as *peptide sequence* expressly.

4. Since the FDR can be computed at different levels of the workflow, sometimes we use the term *psmFDR* for referring the PSM-level FDR, *pepFDR* for the peptide-level FDR, and *protFDR* for the protein-level FDR.

References

1. Kim MS, Pinto SM, Getnet D, Nirujogi RS, Manda SS, Chaerkady R, Madugundu AK, Kelkar DS, Isserlin R, Jain S, Thomas JK, Muthusamy B, Leal-Rojas P, Kumar P, Sahasrabuddhe NA, Balakrishnan L, Advani J, George B, Renuse S, Selvan LD, Patil AH, Nanjappa V, Radhakrishnan A, Prasad S, Subbannayya T, Raju R, Kumar M, Sreenivasamurthy SK, Marimuthu A, Sathe GJ, Chavan S, Datta KK, Subbannayya Y, Sahu A, Yelamanchi SD, Jayaram S, Rajagopalan P, Sharma J, Murthy KR, Syed N, Goel R, Khan AA, Ahmad S, Dey G, Mudgal K, Chatterjee A, Huang TC, Zhong J, Wu X, Shaw PG, Freed D, Zahari MS, Mukherjee KK, Shankar S, Mahadevan A, Lam H, Mitchell CJ, Shankar SK, Satishchandra P, Schroeder JT, Sirdeshmukh R, Maitra A, Leach SD, Drake CG, Halushka MK, Prasad TS, Hruban RH, Kerr CL, Bader GD, Iacobuzio-Donahue CA, Gowda H, Pandey A (2014) A draft map of the human proteome. Nature 509:575–581

2. Wilhelm M, Schlegl J, Hahne H, Gholami AM, Lieberenz M, Savitski MM, Ziegler E, Butzmann L, Gessulat S, Marx H, Mathieson T, Lemeer S, Schnatbaum K, Reimer U, Wenschuh H, Mollenhauer M, Slotta-Huspenina J, Boese JH, Bantscheff M, Gerstmair A, Faerber F, Kuster B (2014) Mass-spectrometry-based draft of the human proteome. Nature 509:582–587

3. Ezkurdia I, Vázquez J, Valencia A, Tress M (2014) Analyzing the first drafts of the human proteome. J Proteome Res 13:3854–3855

4. Ezkurdia I, Calvo E, Del Pozo A, Vázquez J, Valencia A, Tress ML (2015) The potential clinical impact of the release of two drafts of the human proteome. Expert Rev Proteomics 12:579–593

5. Elias JE, Gygi SP (2007) Target-decoy search strategy for increased confidence in large-scale protein identifications by mass spectrometry. Nat Methods 4:207–214

6. Wright JC, Choudhary JS (2016) DecoyPyrat: fast non-redundant hybrid decoy sequence generation for large scale proteomics. J Proteomics Bioinform 9:176

7. Navarro P, Vázquez J (2009) A refined method to calculate false discovery rates for peptide identification using decoy databases. J Proteome Res 8:1792–1796

8. Käll L, Storey JD, MacCoss MJ, Noble WS (2007) Assigning significance to peptides identified by tandem mass spectrometry using decoy databases. J Proteome Res 7:29–34

9. Reiter L, Claassen M, Schrimpf SP, Jovanovic M, Schmidt A, Buhmann JM, Hengartner MO, Aebersold R (2009) Protein identification false discovery rates for very large proteomics data sets generated by tandem

mass spectrometry. Mol Cell Proteomics 8:2405–2417

10. Nesvizhskii AI (2010) A survey of computational methods and error rate estimation procedures for peptide and protein identification in shotgun proteomics. J Proteome 73:2092–2123

11. Granholm V, Navarro JF, Noble WS, Käll L (2013) Determining the calibration of confidence estimation procedures for unique peptides in shotgun proteomics. J Proteome 80:123–131

12. Käll L, Canterbury JD, Weston J, Noble WS, MacCoss MJ (2007) Semi-supervised learning for peptide identification from shotgun proteomics datasets. Nat Methods 4:923–925

13. Keller A, Nesvizhskii AI, Kolker E, Aebersold R (2002) Empirical statistical model to estimate the accuracy of peptide identifications made by MS/MS and database search. Anal Chem 74:5383–5392

14. Al N, Aebersold R (2005) Interpretation of shotgun proteomic data: the protein inference problem. Mol Cell Proteomics 10:1419–1440

15. Qeli E, Ahrens CH (2010) PeptideClassifier for protein inference and targeted quantitative proteomics. Nat Biotechnol 28:647–650

16. Searle BC (2010) Scaffold: a bioinformatic tool for validating MS/MS-based proteomic studies. Proteomics 10:1265–1269

17. Meyer-Arendt K, Old WM, Houel S, Renganathan K, Eichelberger B, Resing KA, Ahn NG (2011) IsoformResolver: a peptide-centric algorithm for protein inference. J Proteome Res 10:3060–3075

18. Prieto G, Aloria K, Osinalde N, Fullaondo A, Arizmendi JM, Matthiesen R (2012) PAnalyzer: a software tool for protein inference in shotgun proteomics. BMC Bioinformatics 13:288

19. Uszkoreit J, Maerkens A, Perez-Riverol Y, Meyer HE, Marcus K, Stephan C, Kohlbacher O, Eisenacher M (2015) PIA: an intuitive protein inference engine with a web-based user interface. J Proteome Res 14:2988–2997

20. Seymour SL, Farrah T, Binz P-A, Chalkley RJ, Cottrell JS, Searle BC, Tabb DL, Vizcaíno JA, Prieto G, Uszkoreit J et al (2014) A standardized framing for reporting protein identifications in mzIdentML 1.2. Proteomics 14:2389–2399

21. Desiere F, Deutsch EW, King NL, Nesvizhskii AI, Mallick P, Eng J, Chen S, Eddes J, Loevenich SN, Aebersold R (2006) The PeptideAtlas project. Nucleic Acids Res 34:D655–D658

22. Farrah T, Deutsch EW, Hoopmann MR, Hallows JL, Sun Z, Huang C-Y, Moritz RL (2012) The state of the human proteome in 2012 as viewed through PeptideAtlas. J Proteome Res 12:162–171

23. Farrah T, Deutsch EW, Omenn GS, Campbell DS, Sun Z, Bletz JA, Mallick P, Katz JE, Malmström J, Ossola R et al (2011) A high-confidence human plasma proteome reference set with estimated concentrations in PeptideAtlas. Mol Cell Proteomics 10:M110.006353

24. Savitski MM, Wilhelm M, Hahne H, Kuster B, Bantscheff M (2015) A scalable approach for protein false discovery rate estimation in large proteomic data sets. Mol Cell Proteomics 14:2394–2404

25. MacCoss MJ, Noble WS, Käll L et al (2016) Fast and accurate protein false discovery rates on large-scale proteomics data sets with percolator 3.0. J Am Soc Mass Spectrom 27:1719–1727

26. Higdon R, Reiter L, Hather G, Haynes W, Kolker N, Stewart E, Bauman AT, Picotti P, Schmidt A, van Belle G et al (2011) IPM: an integrated protein model for false discovery rate estimation and identification in high-throughput proteomics. J Proteome 75:116–121

27. Vizcaíno JA, Deutsch EW, Wang R, Csordas A, Reisinger F, Rios D, Dianes JA, Sun Z, Farrah T, Bandeira N et al (2014) ProteomeXchange provides globally coordinated proteomics data submission and dissemination. Nat Biotechnol 32:223

28. Deutsch EW, Csordas A, Sun Z, Jarnuczak A, Perez-Riverol Y, Ternent T, Campbell DS, Bernal-Llinares M, Okuda S, Kawano S et al (2016) The ProteomeXchange consortium in 2017: supporting the cultural change in proteomics public data deposition. Nucleic Acids Res 45:D1100–D1106

29. Paik Y-K, Jeong S-K, Omenn GS, Uhlen M, Hanash S, Cho SY, Lee H-J, Na K, Choi E-Y, Yan F et al (2012) The Chromosome-Centric Human Proteome Project for cataloging proteins encoded in the genome. Nat Biotechnol 30:221–223

30. Segura V, Medina-Aunon J, Mora M, Martínez-Bartolomé S, Abian J, Aloria K, Antúnez O, Arizmendi J, Azkargorta M, Barceló-Batllori S, Beaskoetxea J, Bech-Serra J, Blanco F, Monteiro M, Cáceres D, Canals F, Carrascal M, Casal J, Clemente F, Colomé N, Dasilva N, Díaz P, Elortza F, Fernández-Puente P, Fuentes M, Gallardo O, Gharbi S, Gil C, González-Tejedo C, Hernáez M, Lombardía M, Lopez-Lucendo M, Marcilla M, Mato J, Mendes M, Oliveira E,

Orera I, Pascual-Montano A, Prieto G, Ruiz-Romero C, Sánchez del Pino M, Tabas-Madrid D, Valero M, Vialas V, Villanueva J, Albar J, Corrales F (2013) Surfing transcriptomic landscapes. A step beyond the annotation of chromosome 16 proteome. J Proteome Res 13:158–172

31. UniProt Consortium (2017) UniProt: the universal protein knowledgebase. Nucleic Acids Res 45:D158–D169

32. Vaudel M, Barsnes H, Berven FS, Sickmann A, Martens L (2011) SearchGUI: an open-source graphical user interface for simultaneous OMSSA and X! Tandem searches. Proteomics 11:996–999

33. Vaudel M, Burkhart JM, Zahedi RP, Oveland E, Berven FS, Sickmann A, Martens L, Barsnes H (2015) PeptideShaker enables reanalysis of MS-derived proteomics data sets. Nat Biotechnol 33:22–24

34. Vizcaíno JA, Côté RG, Csordas A, Dianes JA, Fabregat A, Foster JM, Griss J, Alpi E, Birim M, Contell J et al (2012) The PRoteomics IDEntifications (PRIDE) database and associated tools: status in 2013. Nucleic Acids Res 41:D1063–D1069

Chapter 7

Methods and Algorithms for Quantitative Proteomics by Mass Spectrometry

Rune Matthiesen and Ana Sofia Carvalho

Abstract

Protein quantitation by mass spectrometry has always been a resourceful technique in protein discovery, and more recently it has leveraged the advent of clinical proteomics. A single mass spectrometry analysis experiment provides identification and quantitation of proteins as well as information on posttranslational modifications landscape. By contrast, protein array technologies are restricted to quantitation of targeted proteins and their modifications. Currently, there are an overwhelming number of quantitative mass spectrometry methods for protein and peptide quantitation. The aim here is to provide an overview of the most common mass spectrometry methods and algorithms used in quantitative proteomics and discuss the computational aspects to obtain reliable quantitative measures of proteins, peptides and their posttranslational modifications. The development of a pipeline using commercial or freely available software is one of the main challenges in data analysis of many experimental projects. Recent developments of R statistical programming language make it attractive to fully develop pipelines for quantitative proteomics. We discuss concepts of quantitative proteomics that together with current R packages can be used to build highly customizable pipelines.

Key words Protein quantitation, Peptide quantitation, Liquid chromatography, Mass spectrometry, Label-free quantitation, Stable isotope labeling

1 Introduction

Although quantitative proteomics is considered an advanced and costly mass spectrometry technique, the utility of quantitative proteomics is undoubtable. Quantitative proteomics techniques have become more user friendly, sensitive, and robust, and techniques such as label-free quantitation (LFQ) by liquid chromatography–mass spectrometry (LC-MS) are relatively easy to implement. In fact, the quantitative information is often available in the LC-MS/MS raw data even though the experimental design was not intended for protein quantitation. Nevertheless, LFQ across samples does require a well planned experimental design, especially due to the reproducibility of the LC system, spray stability, variation in chemical modification of amino acids introduced during sample

Rune Matthiesen (ed.), *Mass Spectrometry Data Analysis in Proteomics*, Methods in Molecular Biology, vol. 2051, https://doi.org/10.1007/978-1-4939-9744-2_7, © Springer Science+Business Media, LLC, part of Springer Nature 2020

Table 1
MS-based quantitative methods

Method	SIL	In vivo SIL	Clinical
$^{14}N/^{15}N$	+	+	−
SILAC	+	+	(+)
QconCAT	+	+	+
ID	+	+	−
Mass-tag	+	−	+
Dimethyl labeling	+	−	+
ICAT	+	−	+
$^{16}O/^{18}O$	+	−	+
AQUA	+	−	+
SISCAPA	+	−	+
PSAQ™	+	−	+
iTRAQ/TMT	+	−	+
Spectral counting	−	−	+
XIC	−	−	+
SWATH	−	−	+

preparation (e.g., methionine oxidation), and technical variance from trypsin digestion. To ameliorate the contribution of technical variance and to obtain accurate statistical measurements replicate analysis is recommended [1, 2].

There are a vast number of MS-based protein quantitative methods which can be divided into two main categories: (1) stable isotope labeling strategies and (2) label-free methods (see Table 1). Stable isotope labeling strategies can be further divided into metabolic labeling and chemical labeling [3, 4]. Metabolic labeling has the advantage of early incorporation of isotopes into the sample preparation minimizing experimental variance introduced during protein extraction, protein digestion, peptide enrichment/separation and MS analysis. Enzymatic or chemical labeling is typically incorporated during or after protein digestion. Enzymatic or chemical labeling is often suitable for clinical samples and offer higher degree of multiplexing with the downside of slightly increased technical variance due to later incorporation of stable isotopes.

Quantitative methods are usually divided into relative and absolute quantitation methods. However, such division may not be entirely justified since relative quantitative methods, based on chemical modifications such as iTRAQ (isobaric tags for relative and absolute quantitation) in principle, as the name of the reagent suggests, can be used for absolute quantitation, for example, by spiking into the sample a known quantity of stable isotope labeled peptides or proteins.

Each quantitative method has specific attributes that need to be considered. Thus, quantitative algorithms must correct for potential artifacts associated to each different quantitative method [4]. Consequently, the exact same algorithms cannot be used for all quantitative methods. Lau et al. [3] has reviewed a number of programs, most of which are commercial, for quantitative MS-based proteomics. More recently multiple robust pipelines have become freely available. The aim of this review is to discuss the underlined algorithms of the most common MS-based relative protein quantitative methods (*see* Table 1). A number of in vitro chemical stable labeling methods are reviewed by Julka and Regnier [5] and will not be discussed here.

The methods in grey are reviewed in this chapter. **SIL,** stable isotope labeling. The column named "**In vivo SIL**" indicates whether the labels are introduced in vivo (+ or −). The column named "**Clinical**" indicates if the method can be used for clinical samples (+ or −). *ID* isotope dilution [6], *SILAC* stable isotope labeling with amino acids in cell culture [7], *QconCAT* absolute quantitation by spiking in stable isotope-labeled artificial expressed proteins, which comprise concatenated proteotypic peptides [8–10], *SISCAPA* stable isotope standards with capture by anti-peptide antibody [11, 12], *AQUA* absolute quantification [13], *ICAT* isotope coded affinity tag [14], *iTRAQ* isobaric tag for relative and absolute quantitation [15, 16], *TMT* tandem mass tags [17], *PSAQ* protein standard absolute quantification [18], *SWATH* sequential window acquisition of all theoretical fragment ion mass spectra [19].

The quantitative techniques listed in Table 1 can be combined with a different number of mass spectrometry instruments (LC system, mass analyzer, and detector). For example, selected reaction monitoring (SRM) was omitted from Table 1 since it is a mass spectrometry method that can be combined with most of the methods listed (*see* Chapter 12 for details on SRM). However, the combination of the type of mass spectrometer and a specific quantitation method is a key factor for the successful analysis of proteins. For example, ion trap mass analyzers with low mass cutoff are unsuited for iTRAQ quantitation since reporter ions end outside the recorded mass range. Another frequent flaw is the use of stable-isotope chemical labeling targeting N-terminal and lysine residues for enrichment studies of posttranslational modifications on these sites. For example, modifications such as acetylation, methylation or ubiquitin and ubiquitin like modifiers are complex to study using iTRAQ, tandem mass tags, or stable-isotope dimethyl labeling.

SISCAPA (stable-isotope standards with capture by anti-peptide antibody) [11, 12] and AQUA (absolute quantification) [13] are powerful methods for absolute quantitation, nonetheless, are costly and not suitable to all types of modified peptides. On the other hand, QconCAT (Absolute quantitation by spiking in stable-isotope labeled artificial expressed proteins) [8–10] is cost effective once plasmids of concatenated proteotypic peptides are established. However, like AQUA, QconCAT method disregard to some extent posttranslational modifications. Therefore further discussions on absolute quantitative methods exclude QconCAT, SISCAPA and AQUA methods. Nevertheless, the final data output from these methods is similar to the data produced by stable-isotope labeling with amino acids in cell culture (SILAC) method. Computational methods for SILAC data analysis if proper implemented can therefore be directly applied to QconCAT, SISCAPA, and AQUA.

2   Algorithms for Quantitation

The following section describes concepts and algorithms for specific proteomics quantitative methods, though an extensive overview of common computational and statistical issues for all quantitative techniques (background subtraction, noise filtering, mass calibration, transformation, normalization, scaling, peak detection, missing values, classification, and power estimation) are covered in many other reviews and will not be considered here [4, 20].

2.1   SILAC

SILAC [7] was initially developed for cell in culture and model organism studies though recently it has been further developed to study clinical tissue samples by a method referred to as super SILAC [21]. SILAC is in general viewed as a simple and an inexpensive experimental procedure for quantifying peptides and proteins. From a computational point of view, SILAC is a very simple quantitative method. Labeled and unlabeled peptides are efficiently separated in the LC-MS intensity profile (see Fig. 1). Quantitative ratios are calculated between the integrated intensities of labeled and unlabeled peptides over retention time given that there is no peak overlap from unrelated peptides.

The obtained relative quantitative values can be presented as I_H/I_L or I_L/I_H (zero centered fold change $\in(-\infty,+\infty)$), $\log(I_H)-\log(I_L)$ (log fold change), I_H/I_L if $I_H \geq I_L$ and $-I_L/I_H$ if $I_H < I_L$ (fold change $\in(-\infty,-1)$ U$[1,+\infty])$ or as $I_H/(I_L + I_H)$ where I_L and I_H correspond to the intensity count from the unlabeled and labeled peptide, respectively. This last quantitative value representation can be calculated even if intensity counts are zero which is not the case for I_L/I_H representation. The quantitative values are most accurate if the intensity over several MS scans is integrated

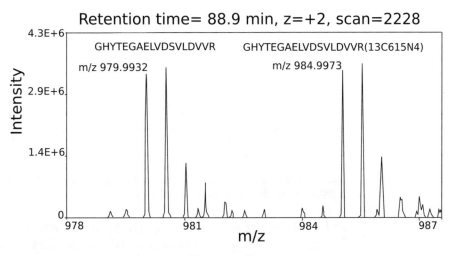

Fig. 1 Survey scan of the unlabeled and 6 × ^{13}C + 4 × ^{15}N labeled peptide GHYTEGAELVDSVLDVVR in the LC-MS profile

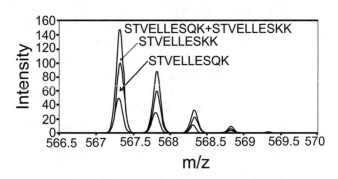

Fig. 2 Overlap between two isotopic distributions from the peptides STVELLESQK and STVELLESKK. The peak overlap shown here should be dealt with while processing the continuous MS data to mass and intensity peak lists

(integration over the retention time dimension). It is worth mentioning that the intensity in the LC-MS profile from other charge states of the peptide can be used to improve the accuracy of the quantitation even though the peptide is identified exclusively in the MS/MS spectrum, corresponding to a specific charge state of the peptide.

The above described simplicity is not always a reality. In practice, analysis of complex protein samples increases the probability of occurring peak overlap in the LC-MS profile. Peak overlap is not restricted to SILAC quantitation, and is observed in general. Such overlap give rise to further problems which require two distinct types of algorithms. The first is exemplified in Fig. 2.

In this example there is an overlap between the isotopic distribution of the peptides STVELLESQK and STVELLESKK, with a mass difference of ~0.03638 Da. Depending on the mass accuracy

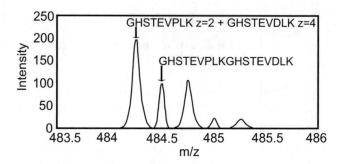

Fig. 3 The peak overlap that be resolved by fitting multiple theoretical isotopic distributions of charge states $z = +1 - 5$

and resolution of the instrument such peaks will reveal one (e.g., Ion-traps) or two (e.g., Orbitrap/FT-ICR) apparent isotopic distributions. Overlap of isotopic peak distributions as shown in Fig. 2 can ultimately be accurately resolved by estimating the peak width and by fitting a multi-Gaussian or a multi-Gaussian/Lorentzian mixture peak model [22]. Additionally, the fitting can be obtained from the corresponding MS/MS spectrum which in some cases can reveal the sequence of several tryptic peptides (aka chimeric spectra). Current quantitative proteomics software should be optimized to better handle chimeric spectra.

Another type of overlap of isotopic distribution can occur when the double charge peptide has approximately the same observed parent ion mass as a peptide with charge state +4 (Fig. 3).

The overlap, in this case, can be resolved by fitting multiple theoretical isotopic distributions to the observed peak pattern. This can be achieved using for example linear regression [23] or the Newton–Gauss unweighted least-squares method. In case of two isobaric peptides, deconvolution of peak overlapping is impossible, and in such cases the quantitation fails. Another down side to consider is the fact that several models can fit the data equally well or even worse the least adequate model fits the data better than the correct model due to measurement errors.

There are an additional number of experimental associated artifacts that can occur depending on the labeling efficiency. In Fig. 1, a small peak can be observed at ~ -0.5 m/z (-1 Da) below the monoisotopic mass for the parent ion of the stable-isotope labeled peptide. This peak can be attributed to incomplete labeling of lysine and arginine with ^{13}C and ^{15}N, or to the activity of transaminases in vivo, that exchange the amino group next to the α-carbon, mainly observed if using $^{13}C_6{}^{15}N_2$ lysine or $^{13}C_6{}^{15}N_4$ arginine. The intensity of -1 Da peak can in some cases surpass 10% of the monoisotopic peak and in general should be included in the integration of the intensity for the labeled peptide's isotopic distributions. Another in vivo-generated artifact is the conversion of $^{13}C_6$ arginine to $^{13}C_5$ proline, giving rise to an additional small

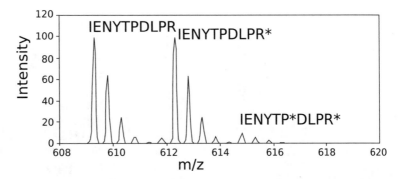

Fig. 4 Survey scan on the elution of the unlabeled and $^{13}C_6$ labeled Arg peptide IENYTPDLPR in the LC-MS profile. In this case $^{13}C_6$ labeled Arg have been in vivo converted to $^{13}C_5$ Pro giving rise to a small peak +2.5 *m/z* (+5 Da) compared to the $^{13}C_6$ labeled Arg peptide IENYTPDLPR. ∗ indicates ^{13}C labeled residues

isotope distribution for peptides that contains one proline residue. This isotopic distribution is displaced +5 Da rightward of the monoisotopic peak of the stable-isotope labeled peptide (*see* Fig. 4). In principle, multiple peaks with +5 Da intervals can be observed depending on the number of proline residues within the peptide. This error can be corrected by including these peaks in the integration procedure.

Alternatively, the probability of ^{13}C-labeled proline can be estimated by dividing the intensity at +5 Da by the total observed intensity of a stable-isotope labeled peptide with one proline. Next, the percentage of these satellite peaks from other proline containing peptides can be calculated using as input the binomial distribution with the estimated proline probability. Thus lowering the probability of introducing an error in the quantitation procedure by inaccurate estimation of the +5 peak intensity (e.g., due to an unaccounted peak overlap). In addition, ^{13}C-labeled proline abundance can be minimized by adding excess proline to labeled and unlabeled growth medium, thus lowering in vivo synthesis of proline. If addition of excess proline is incompatible with the experimental design, the use of mathematical procedures is recommended. Initially, software for SILAC quantitation included programs such as VEMS [24], MSquant (http://msquant. sourceforge.net/), RelEx [25], and ASAPratio [26]. MaxQuant [27] is currently the standard program for SILAC quantitation. More recently R packages such rTANDEM, MSGFplus for database-dependent searches, and MSnbase for quantitation can be used to build a custom pipeline for SILAC.

Thirdly, incomplete labeling of the peptide is a common sample related artifact in cases where the number of cell doublings is insufficient or in case of in vivo synthesizes of arginine or lysine (*see* Fig. 5). The dilution factor can be estimated as $2n$ assuming no in vivo synthesis of arginine and lysine where n is the number of cell

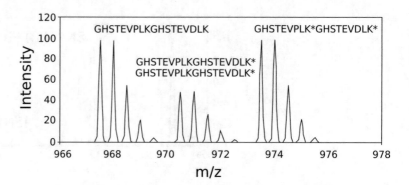

Fig. 5 Survey scan of GHSTEVPLKGHSTEVDLK peptide elution in the LC-MS profile, depicting isotope distributions of the peptide with a different number of $^{13}C_6$ labeled Lys residues ($*$)

doublings [28]. As a result five cell doublings leads to ~0.03% nonlabeled amino acids incorporation in the stable isotope-labeled cell culture. Incomplete labeling of the peptide can only be perceived when peptides with two or more labeled amino acids are detected. Peptides with two or more stable isotope-labeled amino acids can be used to estimate a correction factor, by fitting a binomial distribution to the observed intensity of the unlabeled, single labeled and double-labeled peptide. The estimated probability (incorporation efficiency) is further used in the binomial probability function to correct all the observed intensities.

Protein quantitation of in vivo tissue samples using Super SILAC was first attempted by SILAC labeling Neuro2A cell line as internal standard for an in vivo tissue [29]. However, Geiger et al. [21], encountered several drawbacks by using a single-labeled cell line as reference. The observed quantitative distribution was bimodal and furthermore a large number of peptides from the in vivo tissue mismatched an internal standard peptide from the reference cell line. By increasing the number of reference cell lines to five, the number of peptide matching increased, leading to most accurate quantitation and unimodal distribution of quantitative values.

2.2 ^{18}O Labeling

^{18}O labeling is less widely used than SILAC, label-free or tandem mass tag methods for relative quantitation of proteins. There are two labeling protocols using ^{18}O for MS-based quantitation: (1) In the original protocol labeled proteins are digested in ^{18}O-enriched water [30] while (2) in the alternative approach the standard trypsin digestion step is followed by lyophilization and solubilization of peptides in ^{18}O water labeling [31]. Incomplete labeling is one of the main disadvantages of ^{18}O labeling protocols (*see* Fig. 6). To illustrate this case Fig. 6 depicts the isotopic distribution of peptides in different labeling conditions for two comparative samples (unlabeled and ^{18}O-labeled). There is an overlap between the isotopic

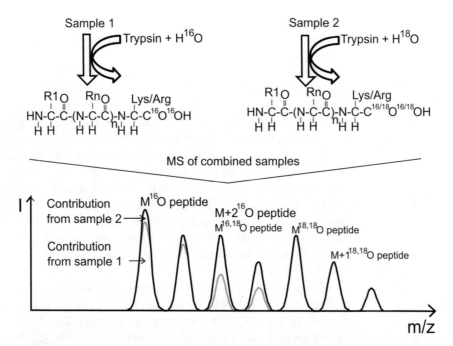

Fig. 6 Experimental design of ^{18}O labeling quantitative approach. The monoisotopic peak ($M\ ^{16}$O peptide) presents a contribution from sample 1, mainly, and partly from sample 2. The peak area difference between the partially labeled (dark) and unlabeled (grey) peaks corresponds to the contribution of sample 2. The peak area difference varies between peptides making ^{18}O labeling quantitation rather complex

distributions of the unlabeled peptides in the two conditions. Incomplete labeled peptides from sample 2 with no ^{18}O incorporated show a perfect overlap with the unlabeled peptides in sample 1, whereas the peptides from sample 2 with one ^{18}O incorporated show partial overlap between the isotopic distribution of peptides in sample 1 and two ^{18}O incorporated peptides in sample 2.

A number of algorithmic methods have been proposed for quantitation using ^{18}O labeling [30, 32–35]. These methods can be divided into two categories which consider different assumptions. Assuming that incorporation of the first ^{18}O and the second ^{18}O atom cannot be correlated or fit to a model, the intensity of the light and heavy form of the peptide can be corrected by Eq. 1 [30]:

$$\left(\frac{^{16}O}{^{18}O}\right) \equiv \frac{I'_m}{I'_{m+4} - \frac{I_{m+4}}{I_m}I'_m + I'_{m+2}\left(1 - \frac{I_{m+2}}{I_m}\right) - I'_m\frac{I_{m+2}}{I_m}\left(1 - \frac{I_{m+2}}{I_m}\right)}$$

(1)

where I'_{m+x} are the observed apparent relative intensities to the total intensity for the peptide ion and I_{m+x} are the theoretical relative intensities to the total intensity for the peptide ion. $m + x$ indicates the monoisotopic mass m and ($m + 1, \ldots, m + 4$) correspond to the second to the fifth peak in the isotopic envelope. The theoretical isotope distributions can be calculated by linear approximations (*see* **Note 1**).

However, for quantitative purposes it is preferable to use accurate calculations and avoid using linear approximations in case the quantitated peptide is sequenced in the MS/MS spectra. In this context the Averagine model can accurately fit the isotopic distributions of un-modified peptides (for more details *see* Chapter 2). However, Averagine is not accurate for peptides with modifications such as glycans, metals, and especially peptides with chlorine atoms incorporated. Alternatively, the R packages enviPat or IsoSpecR can be used to accurately calculate isotopic distributions of such modified peptides. enviPat and IsoSpecR programs are based on the combinatorial method described below, implementing a heuristic approach to obviate combinations that insignificantly contribute to the isotopic distribution.

To calculate theoretical isotopic distributions, it is necessary to understand the different atoms contribution to the isotopic distribution. Biological molecules are mainly composed of carbon (C), hydrogen (H), nitrogen (N), oxygen (O), and sulfur (S). Biological molecules such as proteins and DNA may also bind metals. Natural occurring isotopes of biological compounds occur at almost constant relative abundance (Table 2).

An extensive list of biological relevant isotopes can be found at http://www.ionsource.com/Card/Mass/mass.htm.

The relative isotopic abundance values for isotopes (Table 2) can be used to calculate relative isotopic abundance of different

Table 2
Masses and relative isotopic abundance values of biologically relevant isotopes [36]

Isotope	A	%	Isotope	A + 1	%
^{12}C	12	98.93(8)	^{13}C	13.0033548378(1)	1.07(8)
^{1}H	1.0078250321(4)	99.9885(7)	^{2}H	2.0141017780(4)	0.0115(7)
^{14}N	14.0030740052(9)	99.632(7)	^{15}N	15.0001088984(9)	0.368(7)
^{16}O	15.9949146221(15)	99.757(2)	^{17}O	16.99913150(2)	0.038(1)
^{32}S	31.97207069(12)	94.93(3)	^{33}S	32.97145850(1)	0.76(2)
Isotope	A + 2	%	Isotope	A + 4	%
^{14}C	14.003241988(4)	–	–	–	–
^{3}H	3.0160492675(11)	–	–	–	–
^{18}O	17.9991604(9)	0.205(1)	–	–	–
^{34}S	33.96786683(11)	4.29(3)	^{36}S	35.96708088(3)	0.02(1)

Uncertain digits are shown in parenthesis

biological molecules composed by several different isotopes. The relative isotopic abundance of the monoisotopic mass of a molecule with the composition CxHyNzOvSw can be calculated using the following expression [37, 38].

$$I_m = P_C^x \times P_H^y \times P_N^z \times P_O^v \times P_S^w \tag{2}$$

I_m is the relative abundance of the monoisotopic peak for the molecule. P_C, P_H, P_N, P_O, and P_S are the abundance of the monoisotopic masses of the C, H, N, O, and S elements, and x, y, z, v, w are positive integer values indicating the number of occurrences of the corresponding atom in the biological compound. The expression corresponds to the probability that all the elements in the molecule have the monoisotopic mass. A similar expression can be made for the monoisotopic mass +1

$$\begin{aligned}
I_{m+1} &= \binom{x}{1} P_C^{x-1} P_{C+1} P_H^y P_N^z P_O^v P_S^w \\
&+ P_C^x \binom{y}{1} P_H^{y-1} P_{H+1} P_N^z P_O^v P_S^w + \cdots \\
&+ P_C^x P_H^y P_N^z P_O^v \binom{w}{1} P_S^{w-1} P_{S+1} \tag{3}
\end{aligned}$$

where P_{C+1}, P_{H+1}, P_{N+1}, P_{O+1}, and P_{S+1} are the abundance of the monoisotopic mass +~1 Da of the elements. Once again, the expression corresponds to the probability that one atom in the molecule is the monoisotopic mass +1. Equation 3 can be further expanded to calculate I_{m+2}, I_{m+3} and I_{m+4} by applying the same principals used to expand Eqs. 2–3. Essentially this approach is an approximation, particularly suited for the comparison between theoretical and observed isotopic distributions obtained from mass spectrometers unable to resolve the different elements contribution to the $m + 1$ ion. However, Eq. 3 can be expanded using the same concept considering all possible unique masses rather than only the isotopic abundance for the integer masses m, $m + 1$, $m + 2$, $m + 3$, and $m + 4$. For most common mass spectrometers and for $^{16}O/^{18}O$ labeling this approximation is adequate.

$^{16}O/^{18}O$ ratios calculated using Eq. 1 give a good approximation as long as the peptides in the heavy labeled sample are labeled with either one or two ^{18}O. If the labeling failed so that a peptide in the heavy sample has too high percentage of completely unlabeled peptide then Eq. 1 will give an erroneous quantitation for that peptide. The incorporation efficiency of ^{18}O by trypsin is lower if charged residues are present in -2, -1, $+1$, or $+2$ positions relative to the trypsin cleavage site. However, most peptides incorporate ^{18}O well and since the protein quantitation is based on several peptide quantitative values the low incorporation efficiency is mainly a problem for proteins that are only detected by one or two peptides. Mirgorodskaya and colleagues [39] suggested to indirectly estimate the incorporation rate by analyzing the labeled sample separately and using the experimentally observed peak heights as the expected abundance distribution to overcome erroneous quantitation. They propose a linear matrix equation to calculate the concentration of labeled and unlabeled peptides. Eckel-Passow et al. [33] elaborated on the method proposed by Mirgorodskaya and colleagues. These authors propose a regression model to calculate the peptide-specific incorporation rates of the ^{18}O label, which does not require running the labeled sample independently to obtain the expected distribution of the labeled peptide. In this model the incorporation rate is estimated directly from the multivariable regression model using Eq. 4

$$\hat{\theta}_c = \left(W_c^T W_c\right)^{-1} W_c^T y_c \qquad (4)$$

where θ is a 2×1 vector containing the concentration θ_{c1} (unlabeled) and θ_{c2} (labeled) for peptide c. The hat on top of θ represents the predicted values from the model. The matrix W_c contains the intercept vector and the expected (theoretical) isotopic abundances for the c-th peptide in the unlabeled and labeled samples. Eckel-Passow et al. [33] splits elegantly the matrix W_c into $W_c = X_c S_c$ where.

$$X_c = \begin{bmatrix} 1 & I_m & 0 & 0 \\ 1 & I_{m+1} & 0 & 0 \\ 1 & I_{m+2} & I_m & 0 \\ 1 & I_{m+3} & I_{m+1} & 0 \\ 1 & I_{m+4} & I_{m+2} & I_m \\ \vdots & \vdots & \vdots & \vdots \\ 1 & I_{n-1} & I_{n-3} & I_{n-5} \end{bmatrix} \qquad S_c = \begin{bmatrix} 1 & 0 & 0 \\ 0 & 1 & (1-p)^2 \\ 0 & 0 & 2p(1-p) \\ 0 & 0 & p^2 \end{bmatrix}$$

The first column in X_c is the intercept and the remaining columns are the theoretical isotopic distributions for the peptide c with the labels $^{18}O_0$, $^{18}O_1$, or $^{18}O_2$. The first column in S_c corresponds to the intercept and columns two and three determine the concentration parameters for the labeled and unlabeled sample. p is the purity of the ^{18}O water. This model assumes that it is possible to correlate or model a relation between incorporation of the first ^{18}O with the second which means that the intensity of the light and heavy form of the peptide can be corrected even if the heavy sample contributes to the monoisotopic peak of ^{16}O unlabeled peptide. Vázquez et al. [40] assumes a similar kinetic model for the correlation between the incorporation of the first ^{18}O with the second ^{18}O. However, they use nonlinear, Newton-Gauss unweighted least-squares iteratively to estimate the parameters of the model rather than the multivariate linear regression described by Eckel-Passow et al. [33].

2.3 Primary Amine Labeling

Primary amine labeling constitutes a broad range of chemical labeling strategies that are widely used. It exploits the nucleophilicity of amino groups to displace a leaving group from an activated acid. The N-terminal α-amino group is less nucleophilic than the ε-amino group on lysine residues. This means that the α-amino group has a lower degree of protonation and therefore less derivatized than the ε-amino group at neutral pH, though high or low pH gives equal degree of ionization. The phenolic hydroxyl group on tyrosine can also be derivatized at high pH but are easily hydrolyzed again. Regnier and Julka provide an excellent overview of the chemistry behind primary amine labeling [41].

A major advantage of amine labeling is the increase in the accuracy of protein quantitation since all peptides are labeled in the sample. Coding through derivatization of amines can be achieved at the protein and peptide level. Labeling at the protein level has the advantage to eliminate differential proteolysis of protein samples to be compared since it can be mixed and digested simultaneously. On the other hand, some coding labels prevent trypsin hydrolysis at lysine residues leading to fewer and longer peptides and consequently affecting the accuracy of protein quantitation. Deuterium is often used in amine specific labeling strategies to obtain a differential mass. Deuterium-labeled peptides elute slightly earlier on a reverse phase column than nonlabeled peptides. A solution that can minimize the problem is the introduction of polar groups close to deuterium atoms to inhibit interaction with the column [42]. Using ^{13}C- or ^{18}O-labeled coding is another possibility for solving the differential elution from reverse phase columns; however, for many amine labels only the deuterium form is commercially available [43]. Deuterium coded N-acetoxysuccinimide is a simple and low-cost acetylating agent. Nevertheless, acetylation reduces peptide charge and decrease ionization efficiency, mainly observed for MALDI-MS rather than for ESI-MS [41].

The computational analysis of peptides labeled at amines is straightforward and gives few artifacts. However, the overlap of the isotopic distribution of the unlabeled and labeled peptides is still observed especially if the number of stable-isotope labeled atoms is equal to or less than 4. The overlaps can be taken into account using computational methods described for ^{18}O labeling.

2.3.1 Dimethyl Labeling-Based Quantitation

Stable-isotope labeling by reductive amination uses formaldehyde to globally label the N-terminal α-amino group of peptides and ε-amino group of lysines [44]. This quantitative method is cost-effective and very practical owed to a fast labeling reaction (approximately, 5 min) and high labeling yield without any detectable byproducts. Reductive amination of the N-terminus and lysine side chains is faster than amination of other amino acids like arginine and tryptophan and therefore labeling reaction time is the key

Table 3
Structure and delta mass values of the different dimethyl labeling reagents

Number	Labeling reaction	Mass increment	Mass difference to the lightest label
1		+28.0313	–
2		+30.0439	+2.0126
3		+32.0564	+ 4.0251
4		+34.0631	+ 6.0318
5		+36.0757	+ 8.0444

step for specific labeling [45]. This labeling strategy produces peaks differing by 28 Da relative to its nonderivatized counterpart for each derivatized site when using the nonstable isotope compound (Table 3). The stable isotope labeled compounds have +2 Da, +4 Da, +6 Da, and +8 Da units extra for each derivatized isotopic pair [46–48]. Despite being potentially able to label five samples simultaneously only up to three can be robustly compared using this technique as a difference of less than 4 Da between peptide pairs leads to a significant overlap of the isotopic distribution peaks for the peptide pairs [46, 47]. Moreover, 4 Da unit difference is insufficient to guarantee that peak overlap between the fifth isotopic peak of the lightest isotope and the first isotopic peak of the next highest isotopic ion do not occur. This is mainly a problem for arginine containing canonical peptides which will only have the N-terminal derivatized whereas lysine containing canonical peptides will have labels incorporated at two sites. For example, if

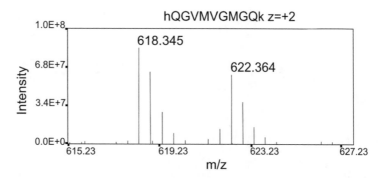

Fig. 7 Survey scan of dimethyl labeled peptide pair of the sequence hQGVMVGMGQk with $z = +2$. Lower case letter indicates the derivatized sites using labeling reagent 3 (*m/z* 618.345) and 5 (*m/z* 622.364) from Table 3

using reagent 3 and 5 from Table 3 the isotope peaks from the peptide hQGVMVGMGQk can be well resolved (*see* Fig. 7).

However, for arginine containing canonical peptides of high mass values the overlap can be significant when combining reagent 3 and 5 from Table 3. Therefore, for quantitative purposes isotope distribution overlap needs to be resolved by adapting the algorithm for O^{16}/O^{18} isotopic pairs described in Subheading 2.2. The use of a correction algorithm can potentially increase the number of experiments to be simultaneously compared to more than three. Deuterium isotope effect constitutes one of the shortcomings for the expansion of the number of experiments to be compared though. Accurate relative quantitation is difficult in the case of isotopic effect on peptide retention. Deuterium isotopes have an effect on the retention process of reverse phase liquid chromatography [43]. It was found that protiated compounds bind to nonpolar moieties in the stationary phase more strongly than deuterated ones [49]. In theory, deuterated peptides should be less retained in comparison to their hydrogen counterpart. In practice one can observe from negligible to significant differences in retention times of an isotopic pair depending on the number of labeled amines on each peptide and the difference in number of isotopes between the isotopologs of a given peptide as demonstrated by several studies [44, 46–48]. In case of significant differences in retention time, the heavy isotopolog is eluted prior to the light one. This needs to be considered when calculating the extracted ion chromatogram (XIC) for quantitation. In most cases the effect on elution time is minor when using reagent 1, 3, and 5 from Table 3 where the difference in number of deuterated atoms is two per label (*see* Fig. 8).

Peptide ratios are in general calculated from the relative intensities from XIC of the monoisotopic peaks in the composite MS spectrum. In general, it is not recommended to use a single scan as the intensity between labeled peptide pairs tend to fluctuate across

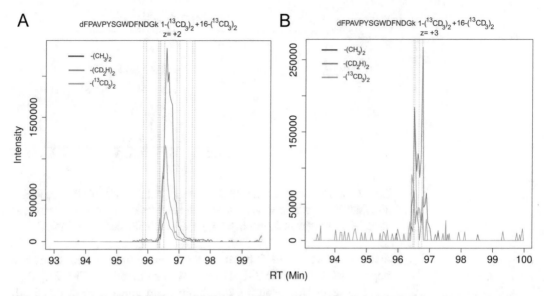

Fig. 8 $XIC_{0.05}$ on dimethyl labeled peptide triplet of the sequence dFPAVPYSGWDFNDGk with $z = +2$ (**a**) and $z = +3$ (**b**). Lower case letter indicates the derivatized sites using labeling reagent 1 (blue), 3 (red) and 5 (green) from Table 3. The dotted vertical lines indicates MS/MS events identifying the peptide

scans. For example, the survey scan of the dimethyl labeled peptide hQGVMVGMGQk in Fig. 7 provides an inaccurate ratio estimate compared to the XIC intensity in Fig. 8. The two different charge states of peptide dFPAVPYSGWDFNDGk from alpha amylase displays similar trends in terms of ratios between samples but the more intense signal in Fig. 8a provides a more accurate ratio estimate due to better data sampling. A curious side note, Fig. 8a provides an argument against spectral counting. In Fig. 8a, the lightest peptide (blue line) is the most intense and also sequenced by MS/MS the most times. However, the peptide from the sample labeled with medium weight compound (red line) has a more than two fold lower intensity overall compared to the peptide from the heaviest labeled sample (green line) yet the spectral counts are three (red line) and two (green line), respectively. The labeled samples correspond to lung cancer and non-lung cancer cases where the light labeled blue peak correspond to a lung cancer sample [50]. Figure 9 displays XIC from a second peptide from alpha amylase confirming the ratios from the peptide in Fig. 8a, b. It is reassuring that two peptides from alpha-amylase confirm a similar significant regulation, consistent even after normalization.

The isotope effect on retention time can be attenuated using different experimental set-ups such as lowering sample complexity to facilitate matching between LC-MS peaks and MS/MS identifications. Complex samples can be separated by a 2D LC approach, particularly if the first dimension of the 2D-LC-MS method is not based on hydrophilicity/hydrophobicity [48]. Instead, low-pH

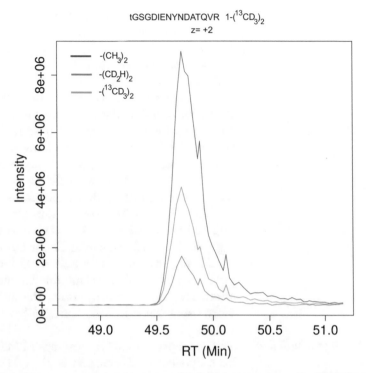

Fig. 9 $XIC_{0.05}$ on dimethyl labeled peptide triplet of the sequence tGSGDIENYN-DATQVR and $z = +2$. Lower case letter indicates the derivatized sites using labeling reagent 1 (blue), 3 (red) and 5 (green) from Table 3

strong cation exchange (SCX) chromatography is an appropriate choice for 1D fractionation, as separation is based on peptide charge and therefore complements the separation based on hydrophobicity. Also enrichment of phosphopeptides can be performed by TiO_2 or immobilized metal affinity chromatography of dimethyl labeled samples prior to LC-MS analysis [48].

Strategies have been developed to circumvent data analysis complexity associated to amine dimethyl labeling methods by decreasing the number of amine labeled groups per peptide. Blocking lysine side chain by guanidation prior to dimethyl labeling generate a single mass tag in the N-terminal of every peptide, creating an uniform 6 Da mass difference between light and heavy isotopically labeled peptide pairs [51]. This assures a minimal overlap of the isotope envelopes from the peptide pairs and with a negligible isotope effect on reversed-phase separation. Recently, the application of dimethyl labeling to the complexity of protein isoforms quantitation has been reported [52]. By labeling intact proteins, the tryptic peptides generated are enriched of peptides with C-terminal arginine due to lysine side chain labeling. This strategy renders longer peptides thereby increasing the probability of peptides that distinguish highly homologous proteins.

Dimethyl stable isotope labeling provides a signal enhancement of the a1 ions in the MS ion series, which are usually hard to detect for most of the nonderivatized peptides. This feature widespread the applicability of dimethyl labeling to de novo sequencing [53]. a1 ions derived from the 20 proteinogenic amino acids gain signal enhancement. For labeled peptides the enhanced signals for the a1 ions combined with accurate mass determination can be used to fingerprint the N-terminal residue. Thereby, overcoming the trouble caused by the frequent absence of b1 ions in MS/MS spectra. The a1 ion signal enhancement has been proposed to improve the confidence of matching accuracy when considered in the scoring algorithm additionally to b- or y-ions [53, 54].

The use of stable-isotope dimethyl labeling has extended to targeted proteomics approaches such as selected reaction monitoring (SRM) [55] (*see* Chapter 12 for more details on SRM).

Finally, data analysis tools should ideally take in consideration potential posttranslational modified lysine residues, such as methylation, acetylation, biotinylation, ubiquitination, and sumoylation, when using stable-isotope dimethyl labeling.

2.3.2 Tandem Mass Tags

The concepts of tandem mass tags (TMT) labeling approaches has been detailed by Thompson et al. [56] and is briefly summarized here. TMT isotopomer labels are isobaric and cannot distinguish different samples at the MS level. This gives increased sensitivity and less sample complexity. The structure of TMT reagents includes (1) a reporter ion, (2) a mass balancer, and (3) a derivatizing agent specific for primary amines [56] or cysteines [57]. These units are linked by labile bonds which lead to intense reporter ions in MS/MS (*see* Fig. 10). The MS/MS spectra, depicted in Fig. 10 was identified in the nuclear fraction of KMH2 cell lines in two different conditions [58] and created using the R package MSnbase [59]. Samples were labeled with reporter 126 and 127 but signals which correspond to reporter 128–130 using a 0.05 Da mass interval were also observed. This suggests that mass accuracy can be optimized to alleviate signals from overlapping fragment ions. Furthermore, large-scale MS/MS spectra analysis can reveal m/z regions with minimum ion fragmentation for future design of new reporters. The isobaric feature is achieved by the mass balancer that compensate for the mass difference of the reporter ions. Commercial iTRAQ reagent sets come with 2–10 different reporter ions for primary amine labeling for the simultaneously analysis of up to 10 samples. Interestingly, a recent method named "improved CILAT (cleavable isobaric labeled affinity tag)" which derivatizes cysteines with TMT and can label up to 12 samples has been proposed [57].

There are two new isotopologs of TMT 127 and 129 reagents that make quantitation with ETD possible. The reporter ions of

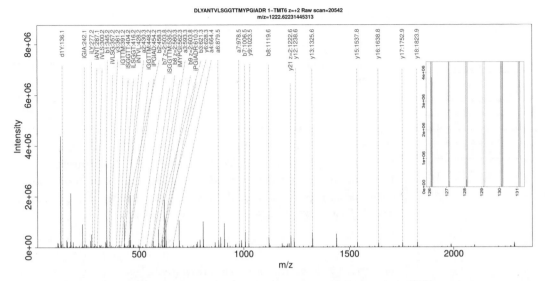

Fig. 10 Raw MS/MS spectrum annotated with fragment ions from DLYANTVLSGGTTMYPGIADR with N-terminal TMT tag. Insert depict a zoom in on the TMT reporter ions where the colors highlight the expected 0.05 Da mass interval for the different reporters

these new isotopes are slightly lighter (6.32 mDa) than the traditionally used TMT 127 and 129 reagents. Using a high mass accuracy/high resolution mass spectrometer (e.g., Orbitrap) the original and the new isotopes can be distinguished therefore expanding the plex number [60, 61]. Beyond 8-plex quantitation, the potential of current chemistry to accumulate even more channels into that same mass range using different types of isotopes could expand TMT to 10-plex and possibly up to 18-plex [60].

2.3.3 iTRAQ Chemical Labeling

Commercially available iTRAQ reagents are supplied together with the datasheet indicating the percentages of each reporter ion reagent that differ by −2, −1, +1, and +2 Da from the reference mass. These percentages f (*see* **Note 2**) need to be considered for correction of the apparent intensity count $I'r$ for each reporter ion Ir, where r indicates a specific reporter ion. Intensity correction can be obtained by solving the following set of linear equations (Eqs. 5–8):

$$\begin{pmatrix} I'_{114.1} = & f_{m,114.1} \times I_{114.1} + f_{m-1,151.1} \times I_{115.1} + f_{m-2,116.1} \times I_{116.1} + f_{m-3,117.1} \times I_{117.1} \\ I'_{115.1} = & f_{m+1,114.1} \times I_{114.1} + f_{m,115.1} \times I_{115.1} + f_{m-1,116.1} \times I_{116.1} + f_{m-2,117.1} \times I_{117.1} \\ I'_{116.1} = & f_{m+2,114.1} \times I_{114.1} + f_{m+1,115.1} \times I_{115.1} + f_{m,116.1} \times I_{116.1} + f_{m-1,117.1} \times I_{117.1} \\ I'_{117.1} = & f_{m+3,114.1} \times I_{114.1} + f_{m+2,115.1} \times I_{115.1} + f_{m+1,116.1} \times I_{116.1} + f_{m,117.1} \times I_{117.1} \end{pmatrix}$$

$$(5)$$

where I represents intensity, f the percentage of the reporter ion, and m the monoisotope mass. The number of linear equations can be adjusted depending on the number of reporter ion intensities that needs to be corrected. The above strategy is the same as presented by Shadforth et al. [62] describing the i-Tracker tool (*see* **Note 3**). However, the above presentation and the way to solve the linear equations are presented in a general fashion. The above equations can be written by matrix notations as:

$$y = X\beta \tag{6}$$

where y describes a vector containing the apparent intensities, X represents a matrix with the percentages of each reporter ion, and β is a parameter vector containing the true intensities of reporter ions to be determined. Note that no error term can be included in Eqs. 5 and 6 of the model (*see* **Note 4**). The parameter vector β can now be estimated by the equation:

$$\hat{\beta} = \left(X^T X\right)^{-1} X^T y \tag{7}$$

and the corrected intensities are now given by:

$$\hat{y} = \hat{\beta} X \tag{8}$$

A number of software tools are available for iTRAQ quantitation such as ProQUANT (Applied Biosytems, Foster City, CA), i-TRACKER/TandTRACK [62, 63] and Multi-Q [64, 65]. More recently a number of powerful R packages for analysis of TMT/iTRAQ data have become available such as isobar [66], MSnbase [59], iTRAQPak [67], OCAP [68], and iPQF [69].

2.4 Label Free Quantitation

LFQ values can in principle, at all times, be extracted from LC-MS runs despite its labeling state. However, the experimental design in LFQ is essential. Different LC-MS runs are directly compared consequently reproducibility in sample preparation is a crucial issue in LC-MS methodologies. LFQ can be obtained by spectral counting, summation of MS/MS fragment ions, extraction of corresponding parent ion intensity from MS scans or a combination of the three. Many quantitative methodologies have been named based on above mentioned three types of extracted intensity values and they differ by the ways of normalizing the quantitative values. For example, normalization can be based on information on the: (1) intensity between runs, (2) number of expected tryptic peptides from a given protein (corresponding to length normalization of DNA/RNA sequencing data) and (3) difference in ionization efficiency of the peptides. The methods further differ on how peptides shared between several proteins are considered. The reader can find a thorough review on different methodologies to calculate protein quantitative values by Blein-Nicolas et al. [70]. Although quantitative values estimated by using certain LFQ methods are sometimes considered absolute, in general these should be regarded as crude estimates due to different peptide ionization efficiency. LFQ methods, overall identify 50–60% more peptides or proteins than methods based on stable isotope labeling since labeling introduce a higher degree of peak complexity caused by MS signal splitting [58]. Noticeable, TMT based methods also produce a lower number of identifications compared to label-free quantitation, however without MS signal splitting. This is likely a consequence of the increased number of LC-MS runs, that enables sequencing of a larger number of peptides and the decreased ionization efficiency for detection. Interestingly, O'Connell et al. have recently demonstrated that TMT labeling provided three times more significant regulated proteins compared to LFQ due to high precision and 25% less missing values [71]. Labeling-based methods require expensive reagents, but on the other hand LFQ call for a larger number of LC-MS runs. In the future, LFQ can become the standard method for high throughput LC-MS if sensitivity of instruments increases and the costs per LC-MS run considerably decreases. For example, the company Evosep has developed LC-MS methodologies that can analyze clinical samples in 5 min. Along, development of computational methods for MS spectra analysis can push it further.

2.4.1 Experimental Considerations for Label-Free Quantitation

In LFQ, technical variation is introduced at each processing step for that reason protein extraction and enzymatic digestion should be tightly controlled. Table 4 resumes most of the factors associated to each sequential step in the LC-MS workflow that contribute to variations within replicas and across samples, covering sample preparation, LC system, MS instrumentation and peptide properties.

Reproducibility of sample preparation and LC-MS runs must be controlled to obtain confident peptide identifications and increased statistical power. A large number of experimental factors

Table 4
Reproducibility associated features in the LFQ workflow

Sample preparation	LC system	MS instrument	Peptide properties
Protein extraction protocols	Peak separation	Mass accuracy	Internal fragment ions
Sample quality	Peak resolution	Spray instability	In-source decay
Enzymatic digestion	Peak width	Dynamic exclusion settings	Ionization efficiency
Buffer pH	Bleeding	Noise in the MS/MS	Fragmentation efficiency
Peptide salt contamination	Tailing	Sampling rate Mass range of detector Cone voltage Detector sensitivity Dynamic range Isolation window of parent ions	

significantly affect reproducibility of LC-MS runs. Some of the aspects in sample preparation that affects the reproducibility of protein identifications and quantitation are in general protein extraction protocols, degree of sample degradation, sample pH which affects peptide charge distribution, enzymatic digestion, salt contamination which causes additional peaks, for example, the generation of Na^+ peaks instead of H^+, and polyethylene glycol (PEG) contamination. In addition, LC factors affecting reproducibility include peak separation, peak resolution, peak widths, bleeding, and tailing. MS instrument related factors affecting reproducibility are mass accuracy, spray instability, dynamic exclusion settings, triggering of MS/MS on noise, sampling rate, mass range of detector, cone voltage, detector sensitivity, dynamic range, isolation window of parent ions, etc. Examples of peptide properties affecting the identification and quantitation results combine internal fragment ions, in-source decay, ionization efficiency, and fragmentation efficiency. Many features are easily manageable if the samples are run in a standardized laboratory on a single instrument within a short time frame. However, if several laboratories are involved in the study using different instruments over a long time frame then the above factors becomes more challenging to control.

Protease efficiency depends on posttranslational modifications and on specific amino acids in the vicinity of the cleavage site. Sample complexity ideally should be kept at a minimum to reduce the number of chimeric spectra from near isobaric coeluting peptides which frequently originate erroneous quantitative values. For full coverage of the complexity of the sample and to alleviate stochastic subsampling it is necessary to acquire two to three LC-MS runs of each biological replica. Ion suppression might lead to false assignment of the main chromatographic peak and increased variability in estimated quantitative values.

The interpretation of quantitative protein values might be obscured by shared peptides between proteins. A specific protein might be represented by the canonical peptides, missed cleaved peptides, modified peptides and all ions can have multiple charge states which need to be taken into account. Some proteins are represented by peptides that ionize poorly than others interfering with the estimation of absolute but not relative quantitative values. Instrument settings should be kept unchanged between runs although it is possible to compare runs with the same instrument settings operated at different time points (*see* Fig. 11a). Regarding spectral counting, the dynamic exclusion setting might obscure the quantitative values and a threshold value of the number of spectral counts must be defined in order to assure an accurate quantitation (*see* Fig. 8).

It is recommended to randomly assign samples to batches in a way that assures approximately an equal number of sample types to each experimental batch thus avoiding experimental bias

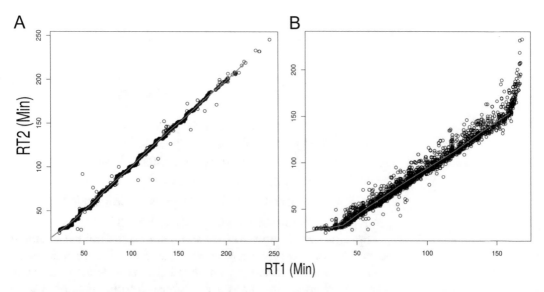

Fig. 11 LC-MS retention time alignment. (**a**) LC-MS alignment of two runs obtained at different time points and different LC-MS column. (**b**) LC-MS alignment of two runs obtained at different time points and with different LC gradients

[50, 72]. Furthermore, within each batch the order of the sample in the mass spectrometer acquisition queue must be considered. For example, if a set of samples from one condition are analyzed first followed by a second set of samples from a second condition, overtime degradation of the LC column flow might introduce an experimental bias. It is therefore preferable to alternate samples from different conditions (e.g., set A and set B with three replicas each, ABABAB scheme is recommended instead of AAABBB scheme). Furthermore, per routine blanks are alternated between samples to avoid carry over from one LC-MS run to the next.

To obtain reproducibility in the quantitation it is necessary to assess the coverage of the sample complexity [50]. Furthermore, a subset of samples can be analyzed at different time points, for example between LC column exchanged to evaluate the robustness of the MS data [50].

2.4.2 Spectral Counting

The main advantage of LFQ spectral counting is the simplicity of implementation and direct association of peak count to identified peptides. Furthermore, it resembles quantitation of DNA and mRNA by DNA sequencing technologies. Spectral counting is often used to obtain relative protein quantitation between samples; nevertheless, it requires a larger number of spectral counts to avoid erroneous quantitation (*see* Fig. 8).

Fu and coworkers suggest a two-step strategy for spectral counting. In the first step spectral index is calculated for all proteins in each sample followed by permutation analysis to establish confidence intervals assessing differential protein expression [73]. The spectral index SpI_i is defined in Eq. 9:

$$SpI_i = \left(\frac{\bar{S}_{Ai}}{\bar{S}_{Ai} + \bar{S}_{Bi}} * \frac{N_{Ai}^D}{N_A^T} \right) - \left(\frac{\bar{S}_{Bi}}{\bar{S}_{Ai} + \bar{S}_{Bi}} * \frac{N_{Bi}^D}{N_B^T} \right) \qquad (9)$$

$\bar{S}_{Ai}$ = Average spectral count for protein i in sample A
$\bar{S}_{Bi}$ = Average spectral count for protein i in sample B
N_{Ai}^D = Number of detections for protein i in sample A
N_A^T = Total number of samples in sample A
N_{Bi}^D = Number of detections for protein i in sample B
N_B^T = Total number of samples in sample B

Spectral index outperformed four other statistical tests namely Student's t-test, G-test, Bayesian t-test, and Significance Analysis of Microarrays when tested on simulated data sets. Spectral index is easy to implement and statistically robust. Permutation of sample annotations is used to obtain statistically significant cutoff values for SpI (e.g., the authors applied 1000 permutations and a 99th percentile cutoff).

2.4.3 Protein Abundance Index (PAI) and Exponential Modified PAI (emPAI) in Spectral Counting Based Absolute Quantitation

$PAI = \frac{N_{obsd}}{N_{obsbl}}$ has been proposed as a simple method to obtain estimate of absolute protein abundance in a sample. N_{obsd} is the number of observed peptides for a given protein and N_{obsbl} is the number of observable peptides. emPAI correspond to exponential modified PAI values defined by $emPAI = 10^{PAI} - 1$. The percentage of deviation of emPAI values from actual protein abundances was estimated to be 63% [74].

2.4.4 APEX

The Java program Absolute Protein Expression (APEX) aims at improving the accuracy of spectral counting by considering the variation of sensitivity in peptide detection by MS techniques [75]. Such considerations are of no use in relative quantitation by spectral counting since the normalization by predicted sensitivity cancels out. However, the objective is to provide absolute quantitative values that are more accurate than for example emPAI. APEX for protein i is defined in Eq. 10:

$$APEX_i = \frac{p_i \left(\frac{n_i}{O_i} \right)}{\sum_{k=1}^N p_k \left(\frac{n_k}{O_k} \right)} * C \qquad (10)$$

n_i is the total observed spectral count for protein i. O_i corresponds to the expected or predicted total count for protein i, and p_i is the protein identification probability. These parameters define the numerator in Eq. 9. The denominator corresponds to the sum of the numerator iterating over all k proteins (k the iteration variable, [1-N]). C is a user defined normalization factor which can for example be the total protein concentration.

LFQ based LC-MS intensity profiling is one of the most frequent used quantitative methods, mainly because of improved reproducibility of chromatography systems and ion spray stability within recent years. LFQ is most adequate for sequential MS analysis of comparative samples using exactly the same experimental settings. The introduction of variations in the experimental procedures affects the reproducibility of the analysis of a set of comparative samples. In the LC separation step several parameters must be thoroughly controlled such as composition of the chromatography buffers or sample loading buffer (especially solutions pH) to prevent differences in relative intensities between the charge states of the peptides and downstream artifacts. Differences in the chromatography buffer or sample loading buffer composition (especially pH needs to be controlled) originates different relative intensities between charge states of peptides which then leads to downstream artifacts. Additionally, if the chromatography column is exchanged, the elution profile of peptides will be difficult to reproduce making it difficult to align and compare LC-MS runs, although not impossible (Fig. 11a). If the mass spectrometer or ion spray needle is changed then it is likely that one can observe a large effect on the intensity values again making the data analysis more challenging.

Most of the sample preparation methods used in LFQ are currently non-gel based due to improvements in sensitivity of current instruments allowing for a larger dynamic range coverage. However, previously, quantitation of in gel digested peptides from histone enriched fractions produce reliable and reproducible results (average coefficient of variance of ~17% *see* **Notes 5** and **6**) [76]. Furthermore, implementation of MS-LFQ in clinical proteomics is relatively straightforward. For example, a frequent strategy for label-free quantitation uses analytical 1D-SDS PAGE followed by 1D-LC-MS in the analysis of biopsy tissue samples. LFQ approaches can identify potential biomarkers which then can be validated by orthogonal quantitative methods using stable isotope labeled samples as internal reference, Western Blot or other quantitative proteomics techniques. Another important aspect to take in consideration is the application of LFQ methods to study aberrant phospho-signaling in diseases such as cancer. Montoya et al. [77] reported 20% coefficient of variation on average for 900 quantified phosphopeptides enriched by TiO_2 prior to LC-MS/MS analysis.

Accurate chromatography alignment is essential for proper quantitation. The mass dimension is normally highly accurate due to the frequent use of lock spray calibration, calibration using buffer contaminants or charge state distributions.

The LC-MS/MS runs contains both LC-MS (which may be used for quantitation) and LC-MS/MS (used for identification) data. This means that the intensity counts obtained in the LC-MS part (survey scans) of the LC-MS/MS runs are less than what would be obtained in LC-MS runs since the sample input is split between MS and MS/MS scans. Therefore one strategy to obtain better sampling is to first identify as many peptides as possible across all sample conditions by LC-MS/MS runs followed by replica LC-MS runs of each individual sample. Other strategies to increase ion counts involves the use of targeted proteomics methodologies (*see* Chapter 12) or SWATH [19]. A frequent used protein quantitative metric based on MS1 intensity is intensity based absolute quantitation (iBAQ) [78]. iBAQ is defined as sum of the MS1 intensities of all the peptides associated to a protein divided by the number of theoretically observable peptides. For an extensive review of quantitative metrics *see* review by Blein-Nicolas and Zivy [70].

3 Data Quality

Estimation of data quality is essential for reliable protein quantitation. A standard procedure in data quality assessment involves MS analysis of BSA standards for the detection of flaws either in the LC or MS instrumentation. Alternatively, visual reports of LC-MS data can be generated for quality monitoring. For example, chromatography and electrospray instability are observable in total ion current (TIC) chromatograms. Extracted ion current (XIC) chromatograms display the elution profiles of specific peptides and are therefore informative about peak width, bleeding and tailing. MS scans can reveal salt adducts (e.g., Na^+ [M + H + 22] + and K^+ [M + H + 38]+) where the parent ion masses are shifted. Other types of contaminations can cause peak series (*see* **Note 7** and Table 5). Essentially, all instrument software provides functions for visualization of MS scans, XIC and TIC chromatograms from raw data.

Comparison of LC-MS data in terms of reproducibility requires the implementation of specific checkpoints. For example, a histogram displaying the distribution of the number of missed cleavage sites in the identified peptides can pin point cases of incomplete enzymatic digestion. Additionally, the level of protein oxidation in the different samples can be inferred from the frequency of methionine sulfoxide and methionine sulfone. Besides, the frequency of cysteine-containing peptides indicates the presence of nonderivatized cysteine residues. Finally, comparison of the overall charge state distributions of the peaks in the survey scan can also pinpoint outliers to be eliminated in the downstream comparison of samples. Quality control measures should be carefully assessed, since consistent differences of a specific sample group might be caused by intrinsic technical properties of the sample group rather than from biological origin.

3.1 Background Subtraction and Noise Filtering

For LC-MS data, background subtraction is basically no longer needed. Nevertheless, noise filtering is still needed mainly to correctly identify chromatographic peak tops. Thus, estimated peak parameters are used as preliminary parameters of peak fitting algorithms to guide the search for optimal fit (e.g., a Gaussian mixture function).

Table 5
List of possible masses and ion series caused by sample contamination in LC-MS analysis

Observation	Probable cause
391.284286	Protonated dioctyl phthalate $(M + H)^+$
413.2667792	Sodium adduct of dioctyl phthalate $(M + Na)^+$
798.5878446	Ammoniated dimer of dioctyl phthalate $(2\ M + NH_4)^+$
803.5437887	Sodiated dimer of dioctyl phthalate $(2\ M + Na)^+$
44.02621475 series	PEGs and other ethoxylated polymers
615.4037494	CHAPS $(MH)^+$
1229.800222	CHAPS $(2\ M + H)^+$

$[Si(CH3)20]_n$	MH^+-CH_4	MH^+	MH^++NH_3
$n = 5$	355.069933	371.101233	388.127782
$n = 6$	429.088724	445.120024	462.146573
$n = 7$	503.107515	519.138815	536.165365
$n = 8$	577.126307	593.157607	610.184156
$n = 9$	651.145098	667.176398	684.202947
$n = 10$	725.16389	741.19519	758.221739

3.2 Mass and Retention Calibration

Mass calibration of LC-MS data has also become less important since instrument software are able to perform lock spray calibration and contaminant-based calibration. Furthermore, database search engines can use a preliminary search result, obtained by applying nonconstrained mass errors as input parameters, to recalibrate the MS data. Alternatively, charge state pairs can be used to recalibrate m/z in survey scans as well as the most common contaminants. The accurate calibration of the m/z dimension implies that the alignment of LC-MS data is reduced to a simple one dimensional problem, the retention time alignment. LC-MS runs can be aligned by calculating similarity measure, (e.g., correlation) between survey scans, thereby creating a matrix with similarity measures [79]. From this matrix an alignment function (linear or nonlinear) can be fit to the best similarities in the matrix. The drawback of such methods is that for long LC-MS runs this becomes computationally harder. However, if peptides are identified by MS/MS then it is computationally simple to use peptide identifications as landmarks [4, 80]. The retention time given in the MS/MS spectra corresponds to the time at which the MS/MS event was triggered and differs from the maximal elution point (*see* Fig. 8). The identification of the maximal elution point should precede peptide identifications and can be used as landmarks for the alignment. The maximum elution time is determined from all retention times from triggered MS/MS spectra which identify the peptide (using all charge states) and from the extracted ion chromatogram based on the peptides parent ion mass at a specific charge state. In some cases it is actually not possible to identify with 100% certainty the maximum elution point. For example, there are cases in which there is only one triggered MS/MS event located in the tail of the elution peak. In such cases information from similar LC-MS runs can be used to ascertain the correct elution peak by aligning the MS scans of the same peptide identified in other runs. In addition, the main peak in the extracted ion chromatogram can be matched to the mass and retention time of other peptides.

The maximum elution point of each peptide after updating the retention times can be used as landmarks to align LC-MS runs (*see* Fig. 11a). State-of-the-art instruments provides an alignment function which consists of a straight line even though the data has been obtained at different time points and using different chromatography columns as long as the LC gradient and MS instrument settings are maintained (*see* Fig. 11a). Consequently, the reproducibility and quality of the chromatography has reached a standard such that the plots in Fig. 11a are frequently made as a quality control check rather than for aligning the LC-MS runs. In contrast, if changes occur in the LC gradient the alignment of the retention time is difficult typically in the start and at the end of the chromatography run (*see* Fig. 11a).

3.3 Detection of Outlier Runs

Poor quality runs can often be identified as outliers by using cluster analysis. Poor quality MS runs lead to inflated variance estimates. This should be avoided since the nonrobust statistical tests which have the highest power depend on accurate estimates of variance. Removing poor-quality runs by identifying outliers in clusters is therefore justified.

The approaches presented in the literature for identification of outliers can roughly be divided into two main groups: (1) clustering based on raw data spectral properties [80] and (2) clustering based on peptides identified from searching MS/MS spectra against a sequence database [81].

4 Replacement of Missing Values

Missing values in quantitative MS-based proteomics is frequent and basically present in every data set. Being able to handle missing values in quantitative proteomics is therefore essential. In R missing values are normally indicated by NA which is not a string or numeric data type. The R function "is.na" can be used to locate missing values, on the other hand the R function "na.omit" removes rows with missing values. Many missing value estimates are between 0 and the minimum detected value. A simple method to insert missing values is by replacing with the minimum detected value or half of the minimum detected value (*see also* Chapter 13). The R package Msnbase integrates a large number of algorithms for missing value estimation [59]. The topic of missing value imputation in proteomics has been reviewed [82–84]. The book by Gelman and Hill [85] provides R code for different methods as well.

5   Basic Transformation, Centering, Scaling and Normalization

In statistics data transformation generally involves applying a continuous invertible function. The purpose of transforming data is (1) to be able to apply a statistical test that requires symmetric distribution of data, (2) to provide better data visualization (e.g., log transform can be used to archive a more uniform spread of data points) and (3) more transparent interpretation of the data (e.g., mass errors can be transformed to ppm values, *see* Chapter 1). In the literature quantitative genomics data analysis often uses the $\log_2$ transformation. The log transform provides better distribution symmetry for right skewed data and allows easy interpretation. For example, a $\log_2$ difference of 1 corresponds to a twofold difference which facilitates the interpretation of the results.

The power transforms are more powerful since they stabilize variance and make the data more normal distribution-like. A slight modification of the power transforms is the Box-Cox transform (Eq. 11).

$$I_{\text{Box-Cox}} = \begin{cases} \dfrac{I^p - 1}{p} & p \neq 0 \\ \log(I) & p = 0 \end{cases} \tag{11}$$

The optimal p can be determined by maximum-likelihood estimation and for this purpose a function is available in the R package geoR.

The next step is normally centering and scaling which involves column-wise manipulations (multivariate statistics).

After mean-centering all columns will have zero mean

$$x_{ij}(\text{mean centered}) = x_{ij}(\text{original}) - \bar{x}_j \tag{12}$$

If variance scaling is applied together with mean centering then the result is autoscaling

$$x_{ij}(\text{autoscaled}) = \left(x_{ij}(\text{original}) - \bar{x}_j\right)/s_j \tag{13}$$

The R function "scale" can be used for mean-centering, variance scaling and autoscaling.

If the data have not been filtered for outliers, then using the median and median absolute deviation (MAD) is more robust and can also be applied using the R function "scale."

A more complete overview of transformation, centering and scaling methods can be found in Chapter 2 of this book titled "Introduction to multivariate statistical analysis in chemometrics" [86]. This book also provides R code to perform the analysis.

Quantile normalization [87] is frequently used to normalize quantitative proteomics data before significance testing for protein regulation. Msnbase integrates a large number of algorithms for normalization (*see* MSnbase::normalise) [59].

6   Final Comment on Significance Testing

The advantage of using Eq. 12 or 13 together with sample permutation is that it is statistically robust and no assumption of the underlying distribution is needed. The disadvantage is that it requires a fair amount of replicas in order to have enough distinct sample permutations. In general the t-test depends on the stated level of significance α, the shape of the populations' distributions, the spreads of the two populations and the sample sizes. An efficient strategy could be to use a large test sample set to estimate the underlining distributions and further use these distributions to simulate the t-test significance on all future runs. The R function described below can be used to establish the simulated distribution of t values (*see* "Bayesian computation with R" for details [88]).

```
tstatistic=function(x,y) {
m=length(x)
n=length(y)
sp=sqrt(((m-1)*sd(x)^2+(n-1)*sd(y)^2)/(m+n-2))
t=(mean(x)-mean(y))/(sp*sqrt(1/m+1/n))
return(t)}
```

Alternatively, the R package limma [89] is frequently used to estimate significance of regulation.

7   Discussion

The ultimate aim is that protein quantitation obtained in various cell types in a given time frame under specific conditions can support the growing field of system biology. The main bottleneck for a more detailed approach is the cost of the quantitative experiments. Currently the relative quantitative methods discussed here are the most cost effective.

QconCAT is a cost effective approach for absolute quantitation and could be used for more extensive studies; however, PTMs will be difficult to accurately quantify by such approach. The computational methods available for SILAC that are discussed in this chapter can be reused for AQUA and QconCAT since the final output of the data is very similar. Although an extra software layer may be needed for AQUA and QconCAT, to deal with accurate calculation of protein concentration using a standard curve made from dilution series.

8 Notes

1. The theoretical isotope distributions can be calculated by linear approximations as described by Wehofsky et al. [90]. Although these linear approximations are not recommended for $^{16}O/^{18}O$ labeling when the quantitated peptide is known. The linear approximations can be used to analyze peaks in a LC-MS run where the peptide sequence is unknown. In such cases one can at best approximate the isotope distribution based on the observed m/z and charge state of the detected peptide in the LC-MS run. The linear approximations are therefore appropriate and can provide faster relative isotopic abundance values. Wehofsky et al. [90] provides the following approximations:

$$I_{m+1} = -1.25446 + 0.05489 \times I_m \text{ and}$$

$$I_{m+2} = 0.13977 + 0.00613 \times I_m + 1.49147E^{-5} \times M^2$$

   Alternatively, the averagine table from the R package prot-Viz (protViz::averagine) can be used.

2. The percentage values $f_{m \pm i,x}$ are provided by the manufacture of the iTRAQ reagent. The percentage for the monoisotopic peak can be calculated as,

$$f_{m,x} = 100\% - \sum_{i}^{n} f_{m \pm i,x}$$

   where i is an nonzero integer and x an specific iTRAQ reporter ion.

3. The above strategy is the same as presented by Shadforth et al. [62] describing the i-Tracker tool. However, the equations provided by Shadforth et al. only works for iTRAQ with 4 reporter ions. The equations provided here are general and can be used for iTRAQ with any number of reporter ions.

4. It is not possible to model an error term in the above case since the number of parameters equals the number of linear equations (*see* Eq. 5) which means that the system will be

underdetermined. There is no unique solution to an under-determined system. However, in some cases it is possible to setup more equations than presented here and it is then worthwhile to add a unit vector as a column to the matrix x in Eq. 6 which will model an error term.

5. Coefficient of variance is defined as the ratio between the standard deviation σ and the mean μ: $c_v = \sigma/\mu$.

6. Especially the cutting of gel pieces have to be done in a careful manner making equal sized squares.

7. Normally samples are cleaned up on Stage-tip or Zip-tip before LC-MS basically eliminating salt adducts. However, when analyzing data from various sources mistakes happen, and it is not unlikely that a data set which originated from a salt-contaminated sample end up in the final batch of data for analysis. In other words, be careful.

Acknowledgments

R.M. is supported by Fundação para a Ciência e a Tecnologia (CEEC position, 2019–2025 investigator). iNOVA4Health—UID/Multi/04462/2013, a program financially supported by Fundação para a Ciência e Tecnologia/Ministério da Educação e Ciência, through national funds and cofunded by FEDER under the PT2020 Partnership Agreement is acknowledged. A.S.C. is supported by Fundação para a Ciência e a Tecnologia (BPD/85569/2012). This work is also funded by FEDER funds through the COMPETE 2020 Programme and National Funds through FCT—Portuguese Foundation for Science and Technology under the projects number PTDC/BTM-TEC/30087/2017 and PTDC/BTM-TEC/30088/2017.

References

1. Molloy MP, Brzezinski EE, Hang J, McDowell MT, VanBogelen RA (2003) Overcoming technical variation and biological variation in quantitative proteomics. Proteomics 3 (10):1912–1919

2. Karp NA, Lilley KS (2007) Design and analysis issues in quantitative proteomics studies. Proteomics 7(Suppl 1):42–50

3. Lau KW, Jones AR, Swainston N, Siepen JA, Hubbard SJ (2007) Capture and analysis of quantitative proteomic data. Proteomics 7 (16):2787–2799

4. Matthiesen R (2007) Methods, algorithms and tools in computational proteomics: a practical point of view. Proteomics 7(16):2815–2832. https://doi.org/10.1002/pmic.200700116

5. Julka S, Regnier F (2004) Quantification in proteomics through stable isotope coding: a review. J Proteome Res 3:350–363

6. Bronstrup M (2004) Absolute quantification strategies in proteomics based on mass spectrometry. Expert Rev Proteomics 1 (4):503–512

7. Ong SE, Blagoev B, Kratchmarova I, Kristensen DB, Steen H, Pandey A, Mann M (2002) Stable isotope labeling by amino acids in cell culture, SILAC, as a simple and accurate approach to expression proteomics. Mol Cell Proteomics 1(5):376–386

8. Mirzaei H, McBee JK, Watts J, Aebersold R (2008) Comparative evaluation of current peptide production platforms used in absolute quantification in proteomics. Mol Cell Proteomics 7(4):813–823. https://doi.org/10.1074/mcp.M700495-MCP200. [pii]: M700495-MCP200.

9. Pratt JM, Simpson DM, Doherty MK, Rivers J, Gaskell SJ, Beynon RJ (2006) Multiplexed absolute quantification for proteomics using concatenated signature peptides encoded by QconCAT genes. Nat Protoc 1 (2):1029–1043. https://doi.org/10.1038/nprot.2006.129. [pii]: nprot.2006.129.

10. Rivers J, Simpson DM, Robertson DH, Gaskell SJ, Beynon RJ (2007) Absolute multiplexed quantitative analysis of protein expression during muscle development using QconCAT. Mol Cell Proteomics 6(8):1416–1427. https://doi.org/10.1074/mcp.M600456-MCP200. [pii]: M600456-MCP200.

11. Anderson L, Hunter CL (2006) Quantitative mass spectrometric multiple reaction monitoring assays for major plasma proteins. Mol Cell Proteomics 5(4):573–588

12. Anderson NL, Anderson NG, Haines LR, Hardie DB, Olafson RW, Pearson TW (2004) Mass spectrometric quantitation of peptides and proteins using Stable Isotope Standards and Capture by Anti-Peptide Antibodies (SISCAPA). J Proteome Res 3(2):235–244

13. Kirkpatrick DS, Gerber SA, Gygi SP (2005) The absolute quantification strategy: a general procedure for the quantification of proteins and post-translational modifications. Methods 35(3):265–273

14. Han DK, Eng J, Zhou H, Aebersold R (2001) Quantitative profiling of differentiation-induced microsomal proteins using isotope-coded affinity tags and mass spectrometry. Nat Biotechnol 19(10):946–951

15. Aggarwal K, Choe LH, Lee KH (2006) Shotgun proteomics using the iTRAQ isobaric tags. Brief Funct Genomic Proteomic 5(2):112–120

16. Shadforth IP, Dunkley TP, Lilley KS, Bessant C (2005) i-Tracker: for quantitative proteomics using iTRAQ. BMC Genomics 6:145

17. Horvatic A, Guillemin N, Kaab H, McKeegan D, O'Reilly E, Bain M, Kules J, Eckersall PD (2019) Quantitative proteomics using tandem mass tags in relation to the acute phase protein response in chicken challenged with Escherichia coli lipopolysaccharide endotoxin. J Proteome 192:64–77. https://doi.org/10.1016/j.jprot.2018.08.009

18. Brun V, Dupuis A, Adrait A, Marcellin M, Thomas D, Court M, Vandenesch F, Garin J (2007) Isotope-labeled protein standards: toward absolute quantitative proteomics. Mol Cell Proteomics 6(12):2139–2149. https://doi.org/10.1074/mcp.M700163-MCP200. [pii]: M700163-MCP200.

19. Ludwig C, Gillet L, Rosenberger G, Amon S, Collins BC, Aebersold R (2018) Data-independent acquisition-based SWATH-MS for quantitative proteomics: a tutorial. Mol Syst Biol 14(8):e8126. https://doi.org/10.15252/msb.20178126

20. Matthiesen R, Azevedo L, Amorim A, Carvalho AS (2011) Discussion on common data analysis strategies used in MS-based proteomics. Proteomics 11(4):604–619. https://doi.org/10.1002/pmic.201000404

21. Geiger T, Cox J, Ostasiewicz P, Wisniewski JR, Mann M (2010) Super-SILAC mix for quantitative proteomics of human tumor tissue. Nat Methods 7(5):383–385. https://doi.org/10.1038/nmeth.1446. [pii]: nmeth.1446.

22. Matthiesen R (2006) Extracting monoisotopic single-charge peaks from liquid chromatography-electrospray ionization-mass spectrometry. Methods Mol Biol 367:37–48

23. Meija J, Caruso JA (2004) Deconvolution of isobaric interferences in mass spectra. J Am Soc Mass Spectrom 15(5):654–658

24. Matthiesen R (2006) Virtual expert mass spectrometrist v3.0: an integrated tool for proteome analysis. Methods Mol Biol 367:121–138

25. MacCoss MJ, Wu CC, Liu H, Sadygov R, Yates JR 3rd (2003) A correlation algorithm for the automated quantitative analysis of shotgun proteomics data. Anal Chem 75 (24):6912–6921

26. Li XJ, Zhang H, Ranish JA, Aebersold R (2003) Automated statistical analysis of protein abundance ratios from data generated by stable-isotope dilution and tandem mass spectrometry. Anal Chem 75(23):6648–6657

27. Cox J, Mann M (2008) MaxQuant enables high peptide identification rates, individualized p.p.b.-range mass accuracies and proteome-wide protein quantification. Nat Biotechnol 26(12):1367–1372. https://doi.org/10.1038/nbt.1511

28. Blagoev B, Mann M (2006) Quantitative proteomics to study mitogen-activated protein kinases. Methods 40:243–250

29. Ishihama Y, Sato T, Tabata T, Miyamoto N, Sagane K, Nagasu T, Oda Y (2005) Quantitative mouse brain proteomics using culture-derived isotope tags as internal standards. Nat Biotechnol 23(5):617–621. https://doi.org/10.1038/nbt1086. [pii]: nbt1086.

30. Yao X, Freas A, Ramirez J, Demirev PA, Fenselau C (2001) Proteolytic 18O labeling for comparative proteomics: model studies with two serotypes of adenovirus. Anal Chem 73 (13):2836–2842

31. Yao X, Afonso C, Fenselau C (2003) Dissection of proteolytic 18O labeling: endoprotease-catalyzed 16O-to-18O exchange of truncated peptide substrates. J Proteome Res 2(2):147–152

32. Mason CJ, Therneau TM, Eckel-Passow JE, Johnson KL, Oberg AL, Olson JE, Nair KS, Muddiman DC, Bergen HR 3rd (2007) A method for automatically interpreting mass spectra of 18O-labeled isotopic clusters. Mol Cell Proteomics 6(2):305–318

33. Eckel-Passow JE, Oberg AL, Therneau TM, Mason CJ, Mahoney DW, Johnson KL, Olson JE, Bergen HR 3rd (2006) Regression analysis for comparing protein samples with 16O/18O stable-isotope labeled mass spectrometry. Bioinformatics (Oxford, England) 22 (22):2739–2745

34. Ramos-Fernandez A, Lopez-Ferrer D, Vazquez J (2007) Improved method for differential expression proteomics using trypsin-catalyzed 18O labeling with a correction for labeling efficiency. Mol Cell Proteomics 6 (7):1274–1286

35. Halligan BD, Slyper RY, Twigger SN, Hicks W, Olivier M, Greene AS (2005) ZoomQuant: an application for the quantitation of stable isotope labeled peptides. J Am Soc Mass Spectrom 16(3):302–306

36. Coursey J, Schwab D, Dragoset R (2001) Atomic weights and isotopic compositions (version 2.3.1). National Institute of Standards and Technology, Gaithersburg, MD. Available: http://physics.nist.gov/Comp 2003, July 7.

37. Matthiesen R, Mutenda KE (2006) Introduction to proteomics. Methods Mol Biol 367:1–36

38. Snyder A (ed) (2001) Interpreting protein mass spectra. A comprehensive resource. Oxford University Press, Oxford

39. Mirgorodskaya O, Kozmin Y, Titov M, Körner R, Sönksen C, Roepstorff P (2000) Quantitation of peptides and proteins by matrix-assisted laser desorption/ionization mass spectrometry using 18O-labeled internal standards. Rapid Commun Mass Spectrom 14:1226–1232

40. Ramos-Fernández A, López-Ferrer D, Vázquez J (2007) Improved method for differential expression Proteomics using trypsin-catalyzed 18O labeling with a correction for labeling efficiency. Mol Cell Proteomics 6:1274–1286

41. Regnier FE, Julka S (2006) Primary amine coding as a path to comparative proteomics. Proteomics 6(14):3968–3979

42. Zhang R, Sioma CS, Thompson RA, Xiong L, Regnier FE (2002) Controlling deuterium isotope effects in comparative proteomics. Anal Chem 74(15):3662–3669

43. Zhang R, Regnier FE (2002) Minimizing resolution of isotopically coded peptides in comparative proteomics. J Proteome Res 1 (2):139–147

44. Hsu JL, Huang SY, Chow NH, Chen SH (2003) Stable-isotope dimethyl labeling for quantitative proteomics. Anal Chem 75:6843–6852

45. Fu Q, Li L (2005) De novo sequencing of neuropeptides using reductive isotopic methylation and investigation of ESI QTOF MS/MS fragmentation pattern of neuropeptides with N-terminal dimethylation. Anal Chem 77 (23):7783–7795. https://doi.org/10.1021/ac051324e

46. Hsu JL, Huang SY, Chen SH (2006) Dimethyl multiplexed labeling combined with microcolumn separation and MS analysis for time course study in proteomics. Electrophoresis 27:3652–3660

47. Boersema PJ, Aye TT, van Veen TA, Heck AJ, Mohammed S (2008) Triplex protein quantification based on stable isotope labeling by peptide dimethylation applied to cell and tissue lysates. Proteomics 8(22):4624–4632. https://doi.org/10.1002/pmic.200800297

48. Boersema PJ, Raijmakers R, Lemeer S, Mohammed S, Heck AJ (2009) Multiplex peptide stable isotope dimethyl labeling for quantitative proteomics. Nat Protoc 4(4):484–494. https://doi.org/10.1038/nprot.2009.21. [pii]: nprot.2009.21.

49. Turowski M, Yamakawa N, Meller J, Kimata K, Ikegami T, Hosoya K, Tanaka N, Thornton ER (2003) Deuterium isotope effects on hydrophobic interactions: the importance of dispersion interactions in the hydrophobic phase. J Am Chem Soc 125(45):13836–13849. https://doi.org/10.1021/ja036006g

50. Carvalho AS, Cuco CM, Lavareda C, Miguel F, Ventura M, Almeida S, Pinto P, de Abreu TT, Rodrigues LV, Seixas S, Barbara C, Azkargorta M, Elortza F, Semedo J, Field JK, Mota L, Matthiesen R (2017) Bronchoalveolar Lavage Proteomics in Patients with Suspected Lung Cancer. Sci Rep 7:42190. https://doi.org/10.1038/srep42190

51. Ji C, Guo N, Li L (2005) Differential dimethyl labeling of N-termini of peptides after guanidination for proteome analysis. J Proteome Res 4:2099–2108

52. She YM, Rosu-Myles M, Walrond L, Cyr TD (2012) Quantification of protein isoforms in mesenchymal stem cells by reductive dimethylation of lysines in intact proteins. Proteomics 12(3):369–379. https://doi.org/10.1002/pmic.201100308

53. Hsu JL, Huang SY, Shiea JT, Huang WY, Chen SH (2005) Beyond quantitative proteomics: signal enhancement of the a(1) ion as a mass tag for peptide sequencing using dimethyl labeling. J Proteome Res 4:101–108

54. Hsu JL, Chen SH, Li DT, Shi FK (2007) Enhanced a(1) fragmentation for dimethylated proteins and its applications for N-terminal identification and comparative protein quantitation. J Proteome Res 6:2376–2383

55. Aye TT, Low TY, Bjorlykke Y, Barsnes H, Heck AJ, Berven FS (2012) Use of stable isotope dimethyl labeling coupled to selected reaction monitoring to enhance throughput by multiplexing relative quantitation of targeted proteins. Anal Chem 84(11):4999–5006. https://doi.org/10.1021/ac300596r

56. Thompson A, Schafer J, Kuhn K, Kienle S, Schwarz J, Schmidt G, Neumann T, Johnstone R, Mohammed AK, Hamon C (2003) Tandem mass tags: a novel quantification strategy for comparative analysis of complex protein mixtures by MS/MS. Anal Chem 75(8):1895–1904

57. Zeng D, Li S (2009) Improved CILAT reagents for quantitative proteomics. Bioorg Med Chem Lett 19(7):2059–2061. https://doi.org/10.1016/j.bmcl.2009.02.022. [pii]: S0960-894X(09)00156-5.

58. Carvalho AS, Ribeiro H, Voabil P, Penque D, Jensen ON, Molina H, Matthiesen R (2014) Global mass spectrometry and transcriptomics array based drug profiling provides novel insight into glucosamine induced endoplasmic reticulum stress. Mol Cell Proteomics 13(12):3294–3307. https://doi.org/10.1074/mcp.M113.034363

59. Gatto L, Lilley KS (2012) MSnbase-an R/Bioconductor package for isobaric tagged mass spectrometry data visualization, processing and quantitation. Bioinformatics 28(2):288–289. https://doi.org/10.1093/bioinformatics/btr645

60. McAlister GC, Huttlin EL, Haas W, Ting L, Jedrychowski MP, Rogers JC, Kuhn K, Pike I, Grothe RA, Blethrow JD, Gygi SP (2012) Increasing the multiplexing capacity of TMTs using reporter ion isotopologues with isobaric masses. Anal Chem 84(17):7469–7478. https://doi.org/10.1021/ac301572t

61. Werner T, Becher I, Sweetman G, Doce C, Savitski MM, Bantscheff M (2012) High-resolution enabled TMT 8-plexing. Anal Chem 84(16):7188–7194. https://doi.org/10.1021/ac301553x

62. Shadforth I, Crowther D, Bessant C (2005) Protein and peptide identification algorithms using MS for use in high-throughput, automated pipelines. Proteomics 5(16):4082–4095

63. Laderas T, Bystrom C, McMillen D, Fan G, McWeeney S (2007) TandTRAQ: an open-source tool for integrated protein identification and quantitation. Bioinformatics (Oxford, England) 23(24):3394–3396

64. Yu CY, Tsui YH, Yian YH, Sung TY, Hsu WL (2007) The Multi-Q web server for multiplexed protein quantitation. Nucleic Acids Res 35(Web Server issue):W707–W712

65. Lin WT, Hung WN, Yian YH, Wu KP, Han CL, Chen YR, Chen YJ, Sung TY, Hsu WL (2006) Multi-Q: a fully automated tool for multiplexed protein quantitation. J Proteome Res 5(9):2328–2338

66. Breitwieser FP, Muller A, Dayon L, Kocher T, Hainard A, Pichler P, Schmidt-Erfurth U, Superti-Furga G, Sanchez JC, Mechtler K, Bennett KL, Colinge J (2011) General statistical modeling of data from protein relative expression isobaric tags. J Proteome Res 10(6):2758–2766. https://doi.org/10.1021/pr1012784

67. D'Ascenzo M, Choe L, Lee KH (2008) iTRAQPak: an R based analysis and visualization package for 8-plex isobaric protein expression data. Brief Funct Genomic Proteomic 7(2):127–135. https://doi.org/10.1093/bfgp/eln007

68. Wang P, Yang P, Yang JY (2012) OCAP: an open comprehensive analysis pipeline for iTRAQ. Bioinformatics 28(10):1404–1405. https://doi.org/10.1093/bioinformatics/bts150

69. Fischer M, Renard BY (2016) iPQF: a new peptide-to-protein summarization method using peptide spectra characteristics to improve protein quantification. Bioinformatics 32(7):1040–1047. https://doi.org/10.1093/bioinformatics/btv675

70. Blein-Nicolas M, Zivy M (2016) Thousand and one ways to quantify and compare protein abundances in label-free bottom-up proteomics. Biochim Biophys Acta 1864(8):883–895. https://doi.org/10.1016/j.bbapap.2016.02.019

71. O'Connell JD, Paulo JA, O'Brien JJ, Gygi SP (2018) Proteome-wide evaluation of two common protein quantification methods. J Proteome Res 17(5):1934–1942. https://doi.org/10.1021/acs.jproteome.8b00016

72. Goeminne LJE, Gevaert K, Clement L (2018) Experimental design and data-analysis in label-free quantitative LC/MS proteomics: a tutorial with MSqRob. J Proteome 171:23–36. https://doi.org/10.1016/j.jprot.2017.04.004

73. Fu X, Gharib SA, Green PS, Aitken ML, Frazer DA, Park DR, Vaisar T, Heinecke JW (2008) Spectral index for assessment of differential protein expression in shotgun proteomics. J Proteome Res 7(3):845–854. https://doi.org/10.1021/pr070271+

74. Ishihama Y, Oda Y, Tabata T, Sato T, Nagasu T, Rappsilber J, Mann M (2005) Exponentially modified protein abundance index (emPAI) for estimation of absolute protein amount in proteomics by the number of sequenced peptides per protein. Mol Cell Proteomics 4(9):1265–1272. https://doi.org/10.1074/mcp.M500061-MCP200. [pii]: M500061-MCP200.

75. Braisted JC, Kuntumalla S, Vogel C, Marcotte EM, Rodrigues AR, Wang R, Huang ST, Ferlanti ES, Saeed AI, Fleischmann RD, Peterson SN, Pieper R (2008) The APEX Quantitative Proteomics Tool: generating protein quantitation estimates from LC-MS/MS proteomics results. BMC Bioinformatics 9:529. https://doi.org/10.1186/1471-2105-9-529. [pii]: 1471-2105-9-529.

76. Beck HC, Nielsen EC, Matthiesen R, Jensen LH, Sehested M, Finn P, Grauslund M, Hansen AM, Jensen ON (2006) Quantitative proteomic analysis of post-translational modifications of human histones. Mol Cell Proteomics 5(7):1314–1325. https://doi.org/10.1074/mcp.M600007-MCP200

77. Montoya A, Beltran L, Casado P, Rodriguez-Prados JC (2011) Cutillas PR characterization of a TiO(2) enrichment method for label-free quantitative phosphoproteomics. Methods 54 (4):370–378. https://doi.org/10.1016/j.ymeth.2011.02.004. [pii]: S1046-2023(11)00039-9.

78. Schwanhausser B, Busse D, Li N, Dittmar G, Schuchhardt J, Wolf J, Chen W, Selbach M (2011) Global quantification of mammalian gene expression control. Nature 473 (7347):337–342. https://doi.org/10.1038/nature10098

79. Vandenbogaert M, Li-Thiao-Te S, Kaltenbach HM, Zhang R, Aittokallio T, Schwikowski B (2008) Alignment of LC-MS images, with applications to biomarker discovery and protein identification. Proteomics 8(4):650–672. https://doi.org/10.1002/pmic.200700791

80. Schulz-Trieglaff O, Machtejevas E, Reinert K, Schluter H, Thiemann J, Unger K (2009) Statistical quality assessment and outlier detection for liquid chromatography-mass spectrometry experiments. BioData Min 2(1):4. https://doi.org/10.1186/1756-0381-2-4. [pii]: 1756-0381-2-4.

81. Matzke MM, Waters KM, Metz TO, Jacobs JM, Sims AC, Baric RS, Pounds JG, Webb-Robertson BJ (2011) Improved quality control processing of peptide-centric LC-MS proteomics data. Bioinformatics 27(20):2866–2872. https://doi.org/10.1093/bioinformatics/btr479. [pii]: btr479.

82. Wieczorek S, Combes F, Lazar C, Giai Gianetto Q, Gatto L, Dorffer A, Hesse AM, Coute Y, Ferro M, Bruley C, Burger T (2017) DAPAR & ProStaR: software to perform statistical analyses in quantitative discovery proteomics. Bioinformatics 33(1):135–136. https://doi.org/10.1093/bioinformatics/btw580

83. Lazar C, Gatto L, Ferro M, Bruley C, Burger T (2016) Accounting for the multiple natures of missing values in label-free quantitative proteomics data sets to compare imputation strategies. J Proteome Res 15(4):1116–1125. https://doi.org/10.1021/acs.jproteome.5b00981

84. Webb-Robertson BJ, Wiberg HK, Matzke MM, Brown JN, Wang J, McDermott JE, Smith RD, Rodland KD, Metz TO, Pounds JG, Waters KM (2015) Review, evaluation, and discussion of the challenges of missing value imputation for mass spectrometry-based label-free global proteomics. J Proteome Res 14(5):1993–2001. https://doi.org/10.1021/pr501138h

85. Gelman A, Hill J (2007) Data analysis using regression and multilevel/hierarchical models. Cambridge University Press, Cambridge

86. Varmuza K, Filzmoser P (2009) Introduction to multivariate statistical analysis in chemometrics. CRC Press, Taylor & Francis Group, Boca Raton, FL, pp 33487–32742

87. Callister SJ, Barry RC, Adkins JN, Johnson ET, Qian WJ, Webb-Robertson BJ, Smith RD, Lipton MS (2006) Normalization approaches for removing systematic biases associated with

mass spectrometry and label-free proteomics. J Proteome Res 5(2):277–286. https://doi.org/10.1021/pr050300l

88. Albert J (2007) Baysian computation with R. Springer, New York, NY

89. Ritchie ME, Phipson B, Wu D, Hu Y, Law CW, Shi W, Smyth GK (2015) limma powers differential expression analyses for RNA-sequencing and microarray studies. Nucleic Acids Res 43 (7):e47. https://doi.org/10.1093/nar/gkv007

90. Wehofsky M, Hoffmann R, Hubert M, Spengler B (2001) Isotopic deconvolution of matrix-assisted laser desorption/ionization mass spectra for substances-class specific analysis of complex samples. Eur J Mass Spectrom 7:39–46

Chapter 8

Interpretation of Tandem Mass Spectra of Posttranslationally Modified Peptides

Jakob Bunkenborg and Rune Matthiesen

Abstract

Tandem mass spectrometry provides a sensitive means of analyzing the amino acid sequence of peptides and modified peptides by providing accurate mass measurements of precursor and fragment ions. Modern mass spectrometry instrumentation is capable of rapidly generating many thousands of tandem mass spectra and protein database search engines have been developed to match the experimental data to peptide candidates. In most studies there is a schism between discarding perfectly valid data and including nonsensical peptide identifications—this is currently managed by establishing a false discovery rate (FDR) but for modified peptides it calls for an understanding of tandem mass spectrometry data. Manual evaluation of the data and perhaps experimental cross-checking of the MS data can save many months of experimental work trying to do biological follow-ups based on erroneous identifications. Especially for posttranslationally modified peptides there is a need for careful consideration of the data because search algorithms seldom have been optimized for the identification of modified peptides and because there are many pitfalls for the unwary. This chapter describes some of the issues that should be considered when interpreting and validating tandem mass spectra and gives some useful tables to aid in this process.

Key words Proteomics, Posttranslational modifications, Mass spectrometry, Database searching

1 Introduction

Proteins can be viewed as linear biopolymers composed of the only 20 different genetically encoded amino acids, but the functionally active form of a protein is often quite different from that of the linear nascent polypeptide chain. Posttranslational modifications (PTMs) are covalent alterations of the polypeptide chain that change the structure. A bewildering number of changes can occur: intramolecular bonds can be formed by cystine disulfide bridges, sequences can be removed from the polypeptide chain by enzymatic cleavage, and most amino acids can be covalently modified. Many cellular functions are regulated by reversible phosphorylation, acetylation, glycosylation, or other enzymatically catalyzed modifications of proteins. In addition to the biologically significant PTMs it is a well-known and

Rune Matthiesen (ed.), *Mass Spectrometry Data Analysis in Proteomics*, Methods in Molecular Biology, vol. 2051, https://doi.org/10.1007/978-1-4939-9744-2_8, © Springer Science+Business Media, LLC, part of Springer Nature 2020

most often ignored fact that proteins are modified during storage and sample handling. When trying to analyze proteins it is very important to be aware of the less mundane PTMs such as deamidation, oxidation, backbone cleavage, and other common spontaneous modifications as well as those that occur in consequence of the sample handling such as alkylation with acrylamide from SDS-PAGE-separated proteins or carbamylation through the use of urea as a denaturant.

Mass spectrometry (MS) measures the mass-to-charge ratio of charged ions and in combination with liquid chromatography it has become the analytical tool of choice for protein identification and characterization. Soft ionization techniques like ESI and MALDI (*see* Chapter 1) allow large biomolecules to be transferred to the gas-phase and ionized. For identification purposes proteins are typically digested to peptides by a specific protease (e.g., trypsin that cleaves after arginine and lysine residues) and the mass of each peptide is determined. Sequence information for each peptide can then be gained by fragmenting the peptide, for example, using collision-induced dissociation (CID) tandem mass spectrometry (MSMS) where a peptide ion is selected and collided with an inert gas. The mass-to-charge ratios of the resulting fragment ions are then measured. The nomenclature for the peptide fragmentation [1–3] is illustrated in Fig. 1a. The most common tandem MS experiment in proteomics is low-energy CID (less than 100 eV) of protonated peptides in ion traps, triple quadrupoles or QqTOFs where there is very little side chain fragmentation. Normally the low-energy collision-induced cleavage of the peptide backbone occurs at the amide bond giving rise to the y- and b-type ions that contain the C- and N-terminal part of the peptide, respectively (*see* Fig. 1b). *See* [4] for a review on peptide sequencing by MS. Additional ions can be generated by the loss of small neutral molecules like water (a −18 Da ion series usually denoted with a superscripted o—e.g., y_i^o) or ammonia (a −17 Da ion series usually denoted with a superscripted asterisk ∗—e.g., y_i^*). Often it is also possible to find immonium ions from the individual amino acids in the low-m/z region of the spectrum. CID fragmentation of multiply charged ions can also give rise to multiply charged fragment ions and it is not uncommon for multiply charged ions to lose one or several N-terminal amino acid residues as neutral fragments [5].

MS instrumentation is rapidly improving and diversifying but there are still different limitations in mass range, mass accuracy, and fragmentation efficiency for each type of instrument and it is necessary to consider these limitations when analyzing the data. For example, to detect deamidation a mass accuracy lower than 1 Da is required. To distinguish phenylalanine from oxidized methionine call for a mass accuracy lower than 0.033 Da. Other techniques for inducing peptide fragmentation like electron capture dissociation (ECD) [6, 7] and electron transfer dissociation (ETD) [8] have

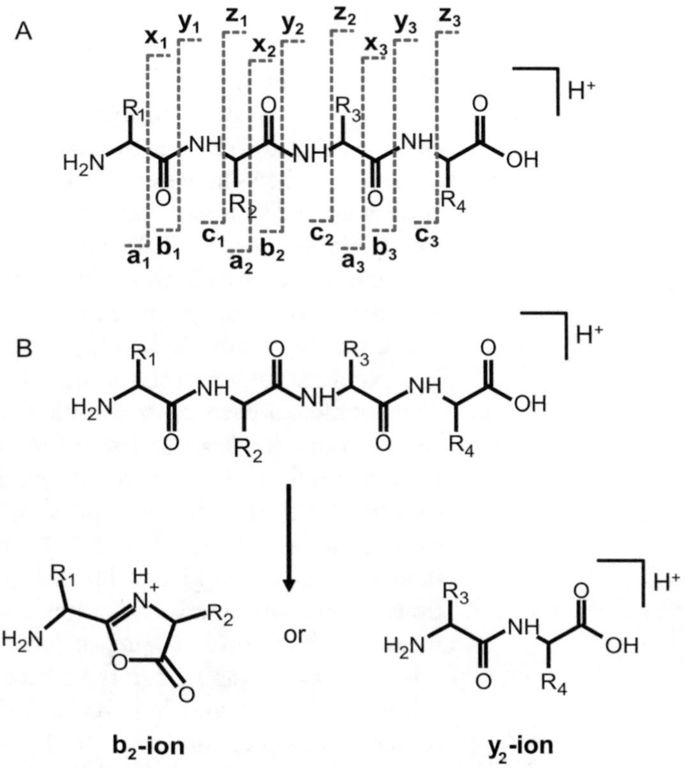

Fig. 1 (**a**) Peptide fragment ion nomenclature. The indices of the amino-terminal containing a, b, and c ions denote the number of residues counted from the N-terminus and the indices of the carboxy-terminal x, y, and z ions are counted from the C-terminus. (**b**) CID fragmentation of the peptide amide bonds to produce N-terminal b-ions and C-terminal y ions. Linear b ions are unstable and can cyclize to form an oxazolone structure involving the carbonyl group of the adjacent residue. The b_1 ion is rarely observed (*see* **Note 7**) because it cannot form the stabilizing oxazolone structure, but it does occur if the N-terminal is derivatized with a carbonyl-containing group (e.g., acetylated)

become more commonplace and provide very powerful tools for analyzing especially labile PTMs as glycosylation [9–11], phosphorylation [8, 12] and γ-carboxyglutamic acid [13]. Many programs is available that can identify the peptide by comparing the experimental MSMS spectrum with the theoretical one calculated for each peptide in a protein database. The mass changes associated with PTMs make mass spectrometry ideally suited for the analysis. Modifications that occur stoichiometrically are often denoted fixed modifications (because all residues of a given type are modified)—as an example the derivatization of cysteines with iodoacetamide occurs with almost 100% efficiency on all cysteine residues. Modifications that occur substoichiometrically (residues are not uniformly modified) are called variable modifications—oxidation of methionine is rarely quantitative unless an oxidizing agent is utilized (i.e., all the residues do not oxidize completely) and methionine sulfoxide is most often included in the analysis as a variable modification of methionine residues.

It is often a difficult problem to assess if the peptide retrieved by the search engine really is the correct sequence because the MSMS data is often far from perfect. Two major limitations apart from data quality usually apply for the identification of PTMs using different search algorithms. Virtually all search engines produce a best-fit solution to a user-defined problem, but it is not always possible to get the correct solution either because the database protein sequence is a variant (or wrong) or because the peptide is modified in an unforeseen way not accounted for by the search parameters. This problem is even worse for many modified peptides where very labile groups can lead to less informative fragmentation patterns from which limited sequence information can be gained. From the mass of the peptide ion it is possible with high mass accuracy to distinguish between PTMs that are nearly isobaric. Fortunately, tandem mass spectra give additional analytical handles on modifications where characteristic neutral losses, composite mass increments in the peptide sequence ions, and diagnostic ions in the low-mass region can lead to the correct interpretation and identification of the peptide. It is beyond the scope of this chapter to present a comprehensive list of all PTMs, but a selection of the most common and well-studied is listed in Table 5. The manual interpretation and validation of tandem mass spectra of posttranslationally modified peptides can often be aided by comparing to the MSMS spectra of the nonmodified sequence, a synthetic version of the modified peptide, or a chemically modified version of the same PTM-peptide (e.g., acetylating the peptide).

Modern mass spectrometry instrumentation directly coupled to liquid chromatography can generate overwhelming amounts of data. Validating the unmodified peptide identifications resulting from database searches of this data is a very time-consuming and at times difficult task, but it is currently necessary unless very stringent identification thresholds are imposed. A very stringent identification threshold will on the other hand lead to many false negatives. The data retrieved by search engines needs to be evaluated and there are many less fortunate examples where the assignments have not been adequately checked (*see* Johnson et al. [14] for a discussion). This chapter aims at giving some helpful tools and rules for interpreting and validating tandem mass spectra of peptides in general and especially of modified peptides.

2   Materials

2.1   Useful Tables for Interpretation of Peptide MSMS Data

Several useful tables are collected below. Table 1 has the exact masses of the most common elements in posttranslational protein modifications—the composite monoisotopic mass of the modified residue can then be calculated once the chemical composition is known.

2.2   Web Resources

A collection of links to software resources to facilitate the process of peptide identification by CID tandem mass spectrometry.

2.2.1   Protein Database Search Engines

Mascot—http://www.matrix-science.com/

X-tandem—http://www.thegpm.org/

SearchGUI—http://compomics.github.io/projects/searchgui.html

Andromeda (part of the MaxQuant package)—http://maxquant.org/

For a more complete list *see* http://en.wikipedia.org/wiki/Mass_spectrometry_software

2.2.2   Protein Analysis Tools

GPMAW—http://www.gpmaw.com

Table 1
List of accurate elemental mass values for the most commonly occurring elements (including stable isotope labeling) in peptides and posttranslationally modified peptides

H-1	1.007825035
H-2	2.014101787
C-12	12.000000000
N-14	14.003074002
O-16	15.994914630
P-31	30.973762000
S-32	31.972070700
C-13	13.003354826
N-15	15.000108970
O-18	17.999160300
Electron	0.000548580
Pos. Charge	−0.000548580

2.2.3 Protein Sequence Databases	nr—ftp://ftp.ncbi.nlm.nih.gov/ Uniprot—http://www.uniprot.org/proteomes
2.2.4 Protein Sequence and PTM Annotations	UniProt—http://www.uniprot.org/ Human Protein Reference Database—http://www.hprd.org/
2.2.5 Protein Modification Resources	Resid—http://pir.georgetown.edu/cgi-bin/resid Unimod—http://www.unimod.org/

3   Methods

3.1 Generating a Theoretical Spectrum

While most search engines return annotated spectra, it can be of immense value to generate theoretical spectra of other possible solutions. Especially with labile modifications as serine O-glycosylation (and to some extent serine and threonine phosphorylation) it can be difficult to pinpoint and assign the exact position of the modified residue. Often there are multiple residues that can be modified in a sequence and distinguishing between the possible modification sites requires careful comparison of the experimental data and the theoretical spectra for each potential modified residue.

The mass of the protonated parent ion $[M + H]^+$ is given by the sum

$$[M + H]^+ = m_N + \Sigma m_i + m_C + m(H^+) \text{ (usually } \Sigma m_i + 19.01784)$$

where m_N is the mass of the N-terminating group (usually a hydrogen), Σm_i is the sum of masses of i amino acid residues, m_C is the mass of the C-terminating group (usually a hydroxyl group), and $m(H^+)$ is the mass of a proton. The monoisotopic peak of a multiple protonated peptide will appear at a mass-to-charge ratio of $[M + nH]^{n+}/n$. The monoisotopic masses m_i of the common amino acids can be found in Tables 2 and 5 lists the masses of the most common PTMs. The commonly observed fragment ion masses for CID tandem mass spectra are then given by where the index is counted from the N-terminus for a-, b-, and c-type ions and from the C-terminus for y- and z- type ions.

$$a_i = m_N + \Sigma m_i - m(e^-) - m(CO) = b_i - m(CO) \text{ (usually } \Sigma m_i - 26.98654)$$

$$b_i = m_N + \Sigma m_i - m(e^-) \text{ (usually } \Sigma m_i + 1.007276)$$

$$y_i = \Sigma m_i + m_C + m(H) + m(H^+) \text{ (usually } \Sigma m_i + 19.01784)$$

For a peptide with n-residues the sum of the two corresponding ions can be calculated as

$$b_i + y_{n-i} = MH^+ + m(H^+) = MH^+ + 1.007276$$

For example, once the b_2-ion has been identified (from the prominent a_2–b_2 pair) the mass of y_{n-2} can be calculated. A list of possible b_2 ions can be found in Table 4. Multiple fragmentations

Table 2
Masses of the common 20 amino acid residues and carbamidomethylated cysteine

Amino acid	Abbreviation	Code	Monomer composition	Monoisotopic mass
Glycine	Gly	G	C_2H_3NO	57.021464
Alanine	Ala	A	C_3H_5NO	71.037114
Serine	Ser	S	$C_3H_5NO_2$	87.032028
Proline	Pro	P	C_5H_7NO	97.052764
Valine	Val	V	C_5H_9NO	99.068414
Threonine	Thr	T	$C_4H_7NO_2$	101.047679
Cysteine	Cys	C	C_3H_5NOS	103.009185
Isoleucine	Ile	I	$C_6H_{11}NO$	113.084064
Leucine	Leu	L	$C_6H_{11}NO$	113.084064
Asparagine	Asn	N	$C_4H_6N_2O_2$	114.042927
Aspartic acid	Asp	D	$C_4H_5NO_3$	115.026943
Glutamine	Gln	Q	$C_5H_8N_2O_2$	128.058578
Lysine	Lys	K	$C_6H_{12}N_2O$	128.094963
Glutamic acid	Glu	E	$C_5H_7NO_3$	129.042593
Methionine	Met	M	C_5H_9NOS	131.040485
Histidine	His	H	$C_6H_7N_3O$	137.058912
Phenylalanine	Phe	F	C_9H_9NO	147.068414
Arginine	Arg	R	$C_6H_{12}N_4O$	156.101111
Cysteine—cbm		C*	$C_5H_8N_2O_2S$	160.030648
Tyrosine	Tyr	Y	$C_9H_9NO_2$	163.063329
Tryptophan	Trp	W	$C_{11}H_{10}N_2O$	186.079313

of the backbone can occur (but less frequently than single fragmentations) and it is not uncommon to observe internal fragments in the lower mass region (less than 700 Da)—especially if there is a proline in the sequence. The theoretical masses of internal ions can be calculated as the sum of residue masses plus the mass of a proton. The information contained in the low-mass region can be used to patchwork the proposed sequence [15]. There is also useful information in the low-mass region from immonium and related fragment ions that are characteristic of specific amino acids (*see* Tables 3 and 5). In addition, some amino acid residues undergo the loss of small neutral molecules like water or ammonia leading to satellite peaks to the major fragment ion series (*see* Tables 3 and 5).

Table 3
Low-mass fragment ions and neutral losses from the common 20 amino acid residues [16, 17]

Amino acid	Code	Immonium and fragment ions	Neutral loss
Glycine	G	30.03	
Alanine	A	44.05	
Serine	S	60.04	18.01 (H_2O)
Proline	P	70.07	
Valine	V	72.08	
Threonine	T	74.06	18.01 (H_2O)
Isoleucine	I	86.10	
Leucine	L	86.10	
Asparagine	N	87.06, 70.03	17.03 (NH_3)
Aspartic acid	D	88.04, 70.03	18.01 (H_2O)
Glutamine	Q	129.10, 101.07, 84.04, 56.05	17.03 (NH_3)
Lysine	K	129.11, 101.11, 84.08, 56.05	17.03 (NH_3)
Glutamic acid	E	102.05, 84.04	18.01 (H_2O)
Methionine	M	104.06	48.00 (CH_4S)
Histidine	H	110.07	
Phenylalanine	F	120.08	
Arginine	R	129.11, 115.09, 112.09, 87.09, 70.07, 60.06	17.03 (NH_3)
Cysteine—cbm	C*	133.04	91.01 (C_2H_5NOS)
Tyrosine	Y	136.08	
Tryptophan	W	159.09, 132.08, 130.07	

Cysteines are derivatized with iodoacetamide to form carbamidomethylcysteine (C*)

The commonly observed c and z + 1 fragment ion masses for ECD/ETD tandem mass spectra can be calculated along the same lines with the residue index running from the N-terminus for c-ions and from the C-terminus for z-ions.

$$c_i = m_N + \Sigma m_i + m(NH_2) + m(H^+) \text{ (usually } \Sigma m_i + 18.03383)$$
$$z + 1_i = \Sigma m_i + m_C - m(NH) + m(H^+) \text{ (usually } \Sigma m_i + 2.99912)$$

Often ECD/ETD spectra have abundant peaks at the charge-reduced species where the precursor ion captures electrons without fragmenting, for example, a triply charged precursor ion at m/z $[M + 3H]^{3+}/3$ could have charge-reduced ions at $[M + 3H]^{2+}\bullet/2$ (the dot indicates that it is an odd electron ion) and at $[M + 3H]^+$ after capturing one or two electrons.

Several software packages offer convenient tools for calculating theoretical masses for peptide digests of proteins, molecular masses

Table 4
Amino acid residue composition and masses of b_2 ions with carbamidomethylated cysteine

115.05	GG	203.10	TT	231.12	MV	258.14	EK	286.15	ER
129.07	AG	205.10	FG	232.08	AC	258.16	RT	286.16	VW
143.08	AA	209.10	AH	233.10	MT	259.09	EE	288.13	TW
145.06	GS	211.14	[I/L]P	235.11	AY	260.11	CV	288.15	MR
155.08	GP	212.10	NP	235.11	FS	260.11	MQ	289.10	CQ
157.10	GV	213.09	DP	235.12	HP	260.14	KM	289.13	CK
159.08	AS	213.16	[I/L]V	237.13	HV	261.09	EM	290.08	CE
159.08	GT	214.12	NV	239.11	HT	261.12	PY	292.08	CM
169.10	AP	214.13	GR	242.15	[I/L]Q	261.16	F[I/L]	292.13	QY
171.11	AV	215.10	DV	242.19	[I/L]K	262.09	CT	292.17	KY
171.11	G[I/L]	215.14	[I/L]T	243.11	NQ	262.12	FN	293.11	EY
172.07	GN	216.10	NT	243.13	E[I/L]	263.09	MM	294.17	HR
173.06	DG	216.10	QS	243.15	KN	263.10	DF	295.11	MY
173.09	AT	216.13	KS	244.09	DQ	263.14	VY	295.14	FF
175.07	SS	217.08	DT	244.09	EN	265.12	TY	298.10	CH
185.09	PS	217.08	ES	244.11	GW	266.12	HQ	300.17	[I/L]W
185.13	A[I/L]	218.06	CG	244.13	DK	266.16	HK	301.13	HY
186.09	AN	219.08	MS	244.14	RS	267.11	EH	301.13	NW
186.09	GQ	219.11	AF	245.08	DE	269.11	HM	302.11	DW
186.12	GK	221.09	GY	245.13	FP	270.19	[I/L]R	304.18	FR
187.07	AD	225.10	HS	245.13	[I/L]M	271.15	NR	308.11	CF
187.07	EG	226.12	PQ	246.09	MN	272.14	DR	311.14	FY
187.11	SV	226.16	KP	247.07	DM	274.12	SW	313.21	RR
189.07	GM	227.10	EP	247.14	FV	274.12	C[I/L]	315.15	QW
189.09	ST	227.18	[I/L][I/L]	248.07	CS	275.08	CN	315.18	KW
195.09	GH	228.13	[I/L]N	249.12	FT	275.13	HH	316.13	EW
195.11	PP	228.13	QV	251.10	SY	276.06	CD	317.14	CR
197.13	PV	228.15	AR	251.15	H[I/L]	276.13	FQ	318.13	MW
199.11	PT	228.17	KV	252.11	HN	276.17	FK	320.17	RY
199.14	VV	229.09	NN	253.09	DH	277.12	EF	321.07	CC
200.10	AQ	229.10	MP	254.16	PR	277.15	[I/L]Y	324.10	CY
200.14	AK	229.12	D[I/L]	256.18	RV	278.11	NY	324.15	HW

(continued)

Table 4
(continued)

201.09	AE	229.12	EV	257.12	QQ	279.10	DY	327.13	YY
201.12	[I/L]S	230.08	DN	257.16	KQ	279.12	FM	334.16	FW
201.12	TV	230.11	QT	257.20	KK	284.14	PW	343.19	RW
202.08	NS	230.15	KT	258.09	CP	285.13	FH	347.12	CW
203.07	DS	231.06	DD	258.11	EQ	285.17	QR	350.15	WY
203.08	AM	231.10	ET	258.12	AW	285.20	KR	373.17	WW

The order of residues does not matter, and [I/L] denotes an isoleucine or leucine residue—these two residues have exactly the same mass. An intense pair of ions separated by 28 Da (a CO group) in the low mass region is usually the signature of an a_2–b_2 pair, and the composition of the b_2 ion can be looked up in this table. An a_2–b_2 pair that does not fit with the masses in this table could be due to modification at the N-terminus. It is not uncommon to observe a number of internal dipeptide from multiple fragmentations that give rise to additional a_2–b_2 pairs

of the peptides and fragment ion masses. One of the most user friendly and versatile packages is GPMAW. (General Protein/Mass Analysis for Windows)—*see* Peri et al. for a brief description [18]. Figure 2 illustrates a typical simple application. After importing the protein sequence into GPMAW and doing a theoretical proteolytic digest, the masses for all ions in a theoretical tandem mass spectrum of an oxidized methionine peptide can be calculated. The experimental spectrum is shown in Fig. 3 as it is returned from a database search.

3.2 Database Searching

Many software tools have been developed to identify the proteins from CID tandem mass spectra of the peptides. Different search engines yield somewhat complementary results and we often search the same data set with two or three different programs [19–21] (*see* Subheading 2.2.1). Several software packages are freely available and offer several tools for database searching and validation. Most can search even very large datasets with many variable modifications, but care must be taken when defining the search parameters and interpreting the outcome. Depending on the organism a database must be selected (*see* Subheading 2.2.3). To avoid erroneous assignments that derive from known protein contaminants (e.g., porcine trypsin is commonly added to digest the proteins but the sequence does not appear in a human database) it has become common to augment the database with an additional FASTA file containing a small set of protein sequences tagged as contaminants [22, 23]. The detailed use and inner workings of mass spectrometry-driven database search engines fall outside the scope of this chapter.

3.3 Assignment and Validation Steps

Most mass spectrometry-based protein database search engines return a score as a measure of how well the theoretical spectrum

Table 5
Posttranslational modifications

Modification	Modified amino acids	Delta composition	Delta mass	Diagnostic ions	Neutral loss
Propionamide	C	$C_3H_5NO_2S$	71.0371	147.059	
N-terminal Pyro-carbamidomethylcysteine	C (N-term)	C_2O	39.9949		
Gamma-carboxylation	E	CO_2	43.9898		43.99
N-terminal Pyro-glutamic acid	E (N-term)	$-H_2O$	−18.0106		
Oxidation (2-Oxohistidine)	H	O	15.9949	126.066	
Methylation	K	CH_2	14.0157	115.123;98.096;84.081	
Dimethylation	K	C_2H_4	28.0313	129.137;84.081	
Acetylation	K	C_2H_2O	42.0106	143.118;126.091;84.081	
Trimethylation	K	C_3H_6	42.0470	143.154;84.081	59.073
N-ubiquitin-lysine (tryptic)	K	$C_4H_6N_2O_2$	114.0429		
Oxidation (methionine sulfoxide)	M	O	15.9949	120.048	63.998
Dioxidation (Methionine sulfone)	M	O_2	31.9898	136.043	
Deamidation	N	$H(-1)N(-1)O$	0.9840		
N-terminal acetylation	N-term	C_2H_2O	42.0106		
N-terminal carbamylation	N-term	CHNO	43.0058		
N-terminal carbamidomethylation	N-term	C_2H_3NO	57.0215		
Hydroxyproline	P	O	15.9949	86.06	
Deamidation	Q	$H(-1)N(-1)O$	0.9840		
N-terminal Pyro-glutamine	Q (N-term)	$-NH_3$	−17.0265		
Methylation	R	CH_2	14.0157	143.129;115.087;112.087; 74.071;70.065	73.064;56.0374;31.0422

(continued)

Table 5
(continued)

Modification	Modified amino acids	Delta composition	Delta mass	Diagnostic ions	Neutral loss
Dimethylation (asymmetric)	R	C_2H_4	28.0313	157.149;115.087;112.087; 88.087;71.060	87.0796;70.0531;45.0579
Dimethylation (symmetric)	R	C_2H_4	28.0313	157.149;115.087;112.087; 88.087;71.060	87.0796;70.0531;31.0422
Trimethylation	R	C_3H_6	42.0470	171.16	
Phosphorylation	S	HO_3P	79.9663		97.977;79.966
O-HexNAc S	S	$C_8H_{13}NO_5$	203.0794	204.087;186.076;168.066; 144.066;138.055;126.055	203.079
Phosphorylation	T	HO_3P	79.9663		97.977;79.966
O-HexNAc T	T	$C_8H_{13}NO_5$	203.0794	204.087;186.076;168.066; 144.066;138.055;126.055	203.079
Kynurenine	W	$C(-1)O$	3.9949	163.087	
Oxidation (hydroxy-tryptophan)	W	O	15.9949	175.087	
Dioxidation (N-formylkynurenine)	W	O_2	31.9898	191.082	
Hydroxytyrosine	Y	O	15.9949	152.071	
Nitration	Y	$H(-1)NO_2$	44.9851	181.061	
Sulfation	Y	O_3S	79.9568		79.957
Phosphorylation	Y	HO_3P	79.9663	216.042	79.966

This short list contains some of the common modifications depending on the biological problem (e.g., histones or serum proteins) and the sample preparation method (e.g., SDS-PAGE separated proteins or an in-solution digestion of urea-denatured proteins). The modifications are added as an additional delta mass. Often an iterative search strategy is used, where posttranslational modifications of the identified proteins are looked up in annotated protein databases (e.g., Swiss-Prot or HPRD—*see* Subheading 2.2.4) and included in a second iteration

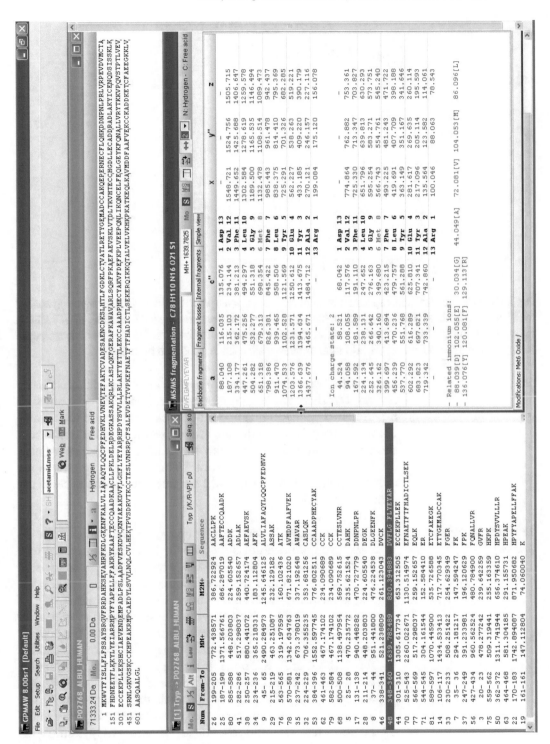

Fig. 2 The program GPMAW is a very versatile tool for protein analysis. In this simple example the protein sequence of human serum albumin is loaded (upper panel) and an in-silico tryptic digestion is performed and the theoretical mass-to-charge ratios of the singly and doubly charged ions are calculated (lower left panel). The theoretical masses of both singly and multiply charged fragment ions can then be calculated for each peptide (lower right panel). The methionine in this example is flagged as being oxidized and the masses are adjusted throughout the calculations

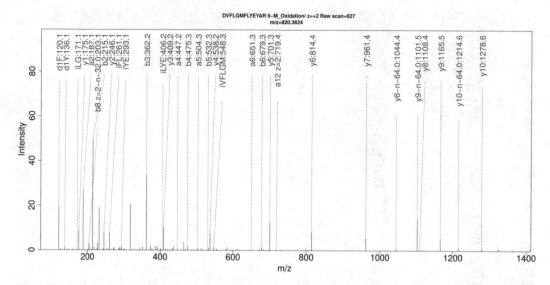

Fig. 3 ESI QqTOF tandem mass spectrum of a doubly charged ion at *m/z* 820.36. Database searching against the IPI human database returns the protein human serum albumin and the peptide DVFLGMFLYEYAR with an oxidized methionine. The y-ion series from y_8 and up displays an additional series of rather intense peaks arising from the neutral loss of CH_3SOH from oxidized methionine residue (*see* **Note 5**)

matches the experimental data. High-scoring peptides are mostly correct, but there is a large grey area where only critical manual or experimental validation can sort out false positive identifications. Identifying peptides that deviate in some way from what is expected should be examined closely with the mantra that "extraordinary results require extraordinary proof." The following questions are mostly empirical but can be used to probe the confidence of the identified peptide.

3.3.1 Peptide Sequence Validation

1. Does the peptide sequence conform to the experiment? In most proteomics studies the proteins have been digested into peptides with a sequence-specific protease prior to analysis. It has been shown that the most commonly used protease, trypsin, cleaves the peptide backbone to the C-terminal side of arginine or lysine residues with a high degree of specificity [24]. Therefore, the amino acid residue preceding the peptide should be arginine or lysine and the peptide C-terminal residue should be K or R. It is possible that there is nonspecific cleavage or that the protein has been processed prior to analysis, but nontryptic peptides in a tryptic protein digest should be examined critically. In many cases there is additional sequence information that can add confidence to seemingly semitryptic peptides where only one end of the peptide follows a tryptic cleavage pattern (for example because of the cleavage of signal-peptides or that the peptide forms the C-terminus of the protein). Likewise, a peptide containing many internal lysine and

arginine peptide is unlikely to survive the incubation with trypsin (unless the peptide is modified so that the cleavage sites are masked).

2. Does the number of basic groups (H, K, R, and N-terminal amino group) correspond to the charge state? The charge state of peptide in ESI depends on several factors and the above rule of thumb should only be taken as a rough estimate. Three major interrelated factors are the basicity of the peptide residues, the conformation of the peptide, and the Coulomb repulsion between multiple protons [25]. In many peptides the maximum charge state can be estimated by the number of basic residues (R, K, H and the N-terminal amino group) but for longer peptides protonation can take place at the next most basic sites (P, Q, and W) [26]. When a parent ion with a low charge state is assigned a potential peptide sequence with several internal basic residues it is necessary to carefully examine the spectra because the assignment could be erroneous. For example, the fragmentation at glycine residues to the C-terminal side is often of low intensity (*see* Subheading 3.3.2, **step 11**) the assigned lysine (128.0950 Da) could be glutamine (128.0586) or the dipeptide AG (128.0586 Da), and arginine (156.1011 Da) could be the dipeptide VG (156.0742 Da).

3. How well does the peptide mass match the experimental mass? The mass accuracy of the instrumentation depends on the type of mass analyzer. A rough estimate would be that the mass accuracy of an ion-trap or triple quadrupole should be better than 1.5 Da; a QqTOF better than 0.15 Da and an FT-ICR/Orbitrap better that 10 ppm. The mass analyzers usually perform a lot better than these values and the experimental setup should be checked if a large proportion of the peptide identifications are made with mass accuracies outside these values (*see* **Note 1**).

4. Are residues consistently modified? An obvious question to raise is if the observed residues are consistent with the experimental procedure? If cysteines have been alkylated with iodoacetamide then, identification of unmodified cystein cause concern (or as being modified by another reagent).

5. Could the peptide derive from contaminating proteins? Including a relevant database of likely contaminants ranging from porcine trypsin, keratins to a list of bovine serum proteins that often contaminate proteins isolated from human cell culture will help eliminate errors [22].

3.3.2 Spectral Features

1. Is there a consecutive series of ions from sequential peptide fragments (e.g., >3 y-ions)? In the case of tryptic peptides there is usually some continuity in the ion-series (with some sequence dependencies—*see* Subheading 3.3.2, **steps 10** and **11**) and for QqTOF MSMS spectra the ion series above the parent ion mass-to-charge ratio usually consists of y-ions. If high m/z b-ions are observed on a QqTOF instrument there must be a basic amino acid residue in the N-terminal fragment—this rule does not apply to ion trap MSMS. A rough rule of thumb for accepting the proposed peptide sequence is if there is a sequence tag of at least three amino acids. This is a somewhat arbitrary criterion and obviously there should not be glaring inconsistencies between the theoretical and experimental spectrum. For example, it should not be possible to extend the sequence tag ion series with an amino acid residue mass to include an intense ion in the series if this next amino acid residue does not fit the retrieved sequence.

2. Is there an intense a_2–b_2 pair of ions separated by 28 Da? The b ions can lose CO to form a-ions and the a_2–b_2 ion pair separated by 28 Da is usually very prominent in the low mass region. By looking for an intense ion pair separated by 28 Da it is possible to guess at the amino acid composition by using Table 4. Additional a_2–b_2 ion pairs can arise from internal fragmentation; *see* [15].

3. Do the observed immonium ions correspond to residues in the peptide (especially V, [I/L], H, F, Y, W)? Depending on the mass range of the instrument it is usually possible to get an idea if the peptide contains any of the listed amino acids. Some amino acid residues give rise to several low-mass fragment ions (*see* Tables 3 and 5) that can increase confidence in determining the amino acid composition of the peptide.

4. Are the y_1-ions observed and in accordance with enzymatic specificity? Tryptic peptides give rise to intense y_1-ions because of the C-terminal position of the basic residues and this C-terminal residue should be consistent with the proposed sequence (*see* **Note 2**)—tryptic peptides have y_1-ions at m/z 147(K) and 175(R). The y_1-ions of chymotryptic peptides are at m/z 132 (I/L), 166(F), 182 (Y), and 205 (W).

5. How well do the fragment ion masses fit the theoretical masses?
   The assignment of fragment ions can be aided by looking at the fragment ion mass differences from the theoretical values. Figure 4 shows an example for the tandem mass spectrum shown in Fig. 3 where the residual mass deviation (right pane) for correctly assigned peaks is small—larger deviations can be caused by overlapping peaks, a fragment peak from another peptide, internal fragment ions or poor ion-statistics,

Mass accuracy in MSMS

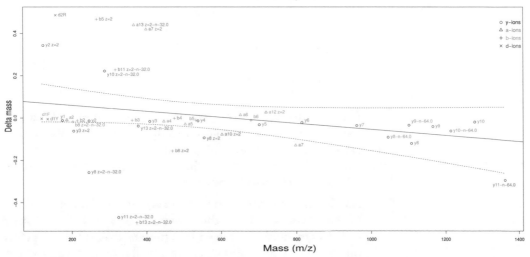

Fig. 4 Displaying the residual masses can often help the assignment of peaks and data validation. The mass accuracy of data from a QqTOF is quite sensitive to temperature fluctuations and a postacquisition recalibration usually brings the mass deviation residuals below 0.05 Da. If the assigned peaks deviate beyond this the assignment should be checked—going back to the raw data prior to data processing as smoothing and centroiding are often helpful to discover overlapping peaks or other causes for poor mass accuracy. Linear regression can help disguising correct ones from incorrect peak assignments. Peak assignments for which the delta mass is outside (red label) or inside (green label) the 95% confidence interval is depicted

but the ions assigned to a_7 and y_8 in Fig. 3 do not fit very well and these assignments should be viewed with some skepticism.

6. Are multiply charged fragment ions observed?

ESI MSMS often gives rise to multiply charged ions and it is common to observe neutral losses of amino acids from the N-terminal end of tryptic peptides [5]. It is important to note the charge state of the ions to avoid mis-assigning a singly charged ion to a multiply charged ion. A multiply charged ion in the tandem mass spectrum with a higher mass than the parent peptide mass suggests that the charge state of the parent ion has been wrongly assigned (e.g., observing a doubly charged fragment ion at $m/z\,750$ with correct isotope pattern for a doubly charged ion when the software reports it has been fragmenting a doubly charged parent ion at $m/z\,600$).

7. Are satellite ions observed from the loss of small neutral molecules? Intense ion series often have a series of associated satellite peaks from the neutral loss of small molecules (*see* Table 3)—especially the loss of water from serine, threonine, aspartic acid, and glutamic acid residues, and ammonia from arginine, lysine, asparagine, and glutamine residues.

8. Can the amino acid sequence be permutated or changed to account for unassigned peaks? If there is a gap in the annotated

y-ion series or an unannotated peak in an otherwise continuous ions series, it is not uncommon that the order of two amino acid residues or composition can be changed to explain the discrepancy. For example, in an ion series containing an asparagine (114.0429 Da) residue one should check that there is not an ion midway to suggest two glycine residues (also 114.0429 Da).

9. If the sequence contains proline—are there internal fragment ions from multiple cleavages? Multiple peptide fragmentations are especially common for peptides with facile cleavage sites and MSMS of especially peptides containing proline or aspartic acid display internal fragment ions.

10. Are intense fragment ions observed from cleavage on the N-terminal side of P and the C-terminal side of D? Low-energy CID fragmentation induced by a mobile proton [27] depends on the charge state of the peptide ion and on how the charges are sequestered by basic residues. Numerous studies have described the facile backbone cleavage N-terminal to proline (*see* for example [28]) because the tertiary amine is more basic than the other backbone atoms. In cases where charges are sequestered (e.g., when the number of protons is lower than the number of arginine residues) there is an enhanced cleavage at aspartic acid residues [27].

11. Are fragment ions from the C-terminal side of P and G of low intensity or absent? Statistical analysis of tandem mass spectra databases has shown that there is a bias against fragmentation to the C-terminal side of proline and glycine residues [29, 30].

3.3.3 Spectral Features of Posttranslational Modifications

1. Can the composite mass of the modified residue be found in a consecutive series of ions? The confidence with which one can identify posttranslationally modified residues depends crucially on the data quality and a direct observation of the composite mass from a modified residue in a consecutive series of ions is stronger evidence than just observing an altered parent ion mass that corresponds to the PTM(s).

2. Are there characteristic neutral losses? In some cases, the posttranslationally modified residue exhibits a characteristic intense loss that increases the confidence in the assignment—often a satellite peak from the parent ion displays the characteristic loss. For example, phosphorylated serine very easily loses a neutral phosphoric acid group and displays an ion series 98 Da less than the expected composite mass ion series—the modified serine can then be identified as a dehydroalanine residue weighing 69 Da.

3. Are there any characteristic low-mass ions? Some posttranslationally modified residues are fairly stable under low-energy

CID conditions and the immonium ion (and other fragment ions) can increase the confidence of the identification; for example, phosphorylated tyrosine residues give rise to an immonium ion at 216.04 Da. Some PTMs give rise to multiple diagnostic ions (*see* Table 5) and observing all the fragment ions is stronger evidence; for example, acetyl lysine gives rise to a set of low-mass ions at m/z 84, 126, and 143.

4. Can the mass increase attributed to the PTM be explained by other means? Since many PTMs are made of the same few elements as amino acids are composed of there are many possibilities of interpreting the data wrongly. When separating proteins on a polyacrylamide gel it is not unusual that nonpolymerized acrylamide reacts with cysteine residues [31, 32]. The propionamide attached to cysteine has the same composition as an alanine residue and depending on the data quality it can be difficult to distinguish the two possibilities of having a propionamide-cysteine or an unmodified cysteine followed by an alanine residue. The solution to this problem is most often to make an informed guess based on the sample handling procedure. If the proteins have been separated by SDS-PAGE, reduced, and alkylated with iodoacetamide the former solution is most likely correct (there should be no free cysteine thiol groups after iodoacetamide treatment). Another example is after alkylation with iodoacetamide there is a side product where the N-terminal amino group has been carbamidomethylated—this modification adds the same mass as a glycine residue (having the same elemental composition). A second problem with iodoacetamide is the dialkylation of peptides leading to a mass increase of 114 that can be confounded with the GlyGly-tag remaining after tryptic proteolysis of ubiquitinated proteins [33].

5. Can missed cleavages be rationalized by neighboring modified residues? Tryptic missed cleavages often occur at amino acid stretches with several adjacent basic residues leading to an additional R or K with a terminal position (either C- or N-terminal). Internal missed cleavages for trypsin are often associated with neighboring acidic groups (aspartate, glutamate, and phosphorylated S, T, or Y).

6. Is the modification consistent with known sequence motifs and target amino acids? Many PTMs only occur on a very small subset of residues determined by the amino acid residue, functional groups, size of the amino acid, or a sequence motif. Looking up information on the various types of modifications can be helpful when analyzing the spectra (*see* **Note 3**). For example, if an asparagine is converted to aspartic acid by

deglycosylation treatment with the glycosidase PNGase F the asparagines can be tentatively assigned as being glycosylated. However, the asparagines should appear in the consensus sequon NXS/T where X can be any amino acid except proline—otherwise the N to D conversion most likely is caused by spontaneous deamidation.

3.3.4   Precursor Ion

The selection of the parent ion and the determination of mass-to-charge ratio and charge state are crucial for the data quality. The problems listed below usually occur as a consequence of software limitations. Visual inspection of the parent ion in the MS survey spectrum can usually identify these problems.

1. Is the charge state correctly assigned? The assignment of charge states depends on the instrumentation and data quality, but even with high-resolution instrumentation the algorithm fails at times leading to an erroneous parent ion mass—manual inspection of the parent ion can often resolve this.

2. Has the correct monoisotopic peak been picked for mass assignment? If the wrong isotope peak has been chosen by the software, the mass assignment is usually off by 1 Da (or 2 Da if the second ^{13}C peak is selected). This can result in peptides being identified as being deamidated, but the correct mass can be assigned manually.

3. Has more than one precursor ion been transmitted for fragmentation simultaneously? A very complex MSMS spectrum can be caused by the simultaneous selection of more than one ion for fragmentation. In some cases, it is possible to identify both sequences from the data, but this usually requires manual interpretation of the spectrum. The search engine Andromeda has included tools for querying data for dual precursor ion fragmentation [34].

4. Is the ion selected for MSMS an in-source generated fragment of another usually more intense ion? It is not uncommon for ions to undergo some in-source fragmentation—either as a loss of ammonia or a backbone cleavage at proline residues. Comparing the elution profiles of the parent ion and the in-source generated fragment in LC-MS (they should have identical elution profiles) is usually helpful for detecting this problem. The data analysis can be handled by comparing the two tandem mass spectra and annotating the in-source generated fragment ion spectrum based on similarity.

3.3.5   Testing and Validating MS Assignments

Often the MSMS data quality of modified peptides leaves a lot to be wished for. Labile modifications can lead to little peptide backbone fragmentation and uninformative spectra, prominent neutral loss series can make the spectra very difficult to interpret, and there are seldom more than one or a few MSMS spectra to base the conclusions on. Faced with these problems there are several possible ways to improve the strength of the identifications and reduce ambiguities.

1. Could another protease be used? By using a several different proteases to process the protein, several different peptides should carry the modification and the consistent observation of a modified residue across different peptide sequences increases the confidence in the assignment.

2. Is it possible to process the samples to prove consistency? For example, to test if the peptide really contains sulfated or phosphorylated residues, the sample can be treated with sulfatase or alkaline phosphatase—if a putatively modified peptide remains unchanged it might be wise to examine the assignments more closely.

3. Is the modification gone if the residue is altered? Site-directed mutagenesis is a powerful but work-intensive strategy for supporting the observations.

4. Synthesize the candidate peptide along with possible ambiguous candidates. In many cases it is possible to synthesize the modified peptide candidate and match the experimental tandem mass spectrum with that of the synthetic version.

5. Complementary fragmentation techniques: By fragmenting the same precursor ion with both CID/HCD and ECD/ETD the complementary fragmentation pathways often yield additional sequence and modification site information. Modifications that are very labile under CID can remain intact under ECD as shown in the example section with an O-GlcNAc modified peptide.

6. In many studies based on cell culture it is possible to grow the cells with stable isotope-labeled amino acids (SILAC). Generating isotope-labeled variants of the modified peptides and swapping labels can provide invaluable information on peptide composition and test assignments.

7. Covalent perturbation proteomics: Often it is possible to chemically modify the entire ensemble of peptides in a predictable uniform stoichiometric manner. This changes the masses of the peptides and the fragment ions in a consistent way that can give an experimental sanity check to the identifications without having to synthesize the peptide candidates. In the example shown in Fig. 8 a candidate lysine-acetylated peptide is contrasted with the covalently perturbed version where all amine groups have been derivatized by reductive dimethylation.

8. Affinity testing—can the peptide be enriched/removed by affinity enrichment/depletion?

9. Are there alternative processing schemes to test assignments? For example, in the case of asparagine glycosylation (N-glycosylation) the entire glycan can be removed by using peptide N-glycosidase F that converts a glycosylated asparagine residue to aspartate in the process (a mass increase of 1 Da). To distinguish this enzymatic process from spontaneous deamidation of asparagines it is possible to perform the deglycosylation in ^{18}O-water, thereby incorporating an ^{18}O atom at transformed asparagines, thereby leaving a trace at the site of glycan attachment (a mass increase of 3 Da) [35]. It is also possible to use a cocktail of endoglycosidases and exoglycosidases that leaves the innermost O-GlcNAc attached to the asparagines (a mass increase of 3 Da) [36, 37]. The spectral complexity of CID/HCD type MSMS is often increased by the remaining sugar residue, but it has the advantage of being direct evidence of glycosylation and having a whole set of diagnostic oxonium fragment ions that can be used for targeted analysis.

3.4 Examples

Protein chemistry and biology are overwhelmingly complex and the modifications listed in Table 5 are only the most common that can be considered. For more extensive lists of modifications Web resources like Resid and Unimod (*see* Subheading 2.2.3) hold a wealth of information. A large body of experimental data has been collected on the fragmentation patterns of histone modifications like monomethylation, dimethylation, and trimethylation and acetylation (*see* **Note 4**). Protein oxidation (*see* **Note 5**) can give rise to a very complex mixture of peptide isoforms and in the case of tryptophan more than six oxidized forms have been reported. Sample preparation plays a crucial role in which side product can be formed. If a gel approach has been used for protein purification the search parameters should reflect that acrylamide can react with cysteines to form propionamide. If the proteins have been digested in the presence of high concentrations of urea there is most likely a lot of carbamylation. If the alkylation of cysteines with iodoacetamide has not been quenched prior to trypsination there will also be carbamidomethylation of the N-terminal amines. In general sample-handling induced modifications are usually sub-stoichiometric and will mostly affect the proteins with the highest concentration. It is unlikely to identify low-abundance proteins solely based on modified peptides (unless a sample handling-related modification is almost quantitative in which case it should be reflected in the peptides assigned to the most abundant proteins).

The assignment and validation of tandem mass spectra of post-translationally modified peptides are not a straightforward task and depends crucially on the MS data quality. This chapter ends with a few examples to give an impression of the data-analysis and how to devise control experiments to validate assignments. To illustrate sample preparation-related PTMs the data derives from a human serum sample that was denatured in 8 M urea, reduced with dithio-threitol, alkylated with iodoacetamide, diluted to 2 M urea, and digested overnight with trypsin at 37 °C.

Oxidation of methionine is a commonly occurring phenomenon both in vivo and in vitro. Figure 3 shows an example where the search engine has returned the peptide sequence DVFLGMFLYEYAR with an oxidized methionine. The y_1-ion fits with a C-terminal arginine and the intense a_2–b_2 pair fits with the residues DV. There is a b-ion series in the low-mass region and a long continuous y ion series and most ions are explained by this sequence. There is an ion at m/z 120 that could be the immonium ion from both phenylalanine and methionine sulfoxide, but the presence of oxidized methionine is confirmed by the intense neutral loss of 64 Da (CH_3SOH). In most cases, it can be helpful to look at the mass accuracy of the fragment ions to increase confidence in the spectral peptide and ion assignment—an example is shown in Fig. 4.

Three very similar spectra of the peptide RHPDYSVVLLLR are displayed in Fig. 5 and assigned to the normal peptide, a carbamoyl-ated, and a carbamidomethylated version. The two basic residues RH at the N-terminus give rise to an intense b-ion series. The peptides differ at the N-terminus as can be seen from the identical y-ion series and the shifted b-ion series. The 14 Da mass difference (*see* **Note 6**) between the spectrum of the carbamoylated and the carbamidomethylated peptide could just as well be explained by a methylation of the N-terminal arginine. The parsimonious inter-pretation is that it is unlikely to have two modifications to the same residue and if the arginine indeed was methylated this peptide containing methylarginine should also be found in a noncarbamoy-lated form.

The parent ion charge state can be misassigned by the software and checking the isotope pattern is often straightforward for high-resolution survey spectra—as illustrated in Fig. 6. As discussed in Subheading 3.3.4 the correct monoisotopic peak should be selected for calculating the peptide mass. Visual inspection of the survey spectrum can point out problems of cofragmentation of multiple ions and it can also give some compositional information especially with SILAC-labeled samples (Table 6).

Crystallins are a family of small proteins found in the lens and the cornea that have a very rich set of PTMs including phosphorylation, lysine acetylation, and O-GlcNAc. Bovine alpha-crystallin is a cheap source of material for method development and two examples of assigning and validating PTMs are given. A tryptic digest of alpha-crystallin was analyzed by reversed-phase LCMS on an Orbitrap and two different validation strategies were applied to the sample. The sample was divided into two parts where one part

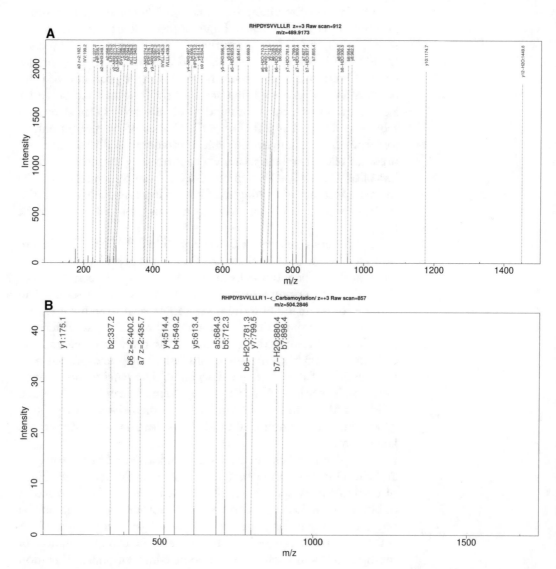

Fig. 5 Three ESI QqTOF tandem mass spectra of triply charged ion at *m/z* 489.92 (**a**), 504.28 (**b**), and 508.95 (**c**). Database searching against the IPI human database returns three different versions of the same peptide from human serum albumin, where the N-terminus has been derivatized with a carbamoyl group from urea (**b**) or a carbamidomethyl group from iodoacetamide (**c**). It is typical to see these unwanted side products of sample handling for the most abundant proteins

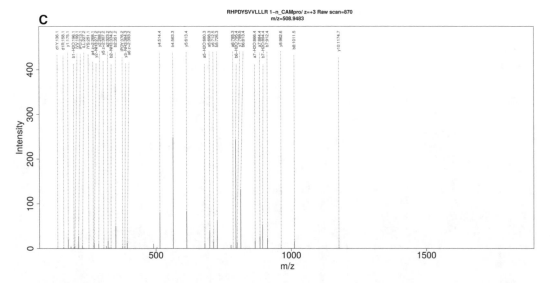

C

RHPDYSVVLLLR 1–n_CAMpro/ z=+3 Raw scan=870
m/z=508.9483

Fig. 5 (continued)

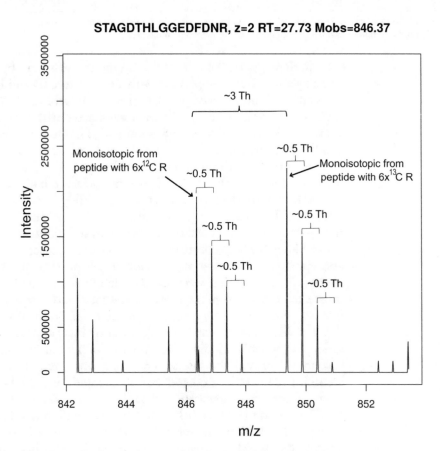

STAGDTHLGGEDFDNR, z=2 RT=27.73 Mobs=846.37

~3 Th

Monoisotopic from
peptide with 6x^{12}C R

~0.5 Th

~0.5 Th

~0.5 Th

~0.5 Th

Monoisotopic from
peptide with 6x^{13}C R

~0.5 Th

~0.5 Th

Fig. 6 Inspection of the MS survey spectrum to check the charge state assignment and peptide composition. A SILAC-labeled sample with two states (Lys/Arg, and Lys-8/Arg-6) was analyzed by LCMS. The excerpt of the survey spectrum displays two doubly charged ions (since the distance between the isotope peaks in each isotope cluster is 0.5 Th) and the two monoisotopic peaks are separated by 3 Th corresponding to a mass difference of 6 Da—therefore the peptide must contain a single arginine residue, no lysine residues and the peptide must be double charged

Table 6
Stable isotope labeled amino acids commonly used for metabolic labeling of proteins using SILAC

Amino acid	Target amino acid	Short name	Composition	Monoisotopic mass
L-lysine	K	K0 or Lys0	$C_6H_{12}N_2O$	128.094963
L-lysine-$^2H_4$	K	K4 or Lys4	$C_6H_8H(2)_4N_2O$	132.120070
L-lysine-$^{13}C_6$	K	K6 or Lys6	$C(13)_6H_{12}N_2O$	134.115092
L-lysine-$^{13}C_6$-$^{15}N_2$	K	K8 or Lys8	$C(13)_6H_{12}N(15)_2O$	136.109162
L-Leucine-$^2H_3$	L	Leu-D3	$C_6H_8H(2)_3NO$	116.102894
L-Tyrosine-$^{13}C_9$	Y	Y9 or Tyr9	$C(13)_9H_9NO_2$	172.093522
L-Arginine	R	Arg0	$C_6H_{12}N_4O$	156.101111
L-arginine-$^{13}C_6$	R	Arg6	$C(13)_6H_{12}N_4O$	162.121240
L-arginine-$^{13}C_6$-$^{15}N_4$	R	Arg10	$C(13)_6H_{12}N(15)_4O$	166.109380

Especially R and K double labeling is widely used because the majority of peptides from a tryptic digestion will have an R or K located at the C-terminus and hence contain at least one labeled amino acid

was covalently perturbed by modifying the amine groups by reductive alkylation. Both samples run with alternating CID and ETD of the same multiply charged precursor ion on an Orbitrap Velos. The CID spectrum shown in Fig. 7a was interrogated with Mascot with HexNAc (S) as variable modification and dimethylation of lysine and the N-terminus retrieved a low-scoring peptide (Mascot score 7) that came from the right protein although it was far below the identification score threshold of 19. The spectrum had the telltale set of low-mass diagnostic ions at m/z 204, 186, 168, 144, 138, and 126 suggesting the presence of an N-acetyl hexosamine—the loss of 203 from several peaks is also indicative of a labile sugar HexNAc. The spectrum is a good counterexample to the concept that low scoring peptides are always of poor quality since almost the entire y-ion series can be identified and the low score most likely derives from the fact that Mascot cannot assign the very intense neutral loss series with the default definition of HexNAc (S). The Mascot score increases significantly by changing the definition of the labile modification so that the neutral loss series are accounted for and the intense oxonium ions ignored. That the identification is indeed correct can be checked by the ETD spectrum of the same precursor ion shown in Fig. 7b—the serine-HexNAc bond that is very labile under CID conditions is left intact during ETD leading to unambiguous assignment of the site of attachment (which in any case was rather straightforward since there was only a single S/T residue in the peptide sequence). The second example shows how assignments in Fig. 8a leading to the identification of the lysine-acetylated peptide can be supported by the covalent perturbation strategy. The amine groups are

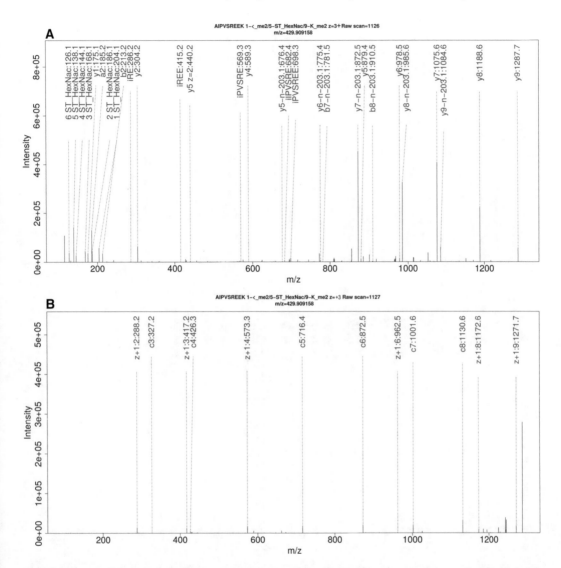

Fig. 7 HCD (**a**) and ETD (**b**) spectrum of the same triply charged precursor ion deriving from a dimethylated tryptic digestion of bovine alpha crystallin acquired on an Orbitrap-Velos ETD. The peptide is identified to be AIPV(HexNAc)SREEK by Mascot with a nonsignificant score of 7. Notice the intense low-mass ions at *m/z* 204, 186, 168, 144, 138, and 126 that are diagnostic of an *N*-acetyl-hexosamine in the HCD spectrum. The vast majority of ions can be meaningfully assigned in the HCD spectrum where the neutral loss series produces the most intense ions. The assignments can be confirmed by the ETD spectrum where the glycan remains attached to the serine residue

dimethylated by reductive alkylation using formaldehyde using standard methods [38]—the mass of both the N-terminus and the C-terminal lysine increases by 28 Da and the y-ion and b-ions are consistently shifted but the internal lysine is protected by the acetylation and has the same mass of 170 Da in both spectra.

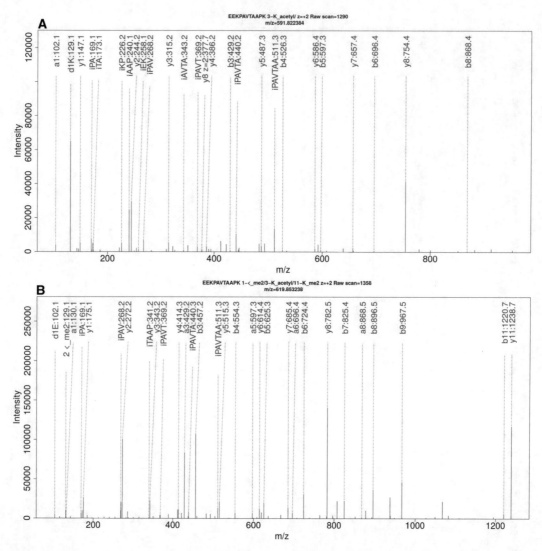

Fig. 8 Identification of a lysine acetylation site in bovine alpha crystallin B-chain and validation by covalent perturbation. A tryptic digest of alpha crystallin was analyzed by LCMS on an Orbitrap Velos and the HCD spectrum of a doubly charged ion with *m/z* 591.82 Th shown in (**a**) led to the identification of a lysine acetylation site in the peptide EEKPAVTAAPK. To validate a part of the tryptic digest was covalently perturbed by reductive dimethylation that changes the mass of all amine groups by 28 Da and reanalyzed by LCMS. Fragmentation of a doubly charged ion at 619.85 Th led to the identification of the same peptide with the N-terminal and the C-terminal lysine amino groups consistently modified and the acetylated lysine residue unchanged shown in (**b**)

4   Notes

1. The most obvious reason for poor mass accuracy is that the calibration has drifted, and the instrument needs to be recalibrated. The mass accuracy in an ion trap depends critically on space charging and one could consider reducing the fill-time/

number of ions per scan or performing a zoom-scan on the ions selected for fragmentation to determine charge state and mass. In a QqTOF the mass accuracy of a well-calibrated instrument depends on having sufficient ion statistics to determine the peak maximum. Some deviation can be caused by other ions or background noise interfering with the isotope pattern peak maxima, but the most common problem is a mass drift in calibration caused by temperature fluctuations (affecting the length of the flight tube). This can be alleviated by postacquisition recalibration.

2. Ion traps that use slow resonance excitation CID to generate tandem mass spectra have a low mass cutoff at approx. one-third of the parent ion mass-to-charge ratio. Therefore, for many ion trap types it is uncommon to detect the low-mass ions.

3. The removal of signalling peptides can be predicted and correlated with the observed sequence. The N-terminus of most proteins is heavily processed, and it is estimated that 80–90% of all proteins are acetylated at the N-terminal amino group. Acetylation usually takes place on methionine or if the penultimate residue has a small radius of gyration (G, A, S, C, T, or V) the methionine is cleaved off by an aminopeptidase and the penultimate residue is acetylated. Hence, detecting an N-terminally acetylated tryptophan should make the alarm bells go off. For a survey of acetylated N-terminal residues—*see* for example [39].

4. Histone modifications have been widely studied and there is many modifications with only the most commonly observed ones listed in Table 5. Differentiating between near isobaric modified residues such as trimethylated lysine and acetyl lysine requires high mass accuracy and careful analysis of the spectra to assign diagnostic marker ions and neutral loss series [40]. Further details of fragmentation pathways and structures for the ions can be found for methylated arginine [41], dimethylated arginine [42], and acetyl lysine [43].

5. For a brief review of protein oxidation—*see* Berlett and Stadtman [44]. The fragmentation behavior of oxidized methionine [45] and oxidized carbamidomethylcysteine [46] is similar. The mono-oxidized form has a fairly intense neutral loss of 64 Da from methionine sulfoxide and 107 Da from mono-oxidized carbamidomethylcysteine, whereas the neutral loss from the dioxidized is of low abundance.

6. There are many modifications that can give rise to 14 Da mass differences. Fairly conservative amino acid conversions (G to A, V to L, N to Q, D to E etc.) and other modifications (carbamylation to carbamidomethylation, carbamidomethylation to

propionamide) give rise to 14 Da mass shifts. Therefore, the data should be scrutinized before reporting methylations.

7. The b_1-ion can be seen for arginine, lysine, histidine, and methionine that can form alternative stabilizing structures by means of their side chains [47].

Acknowledgments

R.M. is supported by Fundação para a Ciência e a Tecnologia (FCT investigator program 2012). iNOVA4Health—UID/Multi/04462/2013, a program financially supported by Fundação para a Ciência e Tecnologia/Ministério da Educação e Ciência, through national funds and cofunded by FEDER under the PT2020 Partnership Agreement is acknowledged.

References

1. Roepstorff P, Fohlman J (1984) Proposal for a common nomenclature for sequence ions in mass spectra of peptides. Biomed Mass Spectrom 11(11):601

2. Biemann K (1990) Appendix 5. Nomenclature for peptide fragment ions (positive ions). Methods Enzymol 193:886–887

3. Johnson RS, Martin SA, Biemann K, Stults JT, Watson JT (1987) Novel fragmentation process of peptides by collision-induced decomposition in a tandem mass spectrometer: differentiation of leucine and isoleucine. Anal Chem 59(21):2621–2625

4. Steen H, Mann M (2004) The ABC's (and XYZ's) of peptide sequencing. Nat Rev Mol Cell Biol 5(9):699–711

5. Salek M, Lehmann WD (2003) Neutral loss of amino acid residues from protonated peptides in collision-induced dissociation generates N- or C-terminal sequence ladders. J Mass Spectrom 38(11):1143–1149

6. Cooper HJ, Hakansson K, Marshall AG (2005) The role of electron capture dissociation in biomolecular analysis. Mass Spectrom Rev 24(2):201–222

7. Zubarev RA (2004) Electron-capture dissociation tandem mass spectrometry. Curr Opin Biotechnol 15(1):12–16

8. Syka JE, Coon JJ, Schroeder MJ, Shabanowitz J, Hunt DF (2004) Peptide and protein sequence analysis by electron transfer dissociation mass spectrometry. Proc Natl Acad Sci U S A 101(26):9528–9533

9. Mirgorodskaya E, Roepstorff P, Zubarev RA (1999) Localization of O-glycosylation sites in peptides by electron capture dissociation in a Fourier transform mass spectrometer. Anal Chem 71(20):4431–4436

10. Hakansson K, Cooper HJ, Emmett MR, Costello CE, Marshall AG, Nilsson CL (2001) Electron capture dissociation and infrared multiphoton dissociation MS/MS of an N-glycosylated tryptic peptide to yield complementary sequence information. Anal Chem 73(18):4530–4536

11. Hogan JM, Pitteri SJ, Chrisman PA, McLuckey SA (2005) Complementary structural information from a tryptic N-linked glycopeptide via electron transfer ion/ion reactions and collision-induced dissociation. J Proteome Res 4(2):628–632

12. Stensballe A, Jensen ON, Olsen JV, Haselmann KF, Zubarev RA (2000) Electron capture dissociation of singly and multiply phosphorylated peptides. Rapid Commun Mass Spectrom 14(19):1793–1800

13. Kelleher NL, Zubarev RA, Bush K, Furie B, Furie BC, McLafferty FW, Walsh CT (1999) Localization of labile posttranslational modifications by electron capture dissociation: the case of gamma-carboxyglutamic acid. Anal Chem 71(19):4250–4253

14. Johnson RS, Davis MT, Taylor JA, Patterson SD (2005) Informatics for protein identification by mass spectrometry. Methods 35(3):223–236

15. Schlosser A, Lehmann WD (2002) Patchwork peptide sequencing: extraction of sequence information from accurate mass data of peptide tandem mass spectra recorded at high resolution. Proteomics 2(5):524–533

16. Falick AM, Hines WM, Medzihradszky KF, Baldwin MA, Gibson BW (1993) Low-mass ions produced from peptides by high-energy collision-induced dissociation in tandem mass-spectrometry. J Am Soc Mass Spectrom 4 (11):882–893

17. Papayannopoulos IA (1995) The interpretation of collision-induced dissociation tandem mass-spectra of peptides. Mass Spectrom Rev 14(1):49–73

18. Peri S, Steen H, Pandey A (2001) GPMAW--a software tool for analyzing proteins and peptides. Trends Biochem Sci 26(11):687–689

19. Perkins DN, Pappin DJ, Creasy DM, Cottrell JS (1999) Probability-based protein identification by searching sequence databases using mass spectrometry data. Electrophoresis 20 (18):3551–3567

20. Matthiesen R, Bunkenborg J, Stensballe A, Jensen ON, Welinder KG, Bauw G (2004) Database-independent, database-dependent, and extended interpretation of peptide mass spectra in VEMS V2.0. Proteomics 4 (9):2583–2593

21. Craig R, Beavis RC (2003) A method for reducing the time required to match protein sequences with tandem mass spectra. Rapid Commun Mass Spectrom 17(20):2310–2316

22. Bunkenborg J, Garcia GE, Paz MIP, Andersen JS, Molina H (2010) The minotaur proteome: avoiding cross-species identifications deriving from bovine serum in cell culture models. Proteomics 10(16):3040–3044

23. Schandorff S, Olsen JV, Bunkenborg J, Blagoev B, Zhang Y, Andersen JS, Mann M (2007) A mass spectrometry-friendly database for cSNP identification. Nat Methods 4 (6):465–466

24. Olsen JV, Ong SE, Mann M (2004) Trypsin cleaves exclusively C-terminal to arginine and lysine residues. Mol Cell Proteomics 3 (6):608–614

25. Pallante GA, Cassady CJ (2002) Effects of peptide chain length on the gas-phase proton transfer properties of doubly-protonated ions from bradykinin and its N-terminal fragment peptides. Int J Mass Spectrom 219 (1):115–131

26. Schnier PD, Gross DS, Williams ER (1995) On the maximum charge-state and proton-transfer reactivity of peptide and protein ions formed by electrospray-ionization. J Am Soc Mass Spectrom 6(11):1086–1097

27. Wysocki VH, Tsaprailis G, Smith LL, Breci LA (2000) Special feature: Commentary - Mobile and localized protons: a framework for understanding peptide dissociation. J Mass Spectrom 35(12):1399–1406

28. Breci LA, Tabb DL, Yates JR, Wysocki VH (2003) Cleavage N-terminal to proline: analysis of a database of peptide tandem mass spectra. Anal Chem 75(9):1963–1971

29. Kapp EA, Schutz F, Reid GE, Eddes JS, Moritz RL, O'Hair RA, Speed TP, Simpson RJ (2003) Mining a tandem mass spectrometry database to determine the trends and global factors influencing peptide fragmentation. Anal Chem 75(22):6251–6264

30. Tabb DL, Smith LL, Breci LA, Wysocki VH, Lin D, Yates JR (2003) Statistical characterization of ion trap tandem mass spectra from doubly charged tryptic peptides. Anal Chem 75 (5):1155–1163

31. Hall SC, Smith DM, Masiarz FR, Soo VW, Tran HM, Epstein LB, Burlingame AL (1993) Mass spectrometric and edman sequencing of lipocortin-I isolated by 2-dimensional SDS PAGE of human-melanoma lysates. Proc Natl Acad Sci U S A 90(5):1927–1931

32. Hamdan M, Bordini E, Galvani M, Righetti PG (2001) Protein alkylation by acrylamide, its N-substituted derivatives and cross-linkers and its relevance to proteomics: a matrix assisted laser desorption/ionization-time of flight-mass spectrometry study. Electrophoresis 22(9):1633–1644

33. Nielsen ML, Vermeulen M, Bonaldi T, Cox J, Moroder L, Mann M (2008) Iodoacetamide-induced artifact mimics ubiquitination in mass spectrometry. Nat Methods 5(6):459–460

34. Cox J, Neuhauser N, Michalski A, Scheltema RA, Olsen JV, Mann M (2011) Andromeda: a peptide search engine integrated into the MaxQuant environment. J Proteome Res 10 (4):1794–1805

35. Gonzalez J, Takao T, Hori H, Besada V, Rodriguez R, Padron G, Shimonishi Y (1992) A method for determination of N-glycosylation sites in glycoproteins by collision-induced dissociation analysis in fast atom bombardment mass spectrometry: identification of the positions of carbohydrate-linked asparagine in recombinant alpha-amylase by treatment with peptide-N-glycosidase F in 18O-labeled water. Anal Biochem 205(1):151–158

36. Hagglund P, Bunkenborg J, Elortza F, Jensen ON, Roepstorff P (2004) A new strategy for identification of N-glycosylated proteins and unambiguous assignment of their glycosylation sites using HILIC enrichment and partial deglycosylation. J Proteome Res 3(3):556–566

37. Hagglund P, Matthiesen R, Elortza F, Hojrup P, Roepstorff P, Jensen ON, Bunkenborg J (2007) An enzymatic deglycosylation scheme enabling identification of core fucosylated N-glycans and O-glycosylation site mapping of human plasma proteins. J Proteome Res 6(8):3021–3031

38. Boersema PJ, Raijmakers R, Lemeer S, Mohammed S, Heck AJR (2009) Multiplex peptide stable isotope dimethyl labeling for quantitative proteomics. Nat Protoc 4 (4):484–494

39. Polevoda B, Sherman F (2003) N-terminal acetyltransferases and sequence requirements for N-terminal acetylation of eukaryotic proteins. J Mol Biol 325(4):595–622

40. Zhang K, Yau PM, Chandrasekhar B, New R, Kondrat R, Imai BS, Bradbury ME (2004) Differentiation between peptides containing acetylated or tri-methylated lysines by mass spectrometry: an application for determining lysine 9 acetylation and methylation of histone H3. Proteomics 4(1):1–10

41. Gehrig PM, Hunziker PE, Zahariev S, Pongor S (2004) Fragmentation pathways of N(G)-methylated and unmodified arginine residues in peptides studied by ESI-MS/MS and MALDI-MS. J Am Soc Mass Spectrom 15 (2):142–149

42. Rappsilber J, Friesen WJ, Paushkin S, Dreyfuss G, Mann M (2003) Detection of arginine dimethylated peptides by parallel precursor ion scanning mass spectrometry in positive ion mode. Anal Chem 75(13):3107–3114

43. Kim JY, Kim KW, Kwon HJ, Lee DW, Yoo JS (2002) Probing lysine acetylation with a modification-specific marker ion using high-performance liquid chromatography/electrospray-mass spectrometry with collision-induced dissociation. Anal Chem 74 (21):5443–5449

44. Berlett BS, Stadtman ER (1997) Protein oxidation in aging, disease, and oxidative stress. J Biol Chem 272(33):20313–20316

45. Lagerwerf FM, vandeWeert M, Heerma W, Haverkamp J (1996) Identification of oxidized methionine in peptides. Rapid Commun Mass Spectrom 10(15):1905–1910

46. Steen H, Mann M (2001) Similarity behween condensed phase and gas phase chemistry: fragmentation of peptides containing oxidized cysteine residues and its implications for proteomics. J Am Soc Mass Spectrom 12 (2):228–232

47. Farrugia JM, O'Hair RAJ, Reid GE (2001) Do all b(2) ions have oxazolone structures? Multistage mass spectrometry and ab initio studies on protonated N-acyl amino acid methyl ester model systems. Int J Mass Spectrom 210 (1-3):71–87

Chapter 9

Solution to Dark Matter Identified by Mass-Tolerant Database Search

Rune Matthiesen

Abstract

Recently a mass-tolerant search approach was proposed which suggested novel types of abundant modification with delta masses that left many scratching their heads. These surprising new findings which were hard to explain were later referred to as dark matter of mass spectrometry-based proteomics. Rewards were promised for those who could solve these intriguing new findings. We propose here simple solutions to the novel delta masses identified by mass-tolerant database search.

Key words Open search, Validation, Database-dependent search, Peptide assignments

1 Introduction

Commonly in database dependent searches of MSMS data a list of potential modifications is provided as input to a search engine. As an alternative, open or blind database search of modifications puts no restriction on which modifications or residue changes the search engine can use to match the observed MSMS spectra. Past research have proposed several methods for open or blind database search of modifications but still none is broadly applied [1, 2]. The search space increases exponential with the numbers of alternative masses to consider for each of the different amino acid residues [3, 4]. This means that from a philosophic point of view open search are doomed to fail since we will have an infinite number of allowed delta masses that combines in a factorial manner. This means by unrestricted modifications of a peptide, the peptide can match any spectrum. Therefore, we need to make assumptions of which modification to allow and the final result improves if these considerations are well prepared. For example, we can restrict which modifications we will consider, we can restrict how many times a modification can occur on a sequence and we can restrict which modifications can co-occur on the same peptides. This was the main

Rune Matthiesen (ed.), *Mass Spectrometry Data Analysis in Proteomics*, Methods in Molecular Biology, vol. 2051, https://doi.org/10.1007/978-1-4939-9744-2_9, © Springer Science+Business Media, LLC, part of Springer Nature 2020

argument for proposing the semiopen search [5]. The objective is therefore to restrict to the right extent to prevent FDR explosion and overfitting but still include the most abundant combinations of modifications. If uncontrolled then, a standard search likely outperform semi open search. Semi open searches are preferably only applied on high resolution MS and MSMS data.

The open database search referred to as mass-tolerant search was proposed by Chick et al. [1]. Chick and coworkers proposed a surprising number of identified peptides with delta mases hard to explain of which some were found frequently by Mass tolerant search. This has led to a number of creative proposals to the chemical composition of these new delta masses whereas others are still unexplained and left as dark matter. Rewards have been put forward for those offering explanations of these novel and apparent delta masses. We demonstrate here that misassigned peptides explain a substantial number of the surprising delta masses. Chick et al. claim to have provided a landmark open search system for modifications, yet not a single delta mass found was assigned to modifications in the supplemental data accompanying the paper. Instead a few simple examples are highlighted in the main text. From the theoretical point of view we identify the following concerns with the proposed mass-tolerant search:

1. A fair number of the delta masses fall in a mass region for which there is no known modification or only artificial modifications such as those based on stable isotope labeling. This question if these delta masses correspond to modifications, amino acid substitutions, a combination of these or rather is a consequence of the fact that erroneous peptide sequences were assigned to the spectra.

2. The delta mass correlation does not consider combination of modifications and attempts to fit directly a delta mass on a single residue when in fact the delta mass is a result of several modifications scattered throughout the peptide sequence. Numerous of the delta masses proposed clearly corresponds to combination of modifications yet the provided output lists assign these delta masses to a single residue. The scoring system therefore cannot assign the optimal assignment.

3. Considering part of the sequence deleted but then still allowing fragment ions from the deleted part to score is problematic and a potential source of artefacts. Positive delta masses raise similar concerns.

4. Many of the abundant modifications proposed lack previous evidence pointing to their abundance.

5. Chick *et al* provide cryptic sequence annotations for all the spectra match solutions proposed by the mass-tolerant search

(e.g., delta mass of −89.031 Da corresponds to cleaved N-terminal methionine followed by acetylation).

6. The delta mass of 5 ppm in the standard search which was used for comparison with the mass tolerant search in the original paper appears too tight for fair comparison. It therefore appears as if mass-tolerant search is significantly superior compared to standard searches.

The above stated concerns together with scientific curiosity led us to investigate the assignments put forward by Chick et al. To demonstrate the validity of these problems a standard search with a well-accepted search engine X!tandem [3] was performed focusing on a single LC-MS run in the data set "b1929_293T_proteinI-D_09A_QE3_122212". The data was filtered with Percolator [6] at a 1% FDR threshold and compared with published results from Chick et al. [1]. The X!tandem standard search was performed specifying methionine oxidation, deamidation and protein N-terminal acetylation. Such a standard search is wildly accepted to produce reliable results with low probability of incorrect assignments for high scoring peptides. X!tandem followed by FDR quality filtering by Percolator at 1% FDR gave 26,272 peptide spectra assignments whereas the mass-tolerant search by Chick et al. was reported to give 28,488 peptide spectra assignments at a 1% FDR threshold. Different FDR filtering methods means that the total numbers of spectra assignments are incomparable. It is possible the mass-tolerant approach identified more true matches by considering possible delta masses, but it is also possible that FDR filtering for mass-tolerant approach ended up with more spectra matches because the precise FDR quality filtering algorithm applied was less stringent. Comparing the scan numbers of the spectra assignment from the two approaches revealed 14,738 spectra which were assigned by both, 13,750 MSMS spectra only assigned by mass-tolerant search and 11,534 only by X!tandem-Percolator standard search. The large number of assignments only found by X!tandem-Percolator standard search, which is a general accepted approach, is a major concern and deviate strongly from the presented result in the paper of Chick et al. We speculate that the few spectra assignments uniquely identified in the standard search by Chick and coauthors are a consequence of too tight a mass error in the specification of the parent ion. Next the agreement of the 14,738 MSMS scan numbers that were assigned by both was evaluated. There were no agreements on 4346 of these which raise even stronger concerns. Interrogating these disagreements revealed 425 were caused by unnecessary complex annotations. For example, the peptide "RPQYSNPPVQGEVMEGADNQGAGEQGR" was annotated by the mass-tolerant database search as "RPQYSNPPVQGEVMEGADNQGAGEQGRPVR with delta mass −352.2229 Da". The delta mass of −352.2229 Da is a result of the introduced "PVR" part of the sequence. This means that mass-tolerant searches converts canonical missed cleaved peptides into peptides that are potentially modified with hard to explain delta masses. One can discuss if this is an error, but it is certainly unnecessary complex annotation of spectra matches and a direct source of confusion. Further, this only explained 425 of the 4346 disagreements. Next the remaining 3921 disagreements were scrutinized. Unfortunately, the large majority (>95%) of these turned out to clearly favor the spectra matches provided by X!Tandem-Percolator approach. We highlight here some examples to clarify the output from mass-tolerant search.

Example 1: X!Tandem: "DLYANTVLSGGTTMYPGIADR 14-Methionine oxidation"

Mass-tolerant search: "DLYANTVLSGGSTMYPGIADR delta mass 30.0092 Da"

In terms of annotated ions in the raw spectra these two solutions are comparable. However, a delta mass of 30.0092 Da is not easy to explain in isolation and the two matches gives equal score when matched against the spectrum. X!Tandem provided a solution which concur with the principals of Occam's razor which is not the case for the mass tolerant search. The comparison reveals that mass-tolerant match corresponds to serine mutated to threonine and methionine oxidation. The X!Tandem match is the most likely solution. Note that this also affects negatively the identified proteins of the mass-tolerant search.

Example 2: X!Tandem: "MPTTQETDGFQVK 1-Methionine oxidation"

Mass-tolerant search: "TPTTQETDGFQVK delta mass 45.9884 Da"

In terms of annotated ions in the raw spectra these two solutions are comparable. However, this delta mass is again hard to explain in isolation but on comparing with the X!Tandem match corresponds to different peptide (threonine to methionine) containing a methionine oxidation. Again, the mass-tolerant search provided a less likely peptide match which in this case also affects protein identification.

Example 3: X!Tandem: "ARPDDEEGAAVAPGHPLAK"

Mass-tolerant search: "RAGDELAYNSSSACASSR delta mass −0.8923 Da"

Comparing the spectrum annotation of the two matches reveals that the X!Tandem solution provides a better solution with more abundant ions matched (*see* Fig. 1). Furthermore, the X!Tandem solution does not provide the extra complication of a unexplained delta mass. It is not entirely clear why this happens but supports some of the concerns listed above in the introduction.

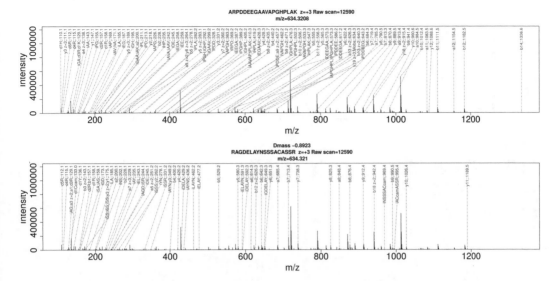

Fig. 1 Raw spectrum annotation of the match from X!Tandem "ARPDDEEGAAVAPGHPLAK" versus mass-tolerant search "RAGDELAYNSSSACASSR delta mass −0.8923 Da". Peaks are annotated with a- , b- , y- , internal (i) and diagnostic (d) ions

Example 4: **X!Tandem**: "IIGATDSSGELMFLMK 12 and 15 Methionine oxidation"

Mass-tolerant search: "AAFGLSEAGFNTACVTK delta mass 1.0094 Da"

Comparing the spectrum annotation of the two matches reveals that the X!Tandem solution provides a solution with more abundant ions matched (*see* Fig. 2). Furthermore, the X!Tandem solution does not provide the extra complication of a unexplained delta mass of 1.0094 Da. Both pipelines provided the same observed parent ion eliminating the possibility that the underlining cause could be misassigned parent ion. It is not entirely clear why this misassignment happens but supports some of the concerns listed above in the introduction.

Example 5: **X!Tandem**: "SLGGGTGSGMGTLLISK 10-Methionine oxidation"

Mass-tolerant search: "SMSDVSAEDVQNLR delta mass 1.0928 Da"

Comparing the spectrum annotation of the two matches reveals that the X!Tandem solution provides a more appropriate solution with more abundant ions matched (*see* Fig. 3). Furthermore, the X!Tandem solution provides a solution that match the parent ion in constrast to the extra complication of an unexplained delta mass of 1.0928 Da. Both pipelines were provided the same observed parent ion eliminating the possibility that the difference could stem from a misassigned parent ion (*see* Fig. 4). It is again unclear why this misassignment happens but supports some of the concerns listed above in the introduction.

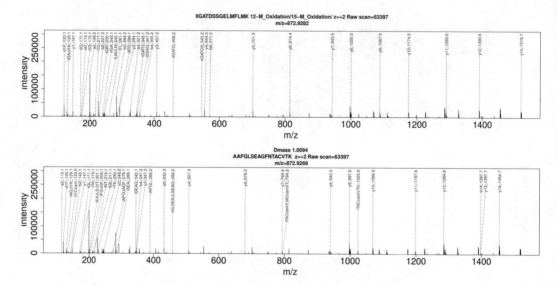

Fig. 2 Raw spectrum annotation of the match from X!Tandem "IIGATDSSGELMFLMK 12 and 15 Methionine oxidation" versus mass-tolerant search "AAFGLSEAGFNTACVTK delta mass 1.0094 Da". Peaks are annotated with a- , b- , y- , internal (i), and diagnostic (d) ions

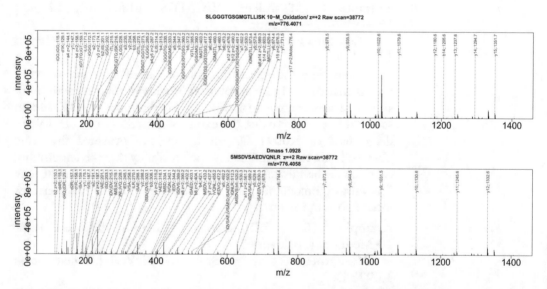

Fig. 3 Raw spectrum annotation of the match from X!Tandem "SLGGGTGSGMGTLLISK 10-Methionine oxidation" versus mass-tolerant search "SMSDVSAEDVQNLR delta mass 1.0928 Da". Peaks are annotated with a- , b- , y- , internal (i), and diagnostic (d) ions

The delta masses from Examples 4 and 5 above was explained by Chick et al. as misidentification of the monoisotopic peak resulting in a detected mass difference of a shift from ^{12}C to ^{13}C (1.0034 Da, *see* Figure 2h of Chick et al.). However, we observe agreement in the parent ion assignment and find a more

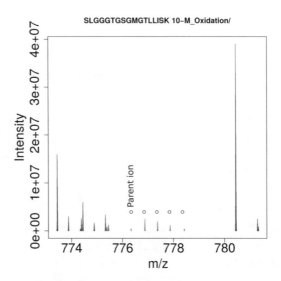

Fig. 4 Zoom in the *m/z* region of the survey scan recorded just before the fragmentation of the peptide spectrum match in Fig. 3. The observed parent ion isotope envelope is highlighted in red

appropriate spectrum match with no uncommon delta masses for the majority of these cases.

The conclusion is that the current state of mass-tolerant search provides a large number of complicated and confusing spectra matches. Further, a substantial number of unexplained delta masses are caused by misassignments with similar spectra macting scores of the correct solution. In some cases Occam's razor distinguise correct from missassigned spectra match. We must scrutinize community efforts attempts to assigning observed delta masses from the output of mass-tolerant search to single chemical modifications, especially considering the types of sequence misassignments described in this chapter. The complicated nature of the outputted spectra annotations by mass-tolerant search question if it is worth fixing the discussed problems.

2 Materials and Software

1. Data set: Raw data file "b1929_293T_proteinI-D_09A_QE3_122212.raw" from ProteomeXchange project "PXD001468" (http://proteomecentral.proteomexchange.org/cgi/GetDataset?ID=PXD001468).

2. UniProt sequence database.

3. RawConvert [7] (http://fields.scripps.edu/rawconv/).

4. X!tandem: https://www.thegpm.org/TANDEM/index.html version Sledgehammer (2013.09.01.1).

5. Percolator (https://github.com/percolator/percolator/releases).

---

3 Methods

3.1 Download the Raw Data File b1929_293T_proteinID_09A_QE3_122212.raw

1. Go to the web page Pride archive [8]: "https://www.ebi.ac.uk/pride/archive/projects/PXD001468"

2. Click on "Download Project Files" in the above right corner.

3. Scroll down to the raw file "b1929_293T_proteinID_09A_QE3_122212.raw" and select for download.

3.2 Download UniProt Human Sequence Database

1. https://www.uniprot.org/uniprot/?query=*&fil=organism%3A%22Homo+sapiens+%28Human%29+%5B9606%5D%22+AND+reviewed%3Ayes

2. *See* Chapter 6 on how to format the reverse sequence database for compatibility with Percolator and Percolator quality filtering of spectra matches.

3.3 Process Raw Data File "b1929_293T_proteinID_09A_QE3_122212.raw" to mgf Format

1. After installing RawConverter the program starts from the start menu in microsft Windows.

2. Press "+" button to select the raw file "b1929_293T_proteinID_09A_QE3_122212.raw".

3. Set experiment type "DDA".

4. Set destination output directory by clicking the button "Browse".

5. Choose MGF as output format.

6. Set the charge states "Options → DDA → Charge States" (*see* **Note 1**).

7. Press the button "Go!"

3.4 Database Dependent Search Using X!Tandem Sledgehammer (See Note 2)

1. Download and compile X!Tandem according to the instructions in the "INSTALL" file.

2. Open the file "input.xml" and change "protein, taxon" from yeast to human, "spectrum, path" to the file path of the mgf file created above in Subheading 3.3 and set the output path "output, path".

3. Open the file "default_input.xml" and modify "spectrum, fragment monoisotopic mass error" from 0.4 to 0.01, "spectrum, parent monoisotopic mass error plus" to 5, "spectrum, parent monoisotopic mass error minus" to 5.

4. Scroll down to "residue, potential modification mass" and set the value to "15.99491462@[M],0.984016@[QN]" without the quotes.

5. Scroll down to "refine, potential N-terminus modifications" and set the value to "42.01056469@[" (the single square bracket indicates the N-terminus, the value again is set without the quotes).

6. Open taxonomy.xml file and change "<taxon label="yeast">" to "<taxon label="human">". Modify the path in the line "<file format="peptide" URL="../fasta/scd.fasta.pro"/>" to contain the path for the UniProt sequence database and delete the next two lines.

7. Execute "tandem.exe input.xml" on the command line (*see* **Notes 3** and **4**).

Finally the output was filtered with Percolator as described in Chapter 6 and spectra matches from X!Tandem were compared in R with the ones provided by Chick et al. (*see* **Note 5**).

4   Notes

1. Charge states from +1 to +6 were specified in RawConverter "Options → DDA → Charge States".

2. There exists a number of graphical interfaces to X!Tandem (e.g., *see* Chapter 6). However, it is relatively easy to modify the xml files directly or make your own script to change the parameter files as indicated in Subheading 3.4 above.

3. The compiled program by the make files on Linux is also named tandem.exe which is confusing, but it also works in Linux-based systems.

4. The file default_input.xml also has the field "spectrum, threads" which is convenient to set as one minus the number of processing units of the computer used.

5. The spectrum identification number in the X! Tandem Percolator search results output correspond to the spectra number in the mgf file. For comparison with Chick et al. [1], the spectra numbers needs conversion to scan numbers.

Acknowledgments

R.M. is supported by Fundação para a Ciência e a Tecnologia (FCT investigator program 2012). iNOVA4Health—UID/Multi/04462/2013, a program financially supported by Fundação para a Ciência e Tecnologia/Ministério da Educação e Ciência, through national funds and cofunded by FEDER under the PT2020 Partnership Agreement is acknowledged. This work is also funded by FEDER funds through the COMPETE 2020 Programme and

National Funds through FCT - Portuguese Foundation for Science and Technology under the projects number PTDC/BTM-TEC/30087/2017 and PTDC/BTM-TEC/30088/2017

References

1. Chick JM, Kolippakkam D, Nusinow DP, Zhai B, Rad R, Huttlin EL, Gygi SP (2015) A mass-tolerant database search identifies a large proportion of unassigned spectra in shotgun proteomics as modified peptides. Nat Biotechnol 33(7):743–749. https://doi.org/10.1038/nbt.3267

2. Tsur D, Tanner S, Zandi E, Bafna V, Pevzner PA (2005) Identification of post-translational modifications by blind search of mass spectra. Nat Biotechnol 23(12):1562–1567. https://doi.org/10.1038/nbt1168

3. Craig R, Beavis RC (2004) TANDEM: matching proteins with tandem mass spectra. Bioinformatics 20(9):1466–1467. https://doi.org/10.1093/bioinformatics/bth092

4. Matthiesen R, Trelle MB, Hojrup P, Bunkenborg J, Jensen ON (2005) VEMS 3.0: algorithms and computational tools for tandem mass spectrometry based identification of post-translational modifications in proteins. J Proteome Res 4(6):2338–2347. https://doi.org/10.1021/pr050264q

5. Carvalho AS, Ribeiro H, Voabil P, Penque D, Jensen ON, Molina H, Matthiesen R (2014) Global mass spectrometry and transcriptomics array based drug profiling provides novel insight into glucosamine induced endoplasmic reticulum stress. Mol Cell Proteomics 13 (12):3294–3307. https://doi.org/10.1074/mcp.M113.034363

6. Spivak M, Weston J, Bottou L, Kall L, Noble WS (2009) Improvements to the percolator algorithm for Peptide identification from shotgun proteomics data sets. J Proteome Res 8 (7):3737–3745. https://doi.org/10.1021/pr801109k

7. He L, Diedrich J, Chu YY, Yates JR (2015) Extracting accurate precursor information for tandem mass spectra by rawconverter. Anal Chem 87:11361–11367

8. Ternent T, Csordas A, Qi D, Gomez-Baena G, Beynon RJ, Jones AR, Hermjakob H, Vizcaino JA (2014) How to submit MS proteomics data to ProteomeXchange via the PRIDE database. Proteomics 14(20):2233–2241. https://doi.org/10.1002/pmic.201400120

Chapter 10

Phosphoproteomics Profiling to Identify Altered Signaling Pathways and Kinase-Targeted Cancer Therapies

Barnali Deb, Irene A. George, Jyoti Sharma, and Prashant Kumar

Abstract

Phosphorylation is one of the most extensively studied posttranslational modifications (PTM), which regulates cellular functions like cell growth, differentiation, apoptosis, and cell signaling. Kinase families cover a wide number of oncoproteins and are strongly associated with cancer. Identification of driver kinases is an intense area of cancer research. Thus, kinases serve as the potential target to improve the efficacy of targeted therapies. Mass spectrometry-based phosphoproteomic approach has paved the way to the identification of a large number of altered phosphorylation events in proteins and signaling cascades that may lead to oncogenic processes in a cell. Alterations in signaling pathways result in the activation of oncogenic processes predominantly regulated by kinases and phosphatases. Therefore, drugs such as kinase inhibitors, which target dysregulated pathways, represent a promising area for cancer therapy.

Key words Mass spectrometry, Cancer signaling, Molecular therapeutics, Targeted therapy

1 Introduction

The eukaryotic system is very multifaceted and synchronized by a plethora of mechanisms conducted at the various levels of cell regulation. Gene regulation is followed by transcription to form mRNA, splicing events, and finally translation of proteins. Proteins are further modified by posttranslational modifications (PTMs) that are mostly reversible in nature. PTMs are generally enzymatic modifications and covalent additions of functional groups to the proteins during its life cycle. The PTMs are mostly governed by the external stimuli generated due to any particular conditions in the biological system. PTMs increase the functional diversity of the proteins and regulate potentially the proteolytic degradation, protein-protein interactions, protein translocations to different cellular component, signaling transduction and metabolism [1]. PTMs also indirectly modulate the other noncovalent protein–protein interactions by regulating their electrostatic and structural properties [2, 3]. Numerous PTMs have been reported so far in

Rune Matthiesen (ed.), *Mass Spectrometry Data Analysis in Proteomics*, Methods in Molecular Biology, vol. 2051, https://doi.org/10.1007/978-1-4939-9744-2_10, © Springer Science+Business Media, LLC, part of Springer Nature 2020

literature (approximately 300) [4]. These PTMs are based on chemical groups, complex groups, polypeptides, cleavage of proteins or even amino acid modification. However, very few PTMs have been extensively studied such as phosphorylation, acetylation, hydroxylation, methylation, glycosylation, AMPylation, SUMOylation, lipidation, ubiquitination, deamidation, thiolation, biotinylation, nitration, sulfation, and so on (Fig. 1a) [4]. The addition of carbohydrate moieties (glycans) to proteins in the endoplasmatic reticulum (N-linked glycans) or in the Golgi (O-linked glycans) is one of the recurrent posttranslational events along with phosphorylation. It has been reported that about 70–80% of the proteins are glycosylated. Number of glycan moieties added to any protein depends on the sites available in the protein sequence [5]. Ubiquitination is another broadly studied PTM. During ubiquitination, addition of ubiquitin tags to the protein (a highly conserved 76-amino acid protein expressed in all cell types), generally serves as a degradation mechanism for the proteins [6].

Phosphorylation is one of the most broadly studied dynamic modifications, which regulates numerous cellular processes such as cellular signaling, cell growth, apoptosis, and differentiation. Protein kinases and phosphatases are the regulatory enzymes that phosphorylate and dephosphorylate proteins respectively through a reversible process [7]. Protein kinases are classified according to the amino acid residue that it phosphorylates [8]. The vast majority of the protein phosphorylation occurs on the serine, threonine and a few tyrosine sites as well [9]. Two-thirds of the proteins encoded by the human genome have been shown to be phosphorylated, and it is expected that more than 90% are actually uncovered to phosphorylation [8]. However, the phosphorylation of any proteome is not completely catalogued and expected to be much higher as inferred from the various experimental data (the number of potential phosphorylation sites in eukaryotic system is estimated to be in the range of 500,000–700,000) [10, 11] (see **Note 1**). The Cell Signaling Technology PhosphoSitePlus (www.phosphosite.org) and the Kinexus PhosphoNET (www.phosphonet.ca) websites both listed over 200,000 known human phosphorylated sites.

Many gene products that regulate the normal cellular functions are altered and may lead to the outrageous forms of cellular growth and proliferation causing cancer [12]. Importantly, dysregulation of protein phosphorylation is a key driver in cancer. The desire to understand the aberrant global phosphorylation events observed in cancer has made phosphorylation one of the best-studied PTMs. Mass spectrometry (MS)-based proteomics has evolved in the past few years as a result of more accurate instrumentation techniques and better methods for sample preparation, quantitation, and data acquisition [13]. Proteomics has enabled the analysis and identification of not only the expressed proteins, but it can also analyze posttranslationally modified proteins (such as phosphorylated

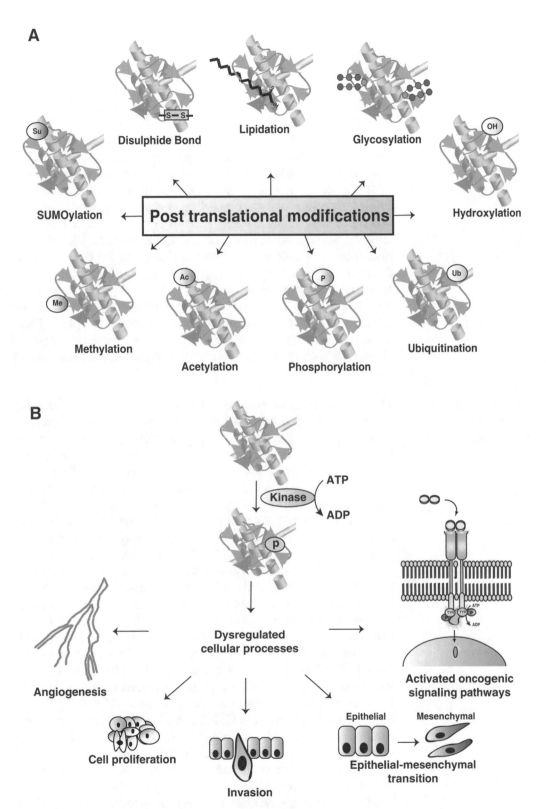

Fig. 1 Posttranslational modifications. (**a**) Widely studied posttranslational modifications. (**b**) Phosphorylation in cancer. Kinases phosphorylate their target proteins and thus lead to dysregulation in the normal cellular processes such as angiogenesis, cell proliferation, invasion, and epithelial–mesenchymal transition

proteins) [14]. As protein expression and modifications regulate the functional properties of cells (e.g., a phosphorylation site can reflect the activity state of a protein), proteomics and phosphoproteomics have been widely used for the identification of potential pharmaceutical targets.

Phosphoproteomics is being extensively studied in this era particularly in cancers for identifying the dysregulated phosphoproteins, altered signaling pathways and to predict the kinases, which could be therapeutically targeted for better treatment outcomes. Phosphorylation-regulated signaling enabled us to study the various aspects of tumor biology such as regulation of cell growth, invasion, angiogenesis, proliferation, and metastasis. Global phosphoproteomics provides a deeper insight into functional consequences of the changes at protein level. The innumerable approaches of targeting an activated kinase in cancers is applicable especially due to an acquisitive kinase inhibitors that are being established by the various pharmaceutical companies [15]. Thus, phosphoproteomics has already emerged as a vital technique in clinical research to assist in diagnosis, prognosis, and treatment of cancers.

1.1 Phosphorylation and Its Biological Importance in Cancer

The process of dynamic phosphorylation regulates protein functions and cell signaling. Mutations in particular kinases as well as phosphatases are the intriguing factors in dysregulation of normal cellular activities. It also disrupts the subtle equilibrium of activation or inactivation of cancer-associated proteins [16]. Cancer is a complex disease, which is governed at various levels including genomics, transcriptomics and proteomics. At the phosphorylation level, the cell undergoes various functional modifications that may affect cell growth proliferation, invasion, angiogenesis, and epithelial–mesenchymal transition (EMT) (Fig. 1b). Altered signaling pathways with subsequent changes in normal cellular mechanisms is a hallmark of cancer cells [12]. Aberrations in kinases and phosphatases have been reported in several tumors, such as gastrointestinal stromal tumors (GISTs) [17, 18], lung cancer [19, 20], hematologic malignancies [21, 22], breast cancer [23, 24], pancreatic cancer [25, 26], and prostate cancer [27]. Numerous variations (about 1000) in the protein kinase expression have already been identified in cancer, which are currently regarded as the potential biomarkers. EGFR for colon and lung cancer, cKIT for GISTs, and human EGFR-related gene (*HER2*) for breast cancer are the most common examples [28, 29]. mTOR is also reported to be activated in renal cancer which promotes angiogenesis and also increases the synthesis of cell cycle proteins by activating VEGF (vascular endothelial growth factor) [30]. Another highly common oncogene Ras is activated in many cancers; however, its activity is very intricate and characterized by several phosphorylation events [31].

1.2   Kinase-Targeted Therapies in Cancer

In chronic myeloid leukemia (CML), most of the patients (95%) have been observed to have a fusion oncogene *BCR-ABL* that is responsible for aberrant activation of the protein-tyrosine kinase. This leads to dysregulation of intracellular signaling with enhanced proliferation and even resistance to apoptosis of hematopoietic stem or progenitor cells, which further leads to proliferation and massive increase in myeloid cell numbers (Fig. 2a) [32]. The comprehensive study of the pathogenetic defects at the molecular level led to the development of imatinib, which inhibits both ABL and BCR-ABL tyrosine kinases. Imatinib specifically inhibits proliferating myeloid cell lines containing BCR-ABL, nonetheless, was minimally harmful to normal cells [33]. The BCR-ABL protein is unique to leukemic cells and expressed at high levels, and its tyrosine kinase activity is essential for its ability to induce leukemia [34]. Imatinib is also effective in the treatment of relapsed/refractory BCR/ABL-positive adult acute lymphocytic leukemia (ALL). 20–40% of these cases have this translocation and complete responses were observed in 60–70% of cases, nevertheless most patients experienced relapse within months of treatment [35]. Imatinib therapy is well tolerated by patients and generally, minimal side effects were observed compared with cytotoxic chemotherapy [36].

In lung cancer pathogenesis, there are a number of cellular pathways observed to be involved, which have advanced to clinical research of relevant investigational drugs. Recently, the involvements of the receptor tyrosine kinases (RTKs) have been studied. The most common example of targeted therapy is the use of inhibitors against EGFR (epidermal growth factor receptor) for the treatment of lung carcinoma [37, 38] (Fig. 2b). The active oncogenic mutations in the *EGFR* have been targeted in lung adenocarcinoma and were observed to show improved treatment outcome [39]. The TCGA study suggested that lung squamous cell carcinoma harbors 7% of cases with *EGFR* amplifications and may respond to the tyrosine kinase inhibitors (TKIs) specific for EGFR such as erlotinib and gefitinib [40]. EGFR is activated through the binding to its ligand, which further leads to the dimerization of the receptor and tyrosine residues phosphorylation of the A-loop of the protein. This leads to as the docking site for several other protein complexes and adaptor molecules that further lead to the activation of several oncogenic signaling cascades [41]. These pathways include the phosphatidylinositol-3-kinase (PI3K)-Akt pathway and the RAS-RAF-MAPK pathway, which are mainly concomitant with cell survival and cell cycle progression respectively. However, despite the effective use of EGFR inhibitor (EGFR-TKI) drugs (such as gefitinib), drug resistance from mutations such as the T790M point mutations are major obstacles in the treatment of non-small cell lung carcinomas. Also, several other alterations such as amplifications in the *MET* oncogenes, overexpression of HGF

A

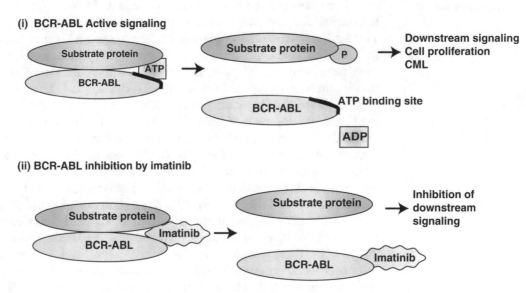

B

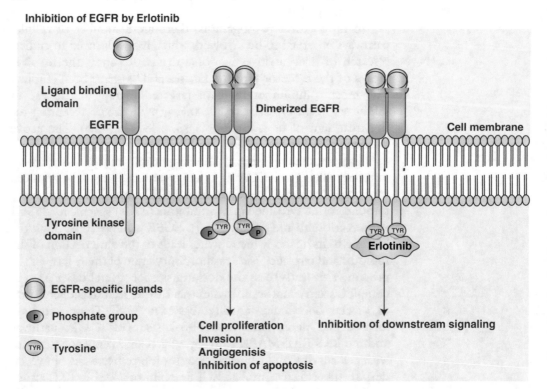

Fig. 2 Kinase inhibitors for targeted therapy. (**a**) (i) Constitutively active signaling due to *BCR-ABL* gene fusion and (ii) mode of action of imatinib targeting the BCR-ABL fusion products. (**b**) Inhibition of EGFR signaling by targeting the receptor kinase by erlotinib

and activation of IGF1R have also been reported to be associated with the acquired resistance to EGFR-TKIs [41, 42]. Owing to the resistance in the first-line drugs (TKIs), several second-line TKIs have been investigated and undergone phase 2 clinical trials such as osimertinib [NCT02769286] [41].

A member of the HER family of receptor tyrosine kinases (EGFR, HER3, and HER4), human epidermal growth factor receptor 2 (HER2) is found to be amplified in breast cancer (20%) and exhibits an aggressive phenotype with poor prognosis [43]. Overexpression of the HER2 oncoprotein is a requisite for the HER2-positive breast cancers and continual activation of its complex downstream signaling pathways. An exceedingly rational therapeutic approach for these cancers is targeting the intracellular tyrosine kinase domain of HER2 with small-molecule tyrosine kinase inhibitors (TKIs) that has been extremely successful in the treatment of other oncogene-driven malignancies. Lapatinib, a highly specific reversible EGFR/HER2 TKI, is approved for the treatment of HER2-positive advanced breast cancer. Lapatinib was approved in 2007 in combination with capecitabine for patients with advanced HER2-positive breast cancer who were previously treated with trastuzumab and chemotherapy [44]. Like lapatinib, neratinib is also another potent EGFR/HER2 kinase inhibitor that binds the ATP pocket and has antitumor effects in cell lines with HER2 overexpression [45].

Several tyrosine kinase inhibitors are undergoing human trials and several are in the pipeline of drug discovery as described in Table 1 [46–49, 52]. The activities of these drugs are restricted to cancers with alterations in kinase targets; hence, broad application of this treatment strategy is challenging. The quick selections of epidemiologically relevant, druggable tyrosine kinase targets need more intervention in the area of high-throughput cancer proteome-based molecular therapeutics. The novel therapeutic approach of anticancer drugs relies on investigation of the phosphorylation pathways and their inhibitors to a great extent. These intensive efforts may cover the silver lining to successful personalized cancer therapy. In future, we expect to have a more precise therapy for cancer with less adverse side effects.

2   Materials

2.1   Cell Lysis and Protein Extraction

Liquid homogenization may be achieved through Dounce homogenizer, Potter-Elvehjem homogenizer, or French press; sonication through an ultrasonicator, freeze-thaw using freezer or dry ice with ethanol; manual grinding with a mortar and a pestle.

Table 1
The list of kinase inhibitors in clinical use or phase trials

Drug	Target receptor	Cancer	Reference
Trastuzumab	HER2	Breast cancer, Pancreatic cancer	Mihaljevic et al. [46]
Pertuzumab	HER2	Breast cancer	Walshe et al. [47]
Lapatinib	HER1/HER2	Breast cancer, HNSCC	Harrington et al. [48]
Canertinib	pan-HER	NSCLC, Ovarian cancer	Janne et al. [49]
Cetuximab	HER1	CRC, HNSCC	Petrelli and Giordano [50]
Bevacizumab	VEGF	CRC, NSCLC	Petrelli and Giordano [50]
Pertuzumab	HER2	Breast cancer, ovarian cancer, NSCLC	Petrelli and Giordano [50]
Gefitinib	HER1	NSCLC	Petrelli and Giordano [50]
Erlotinib	HER1	NSCLC	Petrelli and Giordano [50]
Imatinib	BCR-ABL, c-Kit, PDGFR	CML, GIST	Petrelli and Giordano [50]
Sunitinib	VEGFR, PDGFR, c-Kit, Flt-3	Breast cancer, NSCLC	Petrelli and Giordano [50]
Sorafenib	b-RAF, VEGFR, PDGFR, c-Kit, Flt-3	RCC	Petrelli and Giordano [50]
Lapatinib	HER1, HER2	Breast cancer, NSCLC	Petrelli and Giordano [50]
Idelalisib	PI3Kδ	CLL, FL and SLL	Furman et al. [51]
Vandetanib	HER1, VEGFR, RET	NSCLC, thyroid cancer	Petrelli and Giordano [50]

HNSCC head and neck squamous cell carcinoma, *NSCLC* non-small cell lung carcinoma, *CRC* colorectal cancer, *CML* chronic myeloid leukemia, *GIST* gastrointestinal solid tumors, *RCC* renal cell cancer, *CLL* chronic lymphocytic leukemia, *FL* follicular B cell non-Hodgkin lymphoma, *SLL* and small lymphocytic lymphoma

2.2 Protein Digestion Mostly modified trypsin is used. Some other proteolytic enzymes are also used in isolation or combination such as Glu-C, LysN, Lys-C, Asp-N, or chymotrypsin.

2.3 Phosphopeptide Enrichment TiO_2 beads or/and metal ions—$Fe(3+)$, $Ga(3+)$, $Al(3+)$, $Zr(4+)$, and $Ti(4+)$ or/and anti-Tyr or anti-Thr antibodies.

2.4 Tandem Mass Spectrometry Phosphoproteomic analyses are performed on a mass spectrometer interfaced with a nanoflow liquid chromatography system [53]. Trap column packed and analytical column are used for a linear gradient of acetonitrile and flow of samples into the mass-spectrometer.

2.5 Data Analysis The analysis of the mass spectrometry-acquired data is achieved through the use of various bioinformatics tools such as Proteome Discoverer, motif-X, KinMap, Reactome, KEGG NetPath, NetSlim, WikiPathways, and Cytoscape.

3 Methods

Numerous strategies for sample preparation have been optimized in the recent years. Nevertheless, the basic phosphoproteomics workflow consists of the following steps: protein extraction, protein digestion, phosphopeptide enrichment, mass spectrometry, and data analysis (Fig. 3). Other important modules include stable-isotope encoding of proteins and peptides, high-resolution and multidimensional liquid chromatography (LC), and targeted and discovery-type MS, as well as signal processing algorithms, database searching tools, and downstream biostatistics.

3.1 Cell Lysis and Protein Extraction

Cell lysis is the first step in protein extraction, fractionation, and purification. Many techniques have been developed to obtain the best possible yield and purity for different organisms, sample types (cells or tissue), subcellular structures or specific proteins. Both physical and reagent-based methods may be required to extract high yield of cellular proteins because of the diversity of cell types and cell membrane composition (*see* **Note 2**).

3.2 Protein Digestion

Sequence-specific proteases are utilized to cleave proteins into smaller fragments, or peptides. Trypsin is the most common protease of choice for protein digestion. However, separate or sequential digestion with alternative proteases, such as Glu-C, LysN, Lys-C, Asp-N, or Chymotrypsin, can improve individual protein sequence coverage or generate unique peptide sequences for different MS applications. Digestion of the proteins may be performed in-gel or in-solution (*see* **Note 3**).

3.3 Labeling Techniques

Various labeling methods have enabled researchers to carry out phosphoproteomic analysis and also to monitor dynamic phosphorylation across a number of conditions [15]. The most common labeling strategies used to attain the quantitative measurements include stable isotope labeling by amino acids in cell culture (SILAC; http://www.silac.org) [54–56], isobaric tags for relative and absolute quantitation (iTRAQ) [57] and tandem mass tags (TMT) [58]. However, in case of multiple samples researchers also prefer to perform a label-free quantification (*see* **Note 4**).

3.4 Phosphopeptide Enrichment

The overall low abundance of the reversible phosphorylation process makes it a challenging task to identify the phosphorylated moiety in the complex biological system. Moreover, the overall low abundance of phosphoproteins complicates identification and characterization of this sub-proteome using standard analytical methods. These shortcomings have led to the development of molecular tools to preferentially enrich the phosphoproteome. The enrichment methods are as described below (*see* **Note 5**).

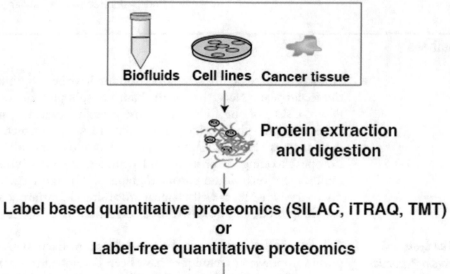

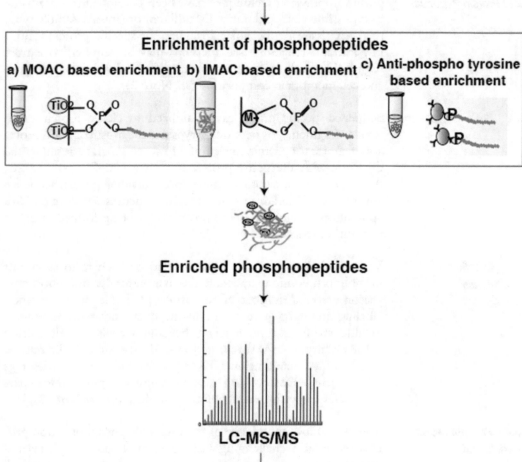

Fig. 3 Workflow for the phosphoproteomics analysis. For sample processing, proteins are extracted and digested using a protease (preferably trypsin). Samples were tagged using the TMT/iTRAQ in quantitative experiment and enriched using the phosphopeptide enrichment protocol (MOAC/IMAC/Anti-pTyr antibody). The samples were run on mass spectrometer for data acquisition

3.4.1 Anti-Phosphotyrosine Antibody-Based Enrichment

The majority of the early enrichment protocol used anti-phosphotyrosine antibodies which were developed using a variety of immunogens [59]. The success of anti-phosphotyrosine antibodies in characterizing tyrosine phosphoproteome in cancers is extended by several mass spectrometry-based studies published in the recent past [60–64]. Rush et al. for the first time reported immunoaffinity-based tyrosine phosphopeptide enrichment and revealed more than 300 distinct tyrosine phosphorylation sites by analyzing protein extracts from human cancer cell lines [65].

3.4.2 Immobilized Metal Affinity Chromatography (IMAC)

IMAC has been the method of choice for phosphopeptide enrichment prior to mass spectrometric analysis for many years and it is still used extensively in many laboratories. Using the affinity of negatively charged phosphate groups towards positively charged metal ions such as $Fe(3+)$, $Ga(3+)$, $Al(3+)$, $Zr(4+)$, and $Ti(4+)$ has made it possible to enrich phosphorylated peptides from protein samples. However, the selectivity of most of the metal ions is limited, when working with highly complex samples, such as whole-cell extracts, results in contamination from nonspecific binding of nonphosphorylated peptides. This problem is majorly caused by highly acidic peptides that also share high binding affinity towards these metal ions [66].

3.4.3 Metal Oxide Affinity Chromatography (MOAC)

The use of MOAC for phosphopeptide enrichment has fully-fledged rapidly in the past few years because of high recoveries and selectivity. Furthermore, the metal oxides are often more stable than traditional silica-based stationary phases [67]. Presently, titanium dioxide (TiO_2) is the most popular metal oxide resin used as a selective affinity support to capture phosphorylated compounds, including peptides [68–70]. At acidic pH, TiO_2 has a positively charged surface that selectively adsorbs phosphorylated species and exhibits outstanding enrichment behavior for phosphopeptides. Notably, when coupled with appropriate solutions for sample loading, column rinsing, and peptide elution, TiO_2 is highly selective to preferentially bind phosphopeptides over acidic peptides [71]. MOAC enrichment is possible with some of the other less preferred metal oxides such as zirconium dioxide (ZrO_2), aluminum hydroxide ($Al(OH)_3$), alumina, and niobium (V) oxide.

Overall, the potential for phosphopeptide enrichment and rapid analysis has improved dramatically over the past years. For the highest protein coverage, future phosphoproteomic techniques will likely employ multiple enrichment techniques along with two-dimensional separations; however, such studies are time consuming.

3.5 Tandem Mass Spectrometry

The processed samples (peptides) are analyzed by mass spectrometry, and presently this approach of shotgun proteomics is the standard choice for researchers, as this technology offers many biological implications [72]. Peptides elude from the reverse phase column at a particular time (retention time) and are ionized and transferred into the gas phase. Further, the selected ions are subjected to tandem mass spectrometry (MS/MS) sequencing to produce fragment ion spectra (MS/MS spectra).

The mass spectrometer scans all peptide ions that were injected into the instrument and archives the MS1 spectrum entailing of mass-to-charge ratios (m/z values) and intensities of all peptide ions. Selected peptide ions ("precursor" ions) detected in the MS1 spectrum were fragmented into fragment ions in the collision cell of the MS instrument. The acquired MS/MS (or MS2) spectrum is a list of m/z values and intensities of all the fragment ions generated by disintegrating an isolated precursor ion. The fragmentation pattern encoded by the MS/MS spectrum allows identification of the amino acid sequence of the peptide that produced it. The fragmentation is most often established through collision induced dissociation (CID). Nonetheless, more than a few alternative mechanisms are also available that are routinely used for specialized presentations and for analyzing the sequence of peptides with PTMs. These fragmentation mechanisms include the electron transfer dissociation (ETD) and higher energy collision dissociation (HCD). Other instruments can be functioned in a mode to acquire automated data-dependent triggering of MS3 acquisition using multistage activation (MSA).

3.6 Data Analysis

After acquiring mass spectrometric data (.raw files), the data was searched against the reference protein sequence database for the identification of proteins from the MS/MS spectra (Fig. 4). For peptide identification, acquired experimental MS/MS spectra is correlated with theoretical spectra predicted for each peptide contained in a protein sequences database (database search approach). It can also be searched against spectra from a spectral library (spectral library searching) or peptide sequences can be extracted directly from the spectra, without referring to a sequence database (de novo sequencing approach) [73]. Furthermore, the following computational analysis of data can be performed.

3.6.1 Motifs Analysis

Determination of motifs from protein PTMs including phosphorylation is required to find out the overrepresented PTMs sites motifs. Identified PTMs from MS/MS spectra from different experiments will provide us the comprehensive dataset of unique PTMs sites. Therefore, in silico prediction of PTMs site motifs and identification of such motifs will enable the researchers to predict several unidentified PTMs. Motif-X algorithm was used for the determination of motifs. It enables the analysis of flanking 12 amino acid residues around the phosphorylation sites against a human protein

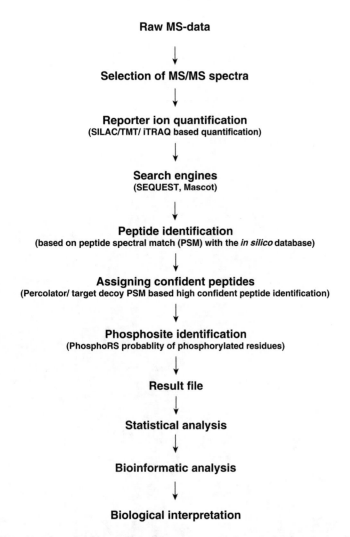

Fig. 4 Workflow for general mass spectrometry-based data analysis. Data acquired from the mass spectrometer were searched against the human reference protein database to identify the peptides. The output can be further analyzed for biological interpretations including pathways and motifs analyses

reference database. Motif-X algorithm selects two datasets; one is the peptide dataset to build motifs, and a background dataset for probability calculations. These two datasets are further converted into position-weight matrices of equal dimensions. Each matrix contains information on the frequency of all residues at the seven positions upstream and downstream of the phosphorylation site [74]. Foreground central residue for phosphorylated tyrosine (pY) was defined. A significance threshold level of 0.000001 and with occurrences of 20 was chosen for the identification of pY site motifs (Fig. 5a) [75].

A

B

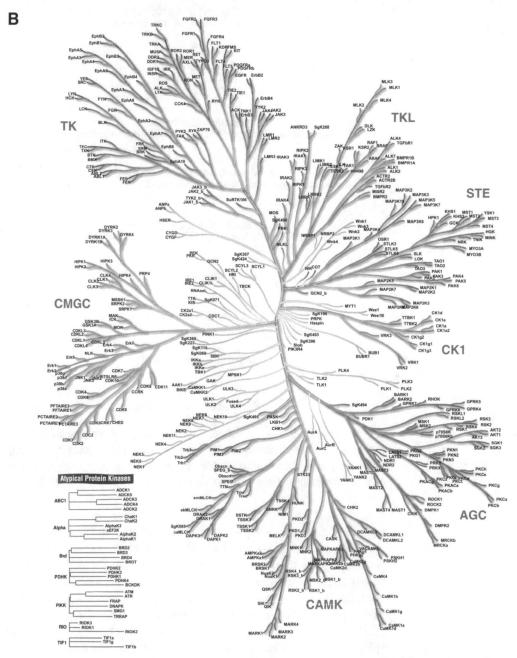

Fig. 5 (**a**) Motifs logos of significantly enriched phosphotyrosine site comprising 12 residues surrounding the phosphorylated tyrosine residue using Motif-X algorithm (Hansen et al.). (**b**) Kinome-dendrogram depicting all family of kinases. The map was built using the KinMap tool. Illustration reproduced courtesy of Cell Signaling Technology, Inc. (www.cellsignal.com)

3.6.2 Enrichment of Kinase Families

An ideal therapeutic target is one that can aid in the elimination of the cancer cells with a high therapeutic index [76]. Kinases have been one of the important therapeutic targets identified in cancer. Functions of several families of kinases have been studied extensively to delineate the biological processes involved in cancer using KinMap [77] (Fig. 5b). In case of a series of diseased condition, including cancer the expression or regulation of phosphorylated proteins is commonly found to be dysregulated [78] (*see* **Note 6**).

3.6.3 Pathway Analysis

Signals that are triggered by ligand do not go through linear pathways but rather through highly interconnected networks. Exploration of such signal transduction networks will benefit our understanding of signaling pathways and provide a systems level view. With the recent developments in high-throughput technologies including proteomics, pathway databases have evolved significantly. There are several human signaling pathway databases including Kyoto Encyclopedia of Genes and Genomes (KEGG), Reactome, NetPath, NetSlim, and WikiPathways. KEGG is a collection of databases dealing with genomes, biological pathways, diseases, drugs, and chemical substances [79]. Reactome is a free, open-source, curated, and peer-reviewed biological pathway database [80]. NetPath and NetSlim open access signaling pathways resources are developed by our group [81, 82]. NetPath provides the reactions annotated from the literature to represent a comprehensive view of individual signaling pathways. NetSlim uses predefined criteria to determine only high-confidence reactions to create a "slim" version of the pathway which can be easily visualized. WikiPathways is a community resource for contributing and maintaining content dedicated to biological pathways [83]. The amount of pathway data generated by researchers which has been stored in these databases has increased exponentially. Therefore, processing and sharing of pathway data in standard data formats is essential for efficient communication of pathway network (*see* **Note 7**). Pathway data in standard formats can be easily used to integrate data available in pathway resources to do computational analyses. The user without a computational knowledge can just download and view these files in visualization software such as Cytoscape [84]. Zhong and colleagues have carried out SILAC-based quantitative phosphoproteomic analysis to identify signaling pathways activated upon stimulation with thymic stromal lymphopoietin (TSLP). In this study, activation of several members of the SRC and TEC family of kinases upon TSLP stimulation were identified. TSLP-induced activation protein phosphatases such as PTPN6 and PTPN11 were described [85]. Furthermore, TSLP signaling pathway was generated using data derived from quantitative phosphoproteomic study (Fig. 6a) [85, 86]. TSLP signaling pathway resource is accessible through NetPath. The pathway data is made

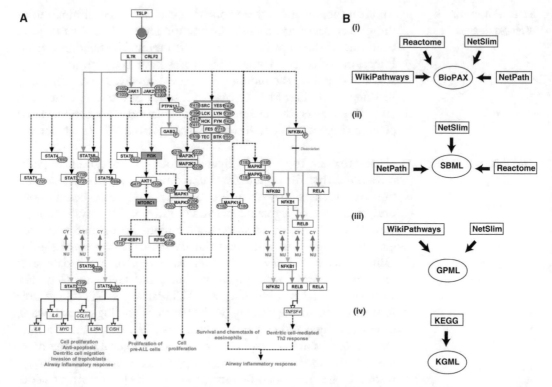

Fig. 6 (**a**) Schematic diagram of thymic stromal lymphopoietin pathway map (http:/www.netpath.org/netslim/ NetSlim_24) (**b**) Depiction of important pathway resources including Reactome, WikiPathways, and NetPath which share data in standard data formats including BioPAX and SBML

available in various standard community data exchange formats such as BioPAX level 3 and SBML version 2.1. TSLP signaling plays an important role in the cell proliferation, inhibition to apoptotic signals and proliferation of pre-ALL cells. IL-33 is another pathway which is available in NetPath [87]. It has been reported that it regulates cell proliferation, tumor invasion and metastasis in colorectal cancer cells through the upregulation of matrix metalloproteinase [88].

4   Notes

1. Global proteomics and phosphoproteomics have proven to provide a comprehensive depiction of the patient and disease stratification. Technological advancements and the reproducible data can build the starting point of several mechanistic studies of the diseased conditions. However, there are some pitfalls in such studies such as lack of robustness of the data and limited coverage of the proteome and phosphoproteome. These could be eliminated through technical standardization and improvisation of bioinformatics approaches. So far the

number of phosphorylated sites identified is approximately 10,000–40,000 [89]. A large number of phosphorylated sites are yet to be captured which may be possible through the implication of more biologically relevant and distinct samples. For example, most of the cell cycle regulating molecules are phosphorylated at various sites through the different stages of cell cycles with maximally phosphorylated proteins in the mitotic stage [90]. One of the other limiting factors in the study of the phosphorylated sites is the lack of knowledge for the biological relevance of the phosphorylations events. Most of the sites which are mapped through the mass-spectrometric studies have not been studied for their functional importance in disease pathogenesis. Most of the phosphorylated sites are known to be poorly conserved throughout species and there is a possibility that the identified sites have no functional inference at all [91]. Large-scale functional studies are required for the implication of the sites in clinical perspective.

2. The first step for a MS-based study is the lysis of cells that would be achieved through liquid homogenization, sonication, freeze–thaw cycles, reagent based or mortar and pestle grinding. Homogenization and grinding are the common types of physical extraction mainly used for the tissue samples. For cell lysates, mostly reagents (such as SDS) are used for efficient extraction.

3. In-solution digestion is mostly recommended for small sample volumes with low to moderate complexity and is more amenable to high-throughput samples. For highly complex samples, in-gel digestion has an inherent advantage because the SDS-PAGE workflow combines protein denaturation with separation, and provides a visual indication of the relative protein abundance in the sample, although peptide recovery is reduced.

4. The labeling technique that was introduced for its application in the cell culture system was through SILAC. It was described as a quantitative proteomic method that was simple, inexpensive, and accurate [92]. SILAC has no chemical manipulation obligation, moreover, the isotopically labeled amino acid and its naturally occurring counterpart has negligible differences chemically [93]. Conversely, the major drawback of this technique is that it cannot be applied for the clinical samples directly. Nevertheless, the successful application of ^{15}N has been useful in eliminating the shortcoming [94]. iTRAQ is another technique which is used at the proteolytic peptide level for four or eight samples that can be quantified simultaneously [95]. Nonetheless, due to the digestion of the proteins, the complexity of the samples increase considerably, making it prerequisite or the samples to undergo an efficient

fractionation step before MS-data acquisition. TMT tags are very frequently used for quantitative proteomics and provides a major advantage as it can be used to tag up to 11 samples (11-plex TMT kits are commercially available) [96]. The heavy isotopes are scattered on carbon or nitrogen atoms and the structurally duplicate TMT tags harbor the equal overall mass. The shortcomings of these tags are that they might impede with a few enrichment techniques for protein modifications such as ubiquitination. In such cases, the enrichment has to be done prior to labeling.

5. It has been described that no single enrichment method can comprehensively cover all phosphorylation events in the cell. The TiO$_2$-based enrichment approach is highly recommended and widely accepted for a wide spectrum of phosphorylation site coverage; however, it may underrepresent the more scarcely phosphorylated residues such as pY. These minimally phosphorylated sites could be captured using the anti-tyrosine antibody-based immunoprecipitation assays. The use of successive or simultaneous enrichment strategies may provide a more comprehensive mapping of the phosphorylated sites.

6. The first report of the involvement of a kinase (v-SRC) in the process of oncogenesis was reported back in the year 1976 which was further revealed as a tyrosine kinase [97–100]. Since then, mutations and other molecular alterations in kinases have been widely studied and many kinase inhibitors are currently in phase trials or used clinically [101, 102]. More than 50% of the aberrantly activated/deactivated protein kinases (due to dysregulation in phosphorylation process) or/and their dysregulated expression may lead to aberrantly activate signaling cascades which further disrupts the regulation of the cell functions and may also give rise to uncontrolled cell growth and proliferation. Hence, systematic and accurate information of protein phosphorylation status and association of the resulting pathophysiological state is immensely important to understand the cellular mechanisms which may lead to the diseased condition. This can be achieved through comparing the changes in the modified status and differences in expression. Moreover, the dysregulation of the otherwise tightly regulated kinases also have been reported to be contributors of oncogenesis [15].

7. Generally used standard data formats include Biological Pathway Exchange (BioPAX) [103], Systems Biology Markup Language (SBML) [104], Proteomics Standards Initiative for Molecular Interaction (PSI-MI) [105], GenMAPP Pathway Markup Language (GPML) [106], and KEGG Markup Language (KGML).

(a) **BioPAX**: It is defined in OWL DL and is represented in the RDF/XML format. BioPAX Level 3 Ontology covers metabolic pathways, molecular interactions, signaling pathways (including molecular states and generics), gene regulation, and genetic interactions. All major databases including Reactome and NetPath provide pathway data in BioPAX level 3 (Fig. 6b (i)).

(b) **SBML**: It is a machine-readable format for representing models of metabolism and cell signaling. It is focused on allowing models to be encoded using XML. SBML models can contain information about sizes, amounts, and kinetics that cannot be expressed with BioPAX. Reactome, NetPath, and NetSlim databases contain data in SBML format (Fig. 6b (ii)).

(c) **PSI-MI**: It is a common data exchange format for an efficient exchange of protein interaction data. The data can be regularly exchanged to create a network of synchronized protein interaction databases including Human Protein Reference Database (HPRD) [107] and NetPath. The PSI-MI format is being developed using a multilevel approach similar to that used by the SBML.

(d) **GPML**: It is an XML-based language, designed to define a pathway consisting of purely graphical elements (lines and shapes) with added biological information (genes, proteins, and data nodes). WikiPathways and NetSlim provide data in GPML format (Fig. 6b (iii)).

(e) **KGML**: It is an exchange format of the KEGG pathway maps, which is converted from internally used KGML+ (KGML + SVG) format. KGML enables automatic drawing of KEGG pathways and provides facilities for computational analysis and modeling of gene/protein networks and chemical networks. KEGG provide data in KGML format (Fig. 6b (iv)).

Acknowledgments

PK is a recipient of the Ramanujan Fellowship awarded by Department of Science and Technology (DST), Government of India. BD is a recipient of INSPIRE Fellowship from Department of Science and Technology (DST), Government of India. IAG is a recipient of junior research fellowship from Council of Scientific and Industrial Research (CSIR), Government of India. JS is a recipient of Bio-CARe Women Scientists award conferred by Department of Biotechnology (DBT), Government of India.

References

1. Duan G, Walther D (2015) The roles of post-translational modifications in the context of protein interaction networks. PLoS Comput Biol 11:e1004049

2. Beltrao P, Bork P, Krogan NJ, Van Noort V (2013) Evolution and functional cross-talk of protein post-translational modifications. Mol Syst Biol 9:714

3. Gavin AC, Bosche M, Krause R, Grandi P, Marzioch M, Bauer A et al (2002) Functional organization of the yeast proteome by systematic analysis of protein complexes. Nature 415:141–147

4. Kumar GK, Prabhakar NR (2008) Post-translational modification of proteins during intermittent hypoxia. Respir Physiol Neurobiol 164:272–276

5. Glavey SV, Huynh D, Reagan MR, Manier S, Moschetta M, Kawano Y et al (2015) The cancer glycome: carbohydrates as mediators of metastasis. Blood Rev 29:269–279

6. Gallo LH, Ko J, Donoghue DJ (2017) The importance of regulatory ubiquitination in cancer and metastasis. Cell Cycle 16:634–648

7. Singh V, Ram M, Kumar R, Prasad R, Roy BK, Singh KK (2017) Phosphorylation: implications in cancer. Protein J 36:1–6

8. Ardito F, Giuliani M, Perrone D, Troiano G, Lo Muzio L (2017) The crucial role of protein phosphorylation in cell signaling and its use as targeted therapy (Review). Int J Mol Med 40:271–280

9. Nghiem HO, Bettendorff L, Changeux JP (2000) Specific phosphorylation of Torpedo 43K rapsyn by endogenous kinase(s) with thiamine triphosphate as the phosphate donor. FASEB J 14:543–554

10. Lemeer S, Heck AJ (2009) The phosphoproteomics data explosion. Curr Opin Chem Biol 13:414–420

11. Ubersax JA, Ferrell JE Jr (2007) Mechanisms of specificity in protein phosphorylation. Nat Rev Mol Cell Biol 8:530–541

12. Hanahan D, Weinberg RA (2000) The hallmarks of cancer. Cell 100:57–70

13. Aebersold R, Mann M (2016) Mass-spectrometric exploration of proteome structure and function. Nature 537:347–355

14. Von Stechow L, Francavilla C, Olsen JV (2015) Recent findings and technological advances in phosphoproteomics for cells and tissues. Expert Rev Proteomics 12:469–487

15. Harsha HC, Pandey A (2010) Phosphoproteomics in cancer. Mol Oncol 4:482–495

16. Lim YP (2005) Mining the tumor phosphoproteome for cancer markers. Clin Cancer Res 11:3163–3169

17. Corless CL, Fletcher JA, Heinrich MC (2004) Biology of gastrointestinal stromal tumors. J Clin Oncol 22:3813–3825

18. Javidi-Sharifi N, Traer E, Martinez J, Gupta A, Taguchi T, Dunlap J et al (2015) Crosstalk between KIT and FGFR3 promotes gastrointestinal stromal tumor cell growth and drug resistance. Cancer Res 75:880–891

19. Rikova K, Guo A, Zeng Q, Possemato A, Yu J, Haack H et al (2007) Global survey of phosphotyrosine signaling identifies oncogenic kinases in lung cancer. Cell 131:1190–1203

20. Sharma SV, Bell DW, Settleman J, Haber DA (2007) Epidermal growth factor receptor mutations in lung cancer. Nat Rev Cancer 7:169–181

21. Zhu N, Xiao H, Wang LM, Fu S, Zhao C, Huang H (2015) Mutations in tyrosine kinase and tyrosine phosphatase and their relevance to the target therapy in hematologic malignancies. Future Oncol 11:659–673

22. Kraus J, Kraus M, Liu N, Besse L, Bader J, Geurink PP et al (2015) The novel beta2-selective proteasome inhibitor LU-102 decreases phosphorylation of I kappa B and induces highly synergistic cytotoxicity in combination with ibrutinib in multiple myeloma cells. Cancer Chemother Pharmacol 76:383–396

23. Jagarlamudi KK, Hansson LO, Eriksson S (2015) Breast and prostate cancer patients differ significantly in their serum Thymidine kinase 1 (TK1) specific activities compared with those hematological malignancies and blood donors: implications of using serum TK1 as a biomarker. BMC Cancer 15:66

24. Hou S, Isaji T, Hang Q, Im S, Fukuda T, Gu J (2016) Distinct effects of beta1 integrin on cell proliferation and cellular signaling in MDA-MB-231 breast cancer cells. Sci Rep 6:18430

25. Paladino D, Yue P, Furuya H, Acoba J, Rosser CJ, Turkson J (2016) A novel nuclear Src and p300 signaling axis controls migratory and invasive behavior in pancreatic cancer. Oncotarget 7:7253–7267

26. Li Z, Lin P, Gao C, Peng C, Liu S, Gao H et al (2016) Integrin beta6 acts as an unfavorable prognostic indicator and promotes cellular malignant behaviors via ERK-ETS1 pathway in pancreatic ductal adenocarcinoma (PDAC). Tumour Biol 37:5117–5131

27. Mehraein-Ghomi F, Church DR, Schreiber CL, Weichmann AM, Basu HS, Wilding G (2015) Inhibitor of p52 NF-kappaB subunit and androgen receptor (AR) interaction reduces growth of human prostate cancer cells by abrogating nuclear translocation of p52 and phosphorylated AR(ser81). Genes Cancer 6:428–444

28. Da Cunha Santos G, Shepherd FA, Tsao MS (2011) EGFR mutations and lung cancer. Annu Rev Pathol 6:49–69

29. Ludwig JA, Weinstein JN (2005) Biomarkers in cancer staging, prognosis and treatment selection. Nat Rev Cancer 5:845–856

30. Dancey JE (2006) Therapeutic targets: MTOR and related pathways. Cancer Biol Ther 5:1065–1073

31. Ahmadian MR (2002) Prospects for anti-ras drugs. Br J Haematol 116:511–518

32. Arora A, Scholar EM (2005) Role of tyrosine kinase inhibitors in cancer therapy. J Pharmacol Exp Ther 315:971–979

33. Druker BJ, Tamura S, Buchdunger E, Ohno S, Segal GM, Fanning S et al (1996) Effects of a selective inhibitor of the Abl tyrosine kinase on the growth of Bcr-Abl positive cells. Nat Med 2:561–566

34. Savage DG, Antman KH (2002) Imatinib mesylate--a new oral targeted therapy. N Engl J Med 346:683–693

35. Druker BJ, Sawyers CL, Kantarjian H, Resta DJ, Reese SF, Ford JM et al (2001) Activity of a specific inhibitor of the BCR-ABL tyrosine kinase in the blast crisis of chronic myeloid leukemia and acute lymphoblastic leukemia with the Philadelphia chromosome. N Engl J Med 344:1038–1042

36. Kantarjian H, Sawyers C, Hochhaus A, Guilhot F, Schiffer C, Gambacorti-Passerini C et al (2002) Hematologic and cytogenetic responses to imatinib mesylate in chronic myelogenous leukemia. N Engl J Med 346:645–652

37. Antonicelli A, Cafarotti S, Indini A, Galli A, Russo A, Cesario A et al (2013) EGFR-targeted therapy for non-small cell lung cancer: focus on EGFR oncogenic mutation. Int J Med Sci 10:320–330

38. Paez JG, Janne PA, Lee JC, Tracy S, Greulich H, Gabriel S et al (2004) EGFR mutations in lung cancer: correlation with clinical response to gefitinib therapy. Science 304:1497–1500

39. Yamamoto H, Toyooka S, Mitsudomi T (2009) Impact of EGFR mutation analysis in non-small cell lung cancer. Lung Cancer 63:315–321

40. Cancer Genome Atlas Research Network (2012) Comprehensive genomic characterization of squamous cell lung cancers. Nature 489:519–525

41. Suda K, Onozato R, Yatabe Y, Mitsudomi T (2009) EGFR T790M mutation: a double role in lung cancer cell survival? J Thorac Oncol 4:1–4

42. Pao W, Miller VA, Politi KA, Riely GJ, Somwar R, Zakowski MF et al (2005) Acquired resistance of lung adenocarcinomas to gefitinib or erlotinib is associated with a second mutation in the EGFR kinase domain. PLoS Med 2:e73

43. Gajria D, Chandarlapaty S (2011) HER2-amplified breast cancer: mechanisms of trastuzumab resistance and novel targeted therapies. Expert Rev Anticancer Ther 11:263–275

44. Chien AJ, Rugo HS (2017) Tyrosine kinase inhibitors for human epidermal growth factor receptor 2-positive metastatic breast cancer: is personalizing therapy within reach? J Clin Oncol 35:3089–3091

45. Wong KK, Fracasso PM, Bukowski RM, Lynch TJ, Munster PN, Shapiro GI et al (2009) A phase I study with neratinib (HKI-272), an irreversible pan ErbB receptor tyrosine kinase inhibitor, in patients with solid tumors. Clin Cancer Res 15:2552–2558

46. Mihaljevic A, Buchler P, Harder J, Hofheinz R, Gregor M, Kanzler S et al (2009) A prospective, non-randomized phase II trial of Trastuzumab and Capecitabine in patients with HER2 expressing metastasized pancreatic cancer. BMC Surg 9:1

47. Walshe JM, Denduluri N, Berman AW, Rosing DR, Swain SM (2006) A phase II trial with trastuzumab and pertuzumab in patients with HER2-overexpressed locally advanced and metastatic breast cancer. Clin Breast Cancer 6:535–539

48. Harrington KJ, El-Hariry IA, Holford CS, Lusinchi A, Nutting CM, Rosine D et al (2009) Phase I study of lapatinib in combination with chemoradiation in patients with locally advanced squamous cell carcinoma of the head and neck. J Clin Oncol 27:1100–1107

49. Janne PA, Von Pawel J, Cohen RB, Crino L, Butts CA, Olson SS et al (2007) Multicenter, randomized, phase II trial of CI-1033, an irreversible pan-ERBB inhibitor, for previously treated advanced non small-cell lung cancer. J Clin Oncol 25:3936–3944

50. Petrelli A, Giordano S (2008) From single- to multi-target drugs in cancer therapy: when

aspecificity becomes an advantage. Curr Med Chem. 15(5):422–432

51. Furman RR, Sharman JP, Coutre SE, Cheson BD, Pagel JM, Hillmen P, Barrientos JC, Zelenetz AD, Kipps TJ, Flinn I, Ghia P, Eradat H, Ervin T, Lamanna N, Coiffier B, Pettitt AR, Ma S, Stilgenbauer S, Cramer P, Aiello M, Johnson DM, Miller LL, Li D, Jahn TM, Dansey RD, Hallek M, O'Brien SM (2014) Idelalisib and rituximab in relapsed chronic lymphocytic leukemia. N Engl J Med. 370 (11):997–1007. https://doi.org/10.1056/NEJMoa1315226

52. Talevi A (2015) Multi-target pharmacology: possibilities and limitations of the "skeleton key approach" from a medicinal chemist perspective. Front Pharmacol 6:205

53. Verma R, Pinto SM, Patil AH, Advani J, Subba P, Kumar M et al (2017) Quantitative Proteomic and Phosphoproteomic Analysis of H37Ra and H37Rv Strains of Mycobacterium tuberculosis. J Proteome Res 16:1632–1645

54. Amanchy R, Zhong J, Hong R, Kim JH, Gucek M, Cole RN et al (2009) Identification of c-Src tyrosine kinase substrates in platelet-derived growth factor receptor signaling. Mol Oncol 3:439–450

55. Amanchy R, Zhong J, Molina H, Chaerkady R, Iwahori A, Kalume DE et al (2008) Identification of c-Src tyrosine kinase substrates using mass spectrometry and peptide microarrays. J Proteome Res 7:3900–3910

56. Harsha HC, Molina H, Pandey A (2008) Quantitative proteomics using stable isotope labeling with amino acids in cell culture. Nat Protoc 3:505–516

57. Ross PL, Huang YN, Marchese JN, Williamson B, Parker K, Hattan S et al (2004) Multiplexed protein quantitation in Saccharomyces cerevisiae using amine-reactive isobaric tagging reagents. Mol Cell Proteomics 3:1154–1169

58. Thompson A, Schafer J, Kuhn K, Kienle S, Schwarz J, Schmidt G et al (2003) Tandem mass tags: a novel quantification strategy for comparative analysis of complex protein mixtures by MS/MS. Anal Chem 75:1895–1904

59. Frackelton AR Jr, Posner M, Kannan B, Mermelstein F (1991) Generation of monoclonal antibodies against phosphotyrosine and their use for affinity purification of phosphotyrosine-containing proteins. Methods Enzymol 201:79–92

60. Sathe G, Pinto SM, Syed N, Nanjappa V, Solanki HS, Renuse S et al (2016) Phosphotyrosine profiling of curcumin-induced signaling. Clin Proteomics 13:13

61. Wu X, Zahari MS, Ma B, Liu R, Renuse S, Sahasrabuddhe NA et al (2015) Global phosphotyrosine survey in triple-negative breast cancer reveals activation of multiple tyrosine kinase signaling pathways. Oncotarget 6:29143–29160

62. Syed N, Barbhuiya MA, Pinto SM, Nirujogi RS, Renuse S, Datta KK et al (2015) Phosphotyrosine profiling identifies ephrin receptor A2 as a potential therapeutic target in esophageal squamous-cell carcinoma. Proteomics 15:374–382

63. Luo W, Slebos RJ, Hill S, Li M, Brabek J, Amanchy R et al (2008) Global impact of oncogenic Src on a phosphotyrosine proteome. J Proteome Res 7:3447–3460

64. Ibarrola N, Kratchmarova I, Nakajima D, Schiemann WP, Moustakas A, Pandey A et al (2004) Cloning of a novel signaling molecule, AMSH-2, that potentiates transforming growth factor beta signaling. BMC Cell Biol 5:2

65. Rush J, Moritz A, Lee KA, Guo A, Goss VL, Spek EJ et al (2005) Immunoaffinity profiling of tyrosine phosphorylation in cancer cells. Nat Biotechnol 23:94–101

66. Thingholm TE, Larsen MR (2016) Phosphopeptide enrichment by immobilized metal affinity chromatography. Methods Mol Biol 1355:123–133

67. Dunn JD, Reid GE, Bruening ML (2010) Techniques for phosphopeptide enrichment prior to analysis by mass spectrometry. Mass Spectrom Rev 29:29–54

68. Selvan LDN, Danda R, Madugundu AK, Puttamallesh VN, Sathe GJ, Krishnan UM et al (2018) Phosphoproteomics of retinoblastoma: a pilot study identifies aberrant kinases. Molecules 23:E1454

69. Barua P, Lande NV, Subba P, Gayen D, Pinto S, Keshava Prasad TS et al (2019) Dehydration-responsive nuclear proteome landscape of chickpea (Cicer arietinum L.) reveals phosphorylation-mediated regulation of stress response. Plant Cell Environ 42:230

70. Gowthami N, Sunitha B, Kumar M, Keshava Prasad TS, Gayathri N, Padmanabhan B et al (2019) Mapping the protein phosphorylation sites in human mitochondrial complex I (NADH: Ubiquinone oxidoreductase): a bioinformatics study with implications for brain aging and neurodegeneration. J Chem Neuroanat 95:13

71. Sugiyama N, Masuda T, Shinoda K, Nakamura A, Tomita M, Ishihama Y (2007)

Phosphopeptide enrichment by aliphatic hydroxy acid-modified metal oxide chromatography for nano-LC-MS/MS in proteomics applications. Mol Cell Proteomics 6:1103–1109

72. Steen H, Mann M (2004) The ABC's (and XYZ's) of peptide sequencing. Nat Rev Mol Cell Biol 5:699–711

73. Nesvizhskii AI (2010) A survey of computational methods and error rate estimation procedures for peptide and protein identification in shotgun proteomics. J Proteomics 73:2092–2123

74. Chou MF, Schwartz D (2011) Biological sequence motif discovery using motif-x. Curr Protoc Bioinformatics Chapter 13:Unit 13:15–24

75. Hansen AM, Chaerkady R, Sharma J, Diaz-Mejia JJ, Tyagi N, Renuse S et al (2013) The Escherichia coli phosphotyrosine proteome relates to core pathways and virulence. PLoS Pathog 9:e1003403

76. Tan WL, Jain A, Takano A, Newell EW, Iyer NG, Lim WT et al (2016) Novel therapeutic targets on the horizon for lung cancer. Lancet Oncol 17:e347–e362

77. Eid S, Turk S, Volkamer A, Rippmann F, Fulle S (2017) KinMap: a web-based tool for interactive navigation through human kinome data. BMC Bioinformatics 18:16

78. Feng X, Lu X, Man X, Zhou W, Jiang LQ, Knyazev P et al (2009) Overexpression of Csk-binding protein contributes to renal cell carcinogenesis. Oncogene 28:3320–3331

79. Kanehisa M, Goto S (2000) KEGG: kyoto encyclopedia of genes and genomes. Nucleic Acids Res 28:27–30

80. Fabregat A, Sidiropoulos K, Garapati P, Gillespie M, Hausmann K, Haw R et al (2016) The Reactome pathway Knowledgebase. Nucleic Acids Res 44:D481–D487

81. Kandasamy K, Mohan SS, Raju R, Keerthikumar S, Kumar GS, Venugopal AK et al (2010) NetPath: a public resource of curated signal transduction pathways. Genome Biol 11:R3

82. Raju R, Nanjappa V, Balakrishnan L, Radhakrishnan A, Thomas JK, Sharma J et al (2011) NetSlim: high-confidence curated signaling maps. Database (Oxford) 2011:bar032

83. Kutmon M, Riutta A, Nunes N, Hanspers K, Willighagen EL, Bohler A et al (2016) WikiPathways: capturing the full diversity of pathway knowledge. Nucleic Acids Res 44: D488–D494

84. Shannon P, Markiel A, Ozier O, Baliga NS, Wang JT, Ramage D et al (2003) Cytoscape: a software environment for integrated models of biomolecular interaction networks. Genome Res 13:2498–2504

85. Zhong J, Kim MS, Chaerkady R, Wu X, Huang TC, Getnet D et al (2012) TSLP signaling network revealed by SILAC-based phosphoproteomics. Mol Cell Proteomics 11:M112 017764

86. Zhong J, Sharma J, Raju R, Palapetta SM, Prasad TS, Huang TC et al (2014) TSLP signaling pathway map: a platform for analysis of TSLP-mediated signaling. Database (Oxford) 2014:bau007

87. Pinto SM, Subbannayya Y, Rex DB, Raju R, Chatterjee O, Advani J et al (2018) A network map of IL-33 signaling pathway. J Cell Commun Signal 12:615–624

88. Liu X, Zhu L, Lu X, Bian H, Wu X, Yang W et al (2014) IL-33/ST2 pathway contributes to metastasis of human colorectal cancer. Biochem Biophys Res Commun 453:486–492

89. Sharma K, D'souza RC, Tyanova S, Schaab C, Wisniewski JR, Cox J et al (2014) Ultradeep human phosphoproteome reveals a distinct regulatory nature of Tyr and Ser/Thr-based signaling. Cell Rep 8:1583–1594

90. Tyanova S, Cox J, Olsen J, Mann M, Frishman D (2013) Phosphorylation variation during the cell cycle scales with structural propensities of proteins. PLoS Comput Biol 9:e1002842

91. Beltrao P, Trinidad JC, Fiedler D, Roguev A, Lim WA, Shokat KM et al (2009) Evolution of phosphoregulation: comparison of phosphorylation patterns across yeast species. PLoS Biol 7:e1000134

92. Ong SE, Blagoev B, Kratchmarova I, Kristensen DB, Steen H, Pandey A et al (2002) Stable isotope labeling by amino acids in cell culture, SILAC, as a simple and accurate approach to expression proteomics. Mol Cell Proteomics 1:376–386

93. Amanchy R, Kalume DE, Pandey A (2005) Stable isotope labeling with amino acids in cell culture (SILAC) for studying dynamics of protein abundance and posttranslational modifications. Sci STKE 2005:pl2

94. Wu CC, Maccoss MJ, Howell KE, Matthews DE, Yates JR III (2004) Metabolic labeling of mammalian organisms with stable isotopes for quantitative proteomic analysis. Anal Chem 76:4951–4959

95. Pierce A, Unwin RD, Evans CA, Griffiths S, Carney L, Zhang L et al (2008) Eight-channel iTRAQ enables comparison of the activity of six leukemogenic tyrosine kinases. Mol Cell Proteomics 7:853–863

96. Stepanova E, Gygi SP, Paulo JA (2018) Filter-based protein digestion (FPD): a detergent-free and scaffold-based strategy for TMT workflows. J Proteome Res 17:1227–1234

97. Stehelin D, Guntaka RV, Varmus HE, Bishop JM (1976) Purification of DNA complementary to nucleotide sequences required for neoplastic transformation of fibroblasts by avian sarcoma viruses. J Mol Biol 101:349–365

98. Spector DH, Varmus HE, Bishop JM (1978) Nucleotide sequences related to the transforming gene of avian sarcoma virus are present in DNA of uninfected vertebrates. Proc Natl Acad Sci U S A 75:4102–4106

99. Brugge JS, Erikson RL (1977) Identification of a transformation-specific antigen induced by an avian sarcoma virus. Nature 269:346–348

100. Hunter T, Sefton BM (1980) Transforming gene product of Rous sarcoma virus phosphorylates tyrosine. Proc Natl Acad Sci U S A 77:1311–1315

101. Wu P, Nielsen TE, Clausen MH (2015) FDA-approved small-molecule kinase inhibitors. Trends Pharmacol Sci 36:422–439

102. Fabbro D, Cowan-Jacob SW, Moebitz H (2015) Ten things you should know about protein kinases: IUPHAR review 14. Br J Pharmacol 172:2675–2700

103. Demir E, Cary MP, Paley S, Fukuda K, Lemer C, Vastrik I et al (2010) The BioPAX community standard for pathway data sharing. Nat Biotechnol 28:935–942

104. Hucka M, Finney A, Sauro HM, Bolouri H, Doyle JC, Kitano H et al (2003) The systems biology markup language (SBML): a medium for representation and exchange of biochemical network models. Bioinformatics 19:524–531

105. Hermjakob H, Montecchi-Palazzi L, Bader G, Wojcik J, Salwinski L, Ceol A et al (2004) The HUPO PSI's molecular interaction format--a community standard for the representation of protein interaction data. Nat Biotechnol 22:177–183

106. Van Iersel MP, Kelder T, Pico AR, Hanspers K, Coort S, Conklin BR et al (2008) Presenting and exploring biological pathways with PathVisio. BMC Bioinformatics 9:399

107. Keshava Prasad TS, Goel R, Kandasamy K, Keerthikumar S, Kumar S, Mathivanan S et al (2009) Human protein reference database--2009 update. Nucleic Acids Res 37: D767–D772

Chapter 11

Mass Spectrometry-Based Characterization of Ub- and UbL-Modified Proteins

Nagore Elu, Benoit Lectez, Juanma Ramirez, Nerea Osinalde, and Ugo Mayor

Abstract

Regulation by ubiquitin (Ub) and ubiquitin-like (UbL) modifiers can confer their substrate proteins a myriad of assignments, such as inducing protein–protein interactions, the internalization of membrane proteins, or their degradation via the proteasome. The underlying code regulating those diverse endpoints appears to be based on the topology of the ubiquitin chains formed.

Experimental characterization of the specific regulation mediated by Ub and UbLs is not trivial. The substoichiometric levels of Ub- and UbL-modified proteins greatly limit their analytical detection in a background of more abundant proteins. Therefore, modified proteins or peptides must be enriched prior to any downstream detection analysis. For that purpose, we recently developed a GFP-tag based isolation strategy. Here we illustrate the usefulness of combining GFP-tag isolation strategy with mass spectrometry (MS) to identify Ub- and UbL-modified residues within the GFP-tagged protein, as well as to uncover the types of Ub and UbL chains formed.

Key words Ubiquitination, Ubiquitin-like, GFP pull down, PTM, Ubiquitin chains

1 Introduction

Ubiquitin (Ub) and ubiquitin-like (UbL) modifiers are a family of highly conserved proteins that share similarity, not only in structure but also in the mechanism of action. Briefly, covalent attachment of Ub and UbL to a target protein is achieved by the sequential intervention of three distinct enzymes: activating E1 enzymes, conjugating E2 enzymes, and E3 ligases (Fig. 1) [1].

Regulation by Ub and UbL modifiers can confer their substrate proteins a myriad of assignments [2], including the promotion of protein–protein interaction, internalization of membrane proteins [3], regulation of transcriptional activity [4], or their degradation via the proteasome [5, 6]. The code regulating those diverse endpoints is based on the topology of the ubiquitin chains that are formed. Ubiquitin is mostly attached covalently to amide groups at

Rune Matthiesen (ed.), *Mass Spectrometry Data Analysis in Proteomics*, Methods in Molecular Biology, vol. 2051, https://doi.org/10.1007/978-1-4939-9744-2_11, © Springer Science+Business Media, LLC, part of Springer Nature 2020

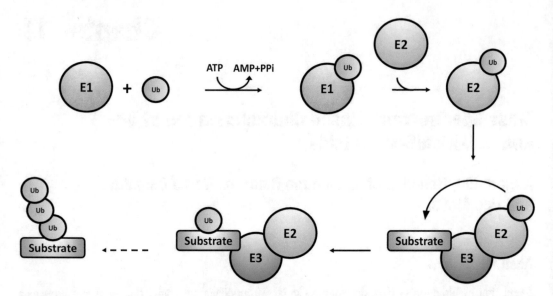

Fig. 1 Ubiquitination pathway by the sequential action of E1 activating enzymes, E2 conjugating enzymes, and E3 ligases. E1 enzymes activate ubiquitin upon ATP consumption, and it is then transferred to the E2 conjugating enzymes. In the last step, E3 ligases transfer the ubiquitin to their substrates (monoubiquitination). Addition of more than one ubiquitin to the substrates results in polyubiquitination. These processes regulate different cellular functions such as protein degradation, signaling or changes in localization

lysine residues on its substrates [5]. Ubiquitination may occur once, resulting in monoubiquitination [7], or simultaneously in several Lys within a protein, leading to polymonoubiquitination [8]. Additionally, as ubiquitin itself has eight available amide groups (seven lysines and its N-terminal amide group), ubiquitin molecules can associate together and form different polyubiquitin chains (M1, K6, K11, K27, K29, K33, K48, K63) [1]. The lysines used throughout the polyubiquitin chain may be the same (homogeneous chains), may be alternated (mixed chains) or even multiple lysines of the same ubiquitin may be modified at the same time (branched chains); consequently, a huge number of distinct conformations can be formed [2]. Those different topological conformations are differently recognized by a number of receptors, conferring the above mentioned variability of destinies to their substrates. For instance, it is well documented that K48 and K11 polyubiquitin chains target proteins to degradation via the proteasome [9, 10], whereas K63 chains have been mainly related to nondegradative roles [11].

Experimental characterization of the specific regulation by Ub and UbLs is not trivial. In most cases, the percentage of molecules modified by those PTM is—for a given protein—in the range of 1–5%. At the same time, in a typical cell extract, several thousand proteins might be ubiquitinated, making it completely impossible to assign the signatures of ubiquitin chain types to a given substrate. Therefore, in order to characterize the PTM signature of a protein of interest, it must be efficiently enriched prior to any downstream detection analysis [12]. For this purpose, we have recently developed a GFP-tag based isolation strategy [13], which can be used either with overexpression or genetic replacement strategies. Here we describe how MS analysis can be used to characterize the Ub- and UbL-modified fraction of those GFP-tagged proteins. Not only for the identification of the ubiquitination sites, but also to delineate the ubiquitin chain types that are formed. It should be highlighted that the procedure presented here can also be combined with any quantitative proteomics approach, including label free strategy, in order to infer the influence a certain ubiquitin ligase or deubiquitinase (DUB) has on the ubiquitination state of the protein of interest. More precisely, by analyzing mass spectrometry samples from control cells and cells in which a certain ubiquitin ligase or DUB is differentially expressed, it is plausible to detect the exact ubiquitination sites and type of ubiquitin chains that are modulated by those enzymes.

1.1 The GFP-Pull Down Strategy

A number of pull down approaches have been developed to validate protein ubiquitination in vivo. However, most of these protocols are carried out using native conditions, which result in the copurification of undesirable interacting proteins that may interfere in the subsequent detection of the ubiquitin signal. Our GFP-tag isolation strategy is based on beads coated with purified recombinant anti-GFP antibody V_HH fragment (Chromotek GmbH, Planegg-Martinsried, Germany) that can withstand very stringent washing conditions [13], therefore, it avoids the copurification of interactors as well as other unspecific proteins. Ubiquitin is bound to its substrates by an isopeptide bond that is kept intact on denaturing environments. Consequently, GFP-tagged proteins, including their ubiquitinated fraction, can be successfully isolated while all noncovalently bound interactors are discarded.

During this protocol, cell cultures are cotransfected with the protein of interest fused to GFP and an epitope-tagged version of ubiquitin (FLAG-ubiquitin) (Fig. 2). In order to isolate the GFP-tagged protein, lysates are incubated with anti-GFP beads in a nondenaturing buffer, supplemented with protease and deubiquitinase inhibitors to prevent protein degradation and deubiquitination, respectively. Beads are then stringently washed to remove nonspecifically bound proteins, and afterward, GFP-tagged protein is eluted from the beads by boiling in Laemmli buffer. Once the GFP-tagged protein is isolated, its nonmodified form can be detected by Western blotting using an antibody against GFP, while its ubiquitinated fraction can be monitored using antibodies against ubiquitin or FLAG. Additionally, the same sample can be analyzed to infer the ubiquitination site(s) [14], as well as chain types [15] in case polyubiquitination occurs.

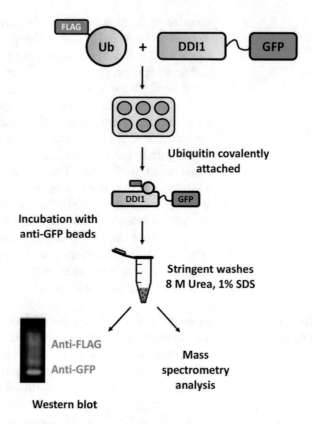

Fig. 2 The GFP-pull down strategy for the characterization of covalent modifications. Schematic illustration of the GFP-pull down protocol for analyzing protein ubiquitination. GFP-tagged proteins and FLAG-tagged ubiquitin are cotransfected into cell cultures. Lysates are then incubated with anti-GFP beads and subjected to stringent washes with 8 M Urea and 1% SDS to discard all interacting proteins. The ubiquitination of purified GFP-tagged protein can be validated by Western blot, and later analyzed by MS to detect ubiquitination sites and ubiquitination chain types

1.2 Mass Spectrometry-Based Characterization of Isolated Ub- and UbL-Modified Proteins

Mass spectrometry (MS) is a powerful tool to identify PTMs, including Ub and UbL, with high confidence from complex proteome mixtures. Nevertheless, reducing the complexity of the sample greatly improves the chance of success. Hence, in order to focus the MS analysis in the Ub- and UbL-modified fraction of the isolated GFP-tagged protein, the nonmodified fraction must be discarded. Therefore, the first step consists on fractionating the protein of interest, which is accomplished by running the eluted GFP-tagged protein by SDS-PAGE (Fig. 3a). Protein bands can be visualized by staining the gel with Blue Coomassie. The main band will correspond to the nonmodified fraction of the GFP-tagged protein, while the bands above are the monoubiquitinated and polyubiquitinated versions of the same protein. Thus, the gel bands corresponding to the modified version of the

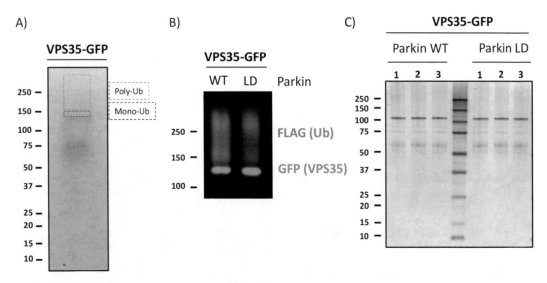

Fig. 3 SDS-PAGE, Western blot, and silver blot analysis prior to MS analysis. MS analysis is used to identify the ubiquitination sites and type of chains formed in VPS35 mediated by its E3 ligase Parkin. (**a**) Eluted GFP-tagged protein (VPS35-GFP) is run on an acrylamide gel and stained with Blue Coomassie. This allows for the identification of unmodified, monoubiquitinated and polyubiquitinated bands of the GFP-tagged protein. Monoubiquitinated and polyubiquitinated bands are used for MS analysis, excluding the unmodified band of the GFP-tagged protein. (**b**) Western blot against anti-FLAG and anti-GFP is used to detect the extension of the ubiquitinated fraction of VPS35 and its nonmodified form, respectively, upon the overexpression of wild-type (WT) and ligase-dead (LD) Parkin. This information is used to facilitate the excision of the polyubiquitinated band. (**c**) Silver blot is used as a quality control to check if the pull down was correctly performed. The equivalent amount of material in all samples is important to get trustworthy mass spectrometry results

GFP-tagged protein should be cut and analyzed, excluding the main unmodified band. It is recommended to run beforehand a Western blot (Fig. 3b) to assess how far the ubiquitinated protein fraction extends, and hence decide the exact point where the gel should be cut. Moreover, performing a silver staining (Fig. 3c) is also helpful in order to check that equivalent amount of material has been eluted in all samples, as this is crucial for appropriate quantitative mass spectrometry results.

Once the gel bands are excised, they are cut into smaller cubes and subjected to in-gel digestion using trypsin. Finally, resulting peptides are extracted from the gel and run into the LC-MS/MS. Acquired raw MS data are then analyzed by a search engine, such as MaxQuant, in order to infer the identity of the ubiquitinated peptides, determine the exact residues that are modified, and elucidate the ubiquitin chain types that have been formed (Fig. 4). For that search, diGly modification on lysines must be selected as variable modifications, since trypsin digestion of ubiquitinated proteins leaves this specific signature attached to the modified residue. If there was a miscleavage, the remnant left in the modified residue would be LRGG, instead of the typical GG. We have compared

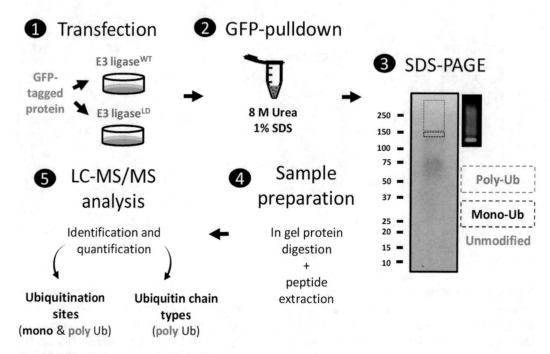

Fig. 4 Schematic illustration of the workflow to identify ubiquitination sites and ubiquitin chain linkages. The protein of interest is GFP-tagged and transfected into a cell culture (1) together with the corresponding E3 ligase, wild type (WT) or ligase dead (LD). Cell lysates are then subjected to the GFP-pull down protocol to isolate the GFP-tagged protein and its Ub-modified forms after stringent washes with 8 M urea and 1% SDS (2). Eluates are then run on a SDS-PAGE and Coomassie Blue stained (3). The most intense band corresponds to the nonmodified form of the protein, while the bands above are the monoubiquitinated (black) and polyubiquitinated (red) forms of the GFP-tagged protein, as verified by Western blot. Monoubiquitinated and polyubiquitinated bands are in-gel digested and the peptides are extracted (4) to later analyze by LC-MS/MS (5) in order to identify and quantify diGly modified peptides to determine specific ubiquitination sites and the ubiquitin chain types. Monoubiquitination band is excluded for the analysis of ubiquitin linkages

searches in which only diGly or both, diGly and LRGG were selected as variable modification. Including LRGG remnant in the search did not show any significant improvement, therefore, in our routine workflow, we do not search for peptides containing this signature.

2   Materials

2.1   GFP-Pull Down

1. Flies, mice, or cell lines expressing a GFP-tagged protein.

2. 1× Phosphate buffered saline (PBS).

3. GFP-Trap-A beads suspension (Chromotek GmbH).

4. Lysis buffer: 50 mM Tris–HCl pH 7.5, 150 mM NaCl, 1 mM EDTA, 0.5% Triton X-100, 1× Protease inhibitor cocktail (Roche Applied Science), 50 mM *N*-ethylmaleimide (NEM).

5. Dilution buffer: 10 mM Tris–HCl pH 7.5, 150 mM NaCl, 0.5 mM EDTA, 1× Protease Inhibitor cocktail, 50 mM *N*-ethylmaleimide (NEM).

6. Washing buffer: 8 M urea, 1% SDS (*see* **Note 1**).

7. 1% Sodium dodecyl sulfate (SDS) in 1× PBS.

8. Elution buffer: 4× Laemmli buffer (250 mM Tris–HCl pH 7.5, 40% Glycerol, 4% SDS, 0.2% Bromophenol blue) and 100 mM Dithiothreitol (DTT).

9. Silver staining kit (Invitrogen).

2.2 Sample Preparation Prior to MS Analysis

1. 4–12% Bolt Bis–Tris Plus precast gels (Invitrogen).

2. Protein standards (Precision plus all blue standards, Bio-Rad #161-0373).

3. Blue Coomassie.

4. 50 mM ammonium bicarbonate (ABC) buffer prepared in MilliQ water.

5. 100% ethanol.

6. 25 mM ABC, 50% ethanol buffer.

7. 10 mM dithiothreitol (DTT) in 50 mM ABC.

8. 55 mM chloroacetamide (CAA) in 50 mM ABC.

9. 12.5 ng/μL trypsin.

10. 100% acetonitrile (ACN).

11. 30% ACN, 3% trifluoroacetic acid (TFA).

12. 0.1% formic acid.

2.3 Software and Other Requirements

1. MaxQuant software (freely available at www.maxquant.org).

2. FASTA sequence database (downloadable from www.uniprot.org).

3   Methods

3.1 The GFP-Pull Down Strategy

1. Prewash beads: resuspend 50 μL of GFP-Trap-A beads suspension per sample in dilution buffer and centrifuge for 2 min at $2700 \times g$ (*see* **Note 2**). Repeat this step three times.

2. Wash cells once in 1× PBS and lyse them using ice-cold Lysis buffer with gentle rolling for 20 min. Use 2 mL of Lysis buffer per 10 cm culture dish (~9×10^6 cells/dish).

3. Centrifuge lysate for 10 min at $16,000 \times g$ at 4 °C and collect supernatant. Keep 30 μL as "input fraction" for Western blot analysis.

4. Mix lysate supernatant with previously washed GFP beads and incubate at room temperature for 3 h with gentle rolling.

5. Centrifuge samples for 2 min at 2700 × g to pellet beads. Keep 30 μL of supernatant as "unbound fraction" for Western blot analysis.

6. Resuspend beads in 1 mL of ice-cold Dilution buffer. Incubate beads for 5 min with gentle rolling and then centrifuge for 2 min at 2700 × g. Discard supernatant.

7. Resuspend beads in 1 mL of Washing buffer. Incubate beads for 5 min with gentle rolling and then centrifuge for 2 min at 2700 × g. Discard supernatant. Repeat this step three times.

8. Resuspend beads in 1 mL of 1% SDS in PBS. Incubate beads for 5 min with gentle rolling and then centrifuge for 2 min at 2700 × g. Discard supernatant.

9. Centrifuge again for 2 min at 2700 × g to fully discard supernatant.

10. Add 25 μL of Elution buffer to the beads and heat at 95 °C for 10 min (*see* **Note 3**).

11. Centrifuge beads at 16,000 × g for 2 min and collect the supernatant.

12. Eluted sample is then checked by silver staining before LC-MS/MS analysis (*see* **Note 4**).

3.2 Sample Preparation Prior to MS Analysis

1. *Protein fractionation*

   (a) Run eluates from GFP-pull down samples in an acrylamide gel under denaturing conditions (*see* **Note 5**).

   (b) Stain the gel using Blue Coomassie.

   (c) Cut the gel bands corresponding to the monoubiquitinated and polyubiquitinated fractions of the GFP-tagged protein (*see* **Note 6**).

   (d) Chop the gel slices into small cubes (*see* **Note 7**).

2. *In-gel protein digestion* (*see* **Notes 8** and **9**). In **steps e–m**, after the corresponding incubation, discard the solution.

   (a) To distain the gel pieces, incubate with 25 mM ABC and 50% ethanol for 20 min. Repeat the same step until no more colorant is released from the gel (*see* **Note 10**).

   (b) Dehydrate unstained gel pieces with 100% ethanol for 10 min. Repeat the same procedure until the gel slices are hard and white (*see* **Notes 11** and **12**).

   (c) Incubate gel with 10 mM DTT in ABC for 45 min at 56 °C to reduce protein disulfide bonds (*see* **Note 13**).

   (d) Dehydrate gel pieces with 100% ethanol for 5 min. Repeat the same procedure until the gel slices are hard and white.

(e) Alkylate cysteine residues by incubation with 55 mM 2-chloroacetamide in ABC for 30 min with no shaking in darkness (*see* **Notes 14** and **15**).

(f) Wash gel with 50 mM ABC for 20 min, and dehydrate with 100% ethanol for 10 min. Repeat the procedure twice.

(g) Repeat dehydration step until gel slices are hard and white.

(h) Carry out protein digestion by covering gel pieces with 12.5 ng/μL trypsin (*see* **Note 16**).

(i) Add 50 mM ABC to the samples and incubate overnight at 37 °C without shaking (*see* **Note 17**).

3. *Peptide extraction* (*see* **Notes 8** and **18**). After the incubations, all solutions must be collected, combined, and saved.

(a) Recover the solution from the digestion and place it in a new tube.

(b) Incubate gel pieces with 30% ACN, 3% TFA for 10 min. Repeat step twice.

(c) Incubate gel pieces with 100% ACN for 10 min. Repeat the same procedure until the gel slices are hard and white.

(d) Dry down collected solution in Speedvac centrifuge (*see* **Notes 19** and **20**).

(e) Resuspend dried peptides in 0.1% formic acid prior to MS analysis.

3.3 MS Data Processing and Bioinformatics Analysis (See Note 21)

1. All MS raw files are collectively processed with a search engine, such as MaxQuant.

2. The settings applied for peptide/protein identification are the ones defined by default in MaxQuant [16]. With the exception of the following parameters:

(a) In addition to Met oxidation and protein N-terminal acetylation, diGly modification on lysines (or LRGG) is set as variable modification to detect Ubi- and UbiL-modified peptides.

(b) Enzyme specificity is set to trypsin, allowing for cleavage N-terminal to Pro and between Asp and Pro.

(c) Enable match between runs option, and set 1.5 min and 20 min for match time window and alignment window, respectively.

3. Open the diGly.txt file generated by MaxQuant using an appropriate program, such as Microsoft Excel.

4. A localization probability of at least 0.75 is required to consider that a certain lysine is confidently ubiquitinated.

5. The types of ubiquitin chains formed are inferred by checking the ubiquitination sites detected on ubiquitin itself.

4 Notes

1. When preparing samples for MS analysis we always prepare fresh urea solutions.

2. The volume of GFP-trap-A beads that is used needs to be optimized for each system. This can be done by incubating equal amounts of protein lysates with different amount of beads. The minimum bead volume required to purify most of the GFP-tagged protein can then be determined by an anti-GFP Western blot to the unbound fraction. We typically use between 20 and 50 µL of beads suspension depending on the system employed. For instance, for a 10 cm cell culture dish we use 50 µL of beads suspension. We recommend pipetting the beads with the end of the tip cut, in order to make wider the hole of the tip and avoid damages to the beads.

3. The volume of Elution buffer that we use is typically half of the beads suspension volume used (i.e., for 50 µL of beads suspension we use 25 µL of Elution buffer). The volume of Elution buffer can be reduced if more concentrated sample is required.

4. We recommend to always run 10% of the eluate in a gel for silver staining (Fig. 3c) to confirm that the pull down process has successfully worked. A similar amount of material between replicas indicates that the process was performed correctly.

5. Run the gel the time necessary so that the unmodified and ubiquitinated versions of the GFP protein are separated.

6. Run samples previously on Western blot (Fig. 3b) to assess how far ubiquitination extends on the gel and to know the exact size at which the polyubiquitination band slice needs to be excised.

7. During the chopping process, gel pieces must be kept wet so that they do not slip off from the tube.

8. Unless otherwise indicated, all incubations are carried out by gentle shaking at room temperature.

9. The volume of reagent added to the gel should be adjusted so that the gel pieces are sufficiently covered/immersed.

10. In order to destain the gel pieces efficiently, it is better to change the destaining buffer every 20 min, instead of increasing the incubation time.

11. In order to dehydrate the gel pieces efficiently, it is better to change the ethanol every 10 min, instead of increasing the incubation time.

12. Be careful when manipulating completely dehydrated gel pieces. They easily jump from the tube.

13. A 100 mM DTT stock can be prepared and stored at −20 °C. When needed, the stock can be diluted 10× with ABC buffer.

14. Iodoacetamide is the reagent of choice for protein alkylation in most proteomics studies. However, it mimics the diglycine tag used for site-specific identification of protein ubiquitination by MS. Therefore, when studying ubiquitination by MS, iodoacetamide must be substituted with chloroacetamide (CAA) [17].

15. CAA is light sensitive so the solution must be stored in darkness.

16. Initially, add trypsin until gel pieces are covered and wait until they swell. If there are still white parts in the gel that have not been completely hydrated, add a bit more of trypsin and wait. Avoid adding excess of trypsin since it may result in autodigestion of the enzyme.

17. During overnight incubation ABC might evaporate, so add excess of ABC in order to make sure that gel pieces will not dry out.

18. Some analytes, such as peptides, have the tendency to bind to untreated tubes. Hence, to minimize sample loss it is recommended to use LoBind eppendorf tubes.

19. Dry until all ACN is removed and resuspend in the desired volume.

20. Dried peptides can be saved at −20 °C for months.

21. All experiments are done in triplicate in order to get conclusive quantitative data.

Acknowledgments

We would like to acknowledge Jabi Beaskoetxea, Kerman Aloria, and Jesus Mari Arizmendi for all their advice and support. The authors are grateful of the technical support provided by UPV/EHU SGIker (ERDF and ESF). This work was supported by Spanish MINECO [grant SAF2016-76898-P].

References

1. Komander D, Rape M (2012) The ubiquitin code. Annu Rev Biochem 81:203–229. https://doi.org/10.1146/annurev-biochem-060310-170328

2. Swatek KN, Komander D (2016) Ubiquitin modifications. Cell Res 26(4):399–422. https://doi.org/10.1038/cr.2016.39

3. Terrell J, Shih S, Dunn R, Hicke L (1998) A function for monoubiquitination in the internalization of a G protein-coupled receptor. Mol Cell 1(2):193–202

4. Pham AD, Sauer F (2000) Ubiquitin-activating/conjugating activity of TAFII250, a mediator of activation of gene expression in Drosophila. Science 289(5488):2357–2360

5. Glickman MH, Ciechanover A (2002) The ubiquitin-proteasome proteolytic pathway: destruction for the sake of construction. Physiol Rev 82(2):373–428. https://doi.org/10.1152/physrev.00027.2001

6. Ciechanover A (2013) Intracellular protein degradation: from a vague idea through the lysosome and the ubiquitin-proteasome system and onto human diseases and drug targeting. Bioorg Med Chem 21(12):3400–3410. https://doi.org/10.1016/j.bmc.2013.01.056

7. Hicke L (2001) Protein regulation by monoubiquitin. Nat Rev Mol Cell Biol 2 (3):195–201. https://doi.org/10.1038/35056583

8. Haglund K, Sigismund S, Polo S, Szymkiewicz I, Di Fiore PP, Dikic I (2003) Multiple monoubiquitination of RTKs is sufficient for their endocytosis and degradation. Nat Cell Biol 5(5):461–466. https://doi.org/10.1038/ncb983

9. Thrower JS, Hoffman L, Rechsteiner M, Pickart CM (2000) Recognition of the polyubiquitin proteolytic signal. EMBO J 19(1):94–102. https://doi.org/10.1093/emboj/19.1.94

10. Wickliffe KE, Williamson A, Meyer HJ, Kelly A, Rape M (2011) K11-linked ubiquitin chains as novel regulators of cell division. Trends Cell Biol 21(11):656–663. https://doi.org/10.1016/j.tcb.2011.08.008

11. Chen ZJ, Sun LJ (2009) Nonproteolytic functions of ubiquitin in cell signaling. Mol Cell 33 (3):275–286. https://doi.org/10.1016/j.molcel.2009.01.014

12. Ramirez J, Lectez B, Osinalde N, Siva M, Elu N, Aloria K, Prochazkova M, Perez C, Martinez-Hernandez J, Barrio R, Saskova KG, Arizmendi JM, Mayor U (2018) Quantitative proteomics reveals neuronal ubiquitination of Rngo/Ddi1 and several proteasomal subunits by Ube3a, accounting for the complexity of Angelman syndrome. Hum Mol Genet 27 (11):1955–1971. https://doi.org/10.1093/hmg/ddy103

13. Lee SY, Ramirez J, Franco M, Lectez B, Gonzalez M, Barrio R, Mayor U (2014) Ube3a, the E3 ubiquitin ligase causing Angelman syndrome and linked to autism, regulates protein homeostasis through the proteasomal shuttle Rpn10. Cell Mol Life Sci 71 (14):2747–2758. https://doi.org/10.1007/s00018-013-1526-7

14. Ramirez J, Martinez A, Lectez B, Lee SY, Franco M, Barrio R, Dittmar G, Mayor U (2015) Proteomic analysis of the ubiquitin landscape in the drosophila embryonic nervous system and the adult photoreceptor cells. PLoS One 10(10):e0139083. https://doi.org/10.1371/journal.pone.0139083

15. Min M, Mevissen TE, De Luca M, Komander D, Lindon C (2015) Efficient APC/C substrate degradation in cells undergoing mitotic exit depends on K11 ubiquitin linkages. Mol Biol Cell 26(24):4325–4332. https://doi.org/10.1091/mbc.E15-02-0102

16. Beer LA, Liu P, Ky B, Barnhart KT, Speicher DW (2017) Efficient quantitative comparisons of plasma proteomes using label-free analysis with MaxQuant. Methods Mol Biol 1619:339–352. https://doi.org/10.1007/978-1-4939-7057-5_23

17. Nielsen ML, Vermeulen M, Bonaldi T, Cox J, Moroder L, Mann M (2008) Iodoacetamide-induced artifact mimics ubiquitination in mass spectrometry. Nat Methods 5(6):459–460. https://doi.org/10.1038/nmeth0608-459

Chapter 12

Targeted Proteomics as a Tool for Quantifying Urine-Based Biomarkers

Sonali V. Mohan, D. S. Nayakanti, Gajanan Sathe, Irene A. George, Harsha Gowda, and Prashant Kumar

Abstract

Mass spectrometry based proteomics approaches are routinely used to discover candidate biomarkers. These studies often use small number of samples to discover candidate proteins that are later validated on a large cohort of samples. Targeted proteomics has emerged as a powerful method for quantification of multiple proteins in complex biological matrix. Parallel reaction monitoring (PRM) and selected reaction monitoring (SRM) are two main methods of choice for quantifying and validating proteins across hundreds to thousands of samples. Over the years, many software tools have become available that enable the users to carry out the analysis. In this chapter, we describe selection of prototypic peptides, sample preparation, generating a response curve, data acquisition and analysis of PRM data using Skyline software for targeted proteomics to quantify candidate markers in urine.

Key words Urine, Targeted proteomics, Parallel reaction monitoring, PRM, Skyline

1 Introduction

Urine, a partial filtrate of blood, offers several advantages for development and detection of protein-based biomarkers as (1) it is an easily accessible body fluid; (2) it can be obtained in large volume; (3) its dynamic range of proteins is smaller compared to other biological fluids [1]. These reasons favor urine to become the most promising source of sample in clinical research for development of novel biomarkers for diagnosis and prognosis. Liquid chromatography coupled to tandem mass spectrometry (MS)-based proteomics has revolutionized the analysis of the entire proteome of cells, tissues, or biofluids [2].

Currently, proteomics can be broadly divided into discovery (shotgun) and targeted MS approaches. Discovery based approaches employ data-dependent acquisition (DDA, also called shotgun proteomics) to identify proteins present in a sample. In DDA, the mass spectrometer performs a survey scan (MS^1) from

Rune Matthiesen (ed.), *Mass Spectrometry Data Analysis in Proteomics*, Methods in Molecular Biology, vol. 2051, https://doi.org/10.1007/978-1-4939-9744-2_12, © Springer Science+Business Media, LLC, part of Springer Nature 2020

which peptide ions with specified charge states and intensities above a threshold value are sequentially selected, isolated and fragmented for product ion scanning (MS^2). This is further coupled with a database search of MS/MS spectra using search algorithms to confidently assign an amino acid sequence to the spectra and thousands of proteins can be quantified in a single run. Targeted MS approaches allow to quantify proteins of interest with high sensitivity and precision across hundreds to thousands of samples. Selected reaction monitoring (SRM), also called multiple reaction monitoring (MRM) and Parallel reaction monitoring (PRM) are the two main methods of choice for targeted quantification of desired proteins in a complex biological matrix. These methodologies generate highly reproducible and quantitatively accurate data of a predetermined set of proteins [3–5]—the a priori knowledge which can come from shotgun proteomics experiments, literature survey, any biochemical experiments, microarray data, transcriptomic data, etc.

1.1 Selected Reaction Monitoring

SRM experiment is generally performed on triple quadrupole-based mass spectrometers and predefined transitions are measured sequentially [6]. In SRM experiments all a priori information on all transition ions is required and all transitions are optimized individually (Fig. 1a).

1.2 Parallel Reaction Monitoring

PRM is performed on high-resolution mass spectrometers like TOF and Orbitrap based platforms [7, 8]. All fragment ions of a predefined precursor ion are measured simultaneously in a single MS/MS scan event (Fig. 1b). It is a highly used method for quantification of multiple proteins in complex biological matrix with an attomole-level detection.

For this book chapter we selected the protein cytokeratin 19 and synthesized the heavy version and the light version of its peptide AALEDTLAETEAR for quantitative analysis using PRM in urine samples.

2 Materials

2.1 Software Tools Available for PRM/ SRM Analysis

Several software applications have been developed over past decade for analyzing the PRM/SRM data. Skyline is standalone software and is used widely by the research community to analyze the PRM and SRM data. Table 1 lists some of the software available for analysis of PRM/SRM data.

2.2 Software Tools Available for Statistical Analysis

Table 2 lists some of the software tools available for carrying out the statistical analysis for the identification of biomarker.

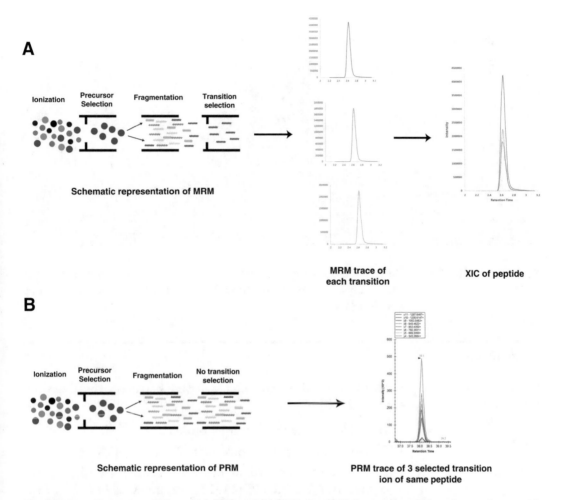

Fig. 1 Schematic representation of (**a**) SRM and (**b**) PRM. For SRM a priori knowledge of every fragment is required while not in the case of PRM. The data on selected transitions is acquired sequentially in SRM experiment while data on all fragment ions is acquired parallely in PRM

Table 1
List of various software available for the analysis of PRM/SRM data

Software	Reference
Skyline	[20]
MultiQuant	Commercial (https://sciex.com/products/software/multiquant-software)
SpectroDive	Commercial (https://biognosys.com/shop/spectrodive)
Pinpoint	Commercial (https://www.thermofisher.com/order/catalog/product/IQLAAEGABSFALDMAXF)

Table 2
List of various software available for the statistical analysis of PRM/SRM data

Software	References
Perseus	[21, 22]
MSstats	[23]
OriginLab	Commercial (https://www.originlab.com/)
SigmaPlot	Commercial (http://www.sigmaplot.co.uk/products/sigmaplot/sigmaplot-details.php)
GraphPad Prism	Commercial (https://www.graphpad.com/scientific-software/prism/)

2.3 Data Availability   Raw files used for generating standard curve and fasta files used for the analysis have been uploaded to PeptideAtlas (Username: PASS01315).

3   Methods

3.1 Selection of Peptides for Targeted Proteomics

Proteotypic peptides are often selected based on the following criteria [9, 10].

- It should unique for the target protein.
- The length of peptide should be 8–25 amino acids.
- It should be identified in MS experiments. The proteomics repositories like PRIDE [11–13], PeptideAtlas [14, 15], ProteomicsDB [16, 17], Human protein map [18] are based on shotgun proteomics data; SRM atlas (http://www.srmatlas.org/) is based on SRM data.
- Any modification including posttranslational modification should preferentially absent for quantifying proteins at the global level (*see* **Note 1**).
- Avoid peptides that contain amino acids that are prone for modification (e.g., methionine or tryptophan (they are prone to oxidation)), and N-terminal glutamine (it has propensity to be converted to pyro-glutamate under acidic conditions).
- Peptides containing glutamine or asparagine residues when followed by glycine may be chemically unstable and convert to glutamate or aspartate through deamidation and hence should be avoided.
- Peptides with aspartic acid residues when followed by glycine undergo dehydration and when followed by proline tend to undergo peptide chain cleavage and should be avoided.

- Peptides having histidine amino acid should be avoided as these residues can give additional charge states.
- Peptides should be neither extremely hydrophobic nor extremely hydrophilic.
- Missed cleavages (KK/RK/KR/RR) should be absent.

3.2 Sample Preparation and Data Acquisition

3.2.1 Sample Preparation

Samples were prepared as described previously [19]. Proteins were extracted using 3 kDa membrane filters and were further reduced, alkylated and digested with trypsin. For generating reverse response curve, digested urine samples were spiked with varying amounts of heavy peptides and constant amount of the light peptide. For quantifying proteins of interest, urine samples were spiked with constant amount of heavy peptides. The samples were cleaned with C-18 stage tip and stored till LC-MS/PRM analysis. Before PRM analysis on Thermo QE HFX mass spectrometer all samples were spiked with peptide ETTVFENLPEK (*see* **Note 2**).

3.2.2 Data Acquisition

Isolation list was exported from skyline and imported to Thermo QE HFX mass spectrometer.

3.2.3 Generation of Reverse Response Curve

1. Open a blank Skyline document, go to settings, peptide settings and select digestion tab to create a background proteome (Fig. 2).

   For creating background proteome, create a FASTA file with all proteins of interest and any peptides used for normalization (*see* **Note 3**).

2. Select background proteome and click "Add."

3. Give name, create the proteome file and add the fasta file.

4. Go to modification tab (Fig. 3).

5. Select all structural modifications present in peptides (e.g., carbamidomethyl cysteine).

6. Select label type—heavy.

7. Select isotope modifications (e.g., Label:13(6)15N(2) (C-term K), Label:13C(6)15N(4) (C-term R)).

8. Select Internal standard type—light.

9. Go to Quantification tab (Fig. 4).

10. Select Regression Fit (linear), Normalization (light/heavy—here we have used light), Method, Regression Weighing, MS level, and Units.

11. Click OK.

12. Go to settings, transition settings and select full scan (Fig. 5).

13. Select Isotope peaks included: Count; Precursor mass analyzer: Orbitrap (for Fusion/QE/Velos data); Peaks: 3 and Resolving power in MS1 filtering.

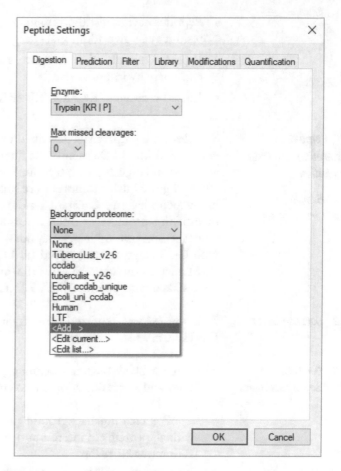

Fig. 2 Descirbes the digestion tab of peptide settings to create a background proteome

14. Select acquisition method: Targeted; Product mass analyzer: Orbitrap (for Fusion/QE/Velos data) and resolving power in MS/MS filtering.

15. Select filter tab (Fig. 6).

16. Enter precursor charge, ion charges (1), ion types (y).

17. Select the charge based on the peptide of interest (*see* **Note 4**). Some peptide may have +2/+3 or both. Type of ions (b or y) depends on whether the study has heavy peptides.

18. Click OK.

19. Go to edit → Insert → Peptides and paste all peptides of interest (Fig. 7).

20. Import raw files acquired for generating the reverse response curve—Go to file → Import → Results (Fig. 8).

21. Select "Add single-injection replicates in files," click OK and select all files to import them in Skyline document.

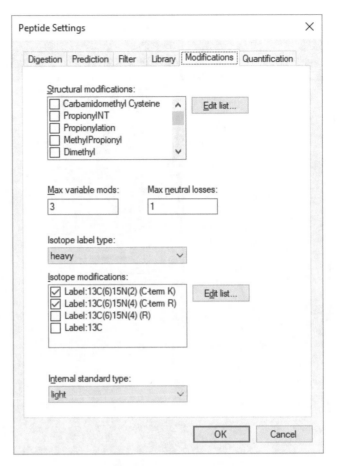

Fig. 3 Describes the modification tab of peptide settings to select all modifications present in the peptides

22. Go to settings and select integrate all.

23. Go to View → Peak Areas → Replicate Comparison.

24. Skyline auto select the peak of interest. Sometimes it may not, in that case select the peak of interest in one file and then right click and select "Apply peak to All." The Peak of interest will get selected automatically in all files.

25. Go to view → Retention Time → Replicate comparison and check for the variation in RT. A huge variation of RT implies that the wrong peak has been annotated as an ion.

26. The rdotp should between 0.99 and 1.

27. Select the peptide which was spiked into the samples to normalize run-to-run variation. In this case we have used peptide ETTVFENLPEK. Check for the variation in peak area the %CV of peak area ≤10 and if not then check the files in which there is a huge variation in peak area and retention time. If the signal is

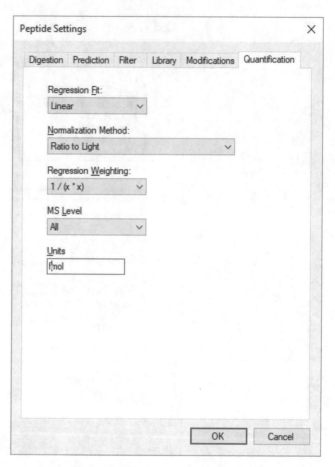

Fig. 4 Describes the quantification tab of peptide settings for creating standard curve

not good for this peptide remove those files from skyline document (Fig. 9).

28. Select a peptide, right click on the peak area graph, click "normalized to" and select "total." The fraction of area coming from all the fragment ions should be equal in all runs, if not then there are interfering peaks in the files where fraction is not equal. Check those files and delete the fragment ion with interfering peaks and/or remove the data file in which signal of peptide is poor (usually these are low amount points of response curve, Fig. 10).

29. Repeat **step 27** for all peptides.

30. Right click on peak area graph, click "normalized to" and select "light" (Fig. 11).

31. Go to view and select document grid. Click on the view (lop left) and select replicates (Fig. 12).

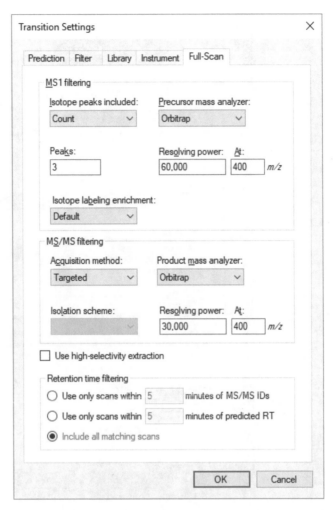

Fig. 5 Describes the scan parameters - it will depend on instrument and parameter used for acquiring the data

32. In sample type, enter standard and in analyte concentration enter the concentration/amount of each file (Fig. 13).

33. Go to view and select calibration curve (Fig. 14).
    Alternatively, the curve can also be made in Excel.

34. Export the data to csv file to create a standard curve with positive slope. Go to file → Export → Report. Export precursor name, total peak area for light and heavy peptides, ratio of heavy to light, retention time, and file name.

35. Plot the curve in Excel with ratio on the y axis and amount on the x axis.

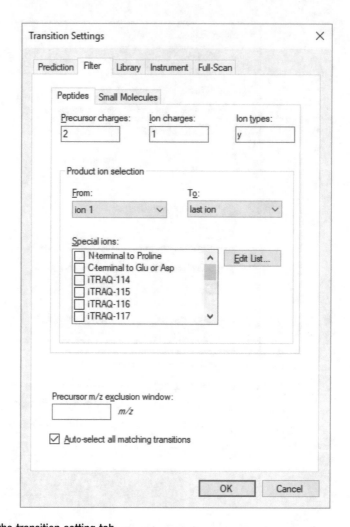

Fig. 6 Describes the transition setting tab

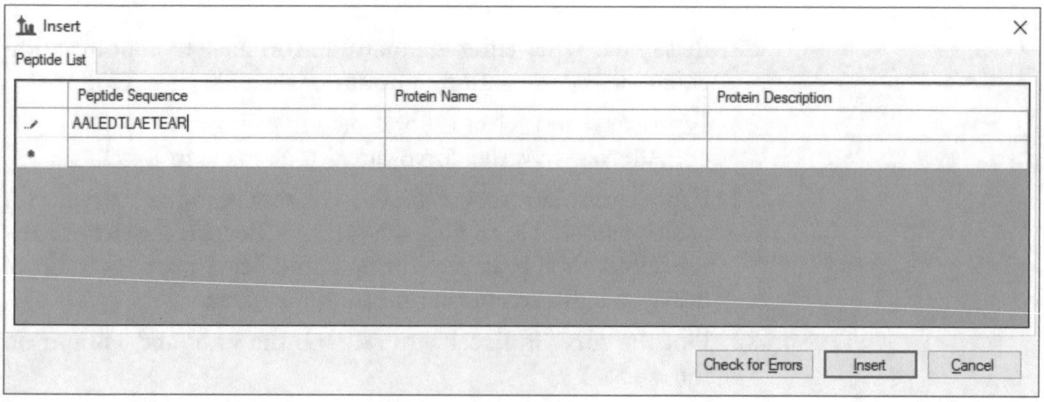

Fig. 7 Shows how to insert peptide sequence in Skyline

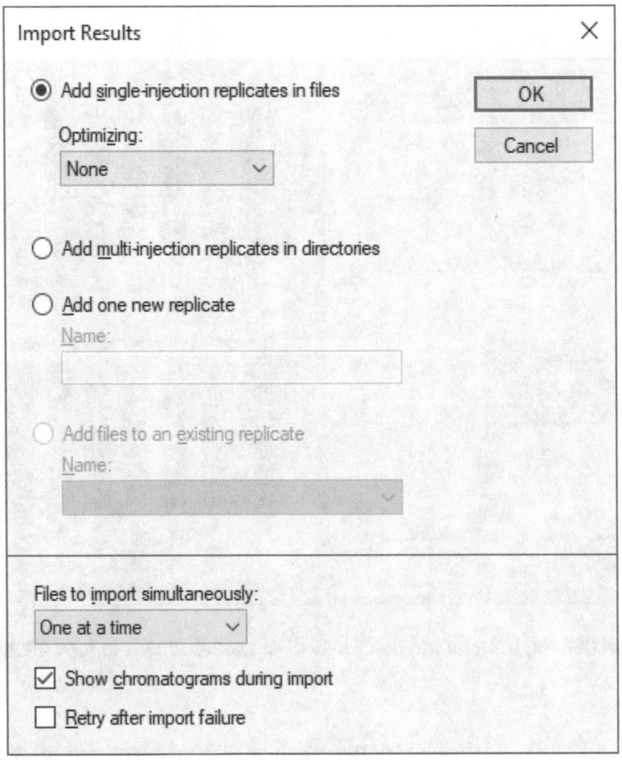

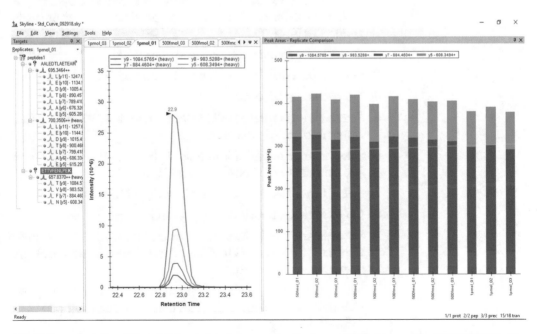

Fig. 8 Describes steps involved in importing the raw files

Fig. 9 Representative XIC and bar graph of area under the curve (AUC) for peptide ETTVFENLPEK. The coefficinet of variatin of AUC of this peptide is less than 5%. This peptide was spiked into the sample just before the mass spec analysis to account for the variation during the analysis

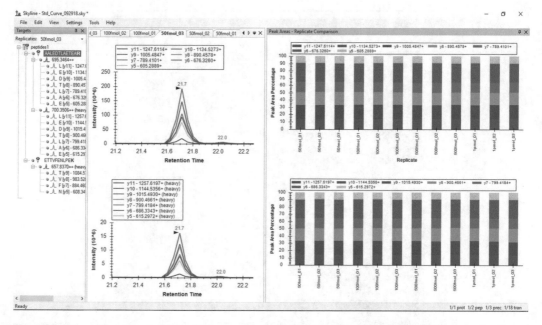

Fig. 10 Fraction of area coming from all fragment ions after normalizing it to total should be same in all runs (Right panel)

3.2.4 Analysis of Samples

1. Open a blank Skyline document, go to settings and select digestion tab to create a background proteome.

   For creating background proteome, create a FASTA file with all proteins of interest and any peptides used for normalization (*see* **Note 3**).

2. Select background proteome and click "Add."

3. Give name, create the proteome file and add the FASTA file.

4. Go to modification tab.

5. Select all structural modifications present in peptides (e.g., carbamidomethyl cysteine).

6. Select isotope modifications (e.g., Label:13(6)15N(2) (C-term K), Label:13C(6)15N(4) (C-term R)).

7. Select Internal standard type—heavy (*see* **Note 5**).

8. Go to settings, transition settings and select full scan.

9. Select Isotope peaks included: Count; Precursor mass analyzer: Orbitrap (for Fusion/QE/Velos data); Peaks: 3 and Resolving power in MS1 filtering.

10. Select acquisition method: Targeted; Product mass analyzer: Orbitrap (for Fusion/QE/Velos data) and resolving power in MS/MS filtering.

11. Click OK.

12. Go to edit → Insert → Peptides and paste all peptides of interest.

A

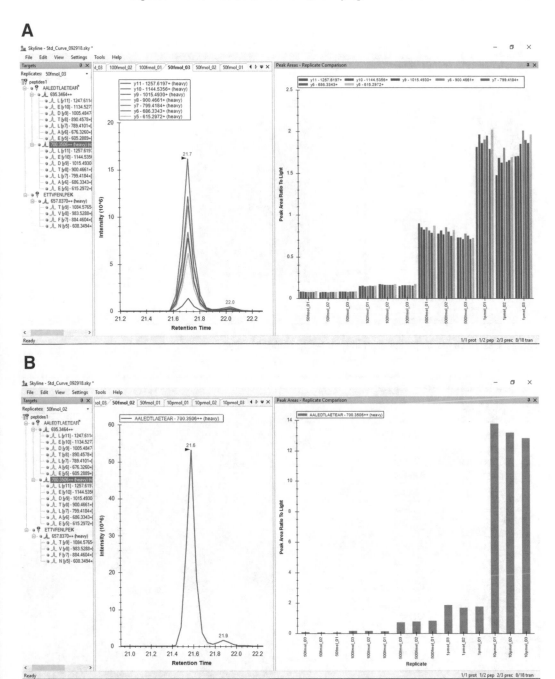

B

Fig. 11 XIC and bar graph to represent the variation in normalized AUC of heavy peptide AALEDTLAETEAR. (**a**) each fragment and (**b**) peptide across different concentrations

13. Import the raw files acquired on mass spectrometer.

14. Select "Add single-injection replicates in files," click OK and select all files to import in Skyline document.

15. Go to settings and select integrate all (Fig. 15) (*see* **Note 6**).

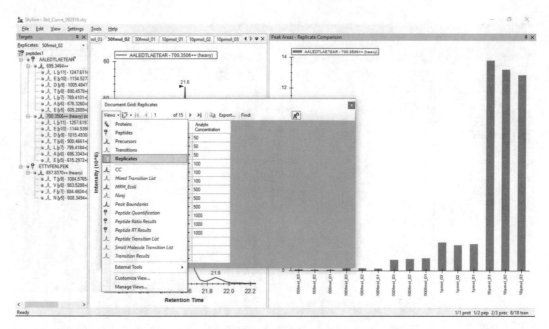

Fig. 12 Describes the steps involved in generating standard curve

Document Grid: Replicates

Views ▾ 📑 ▾ |◀ ◀ | 1   of 15 | ▶ | ▶| | 📋 Export... | Find: A^a

	Replicate	Sample Type		Analyte Concentration
▶	50fmol_03	Standard	▾	50
	50fmol_02	Standard	▾	50
	50fmol_01	Standard	▾	50
	100fmol_03	Standard	▾	100
	100fmol_02	Standard	▾	100
	100fmol_01	Standard	▾	100
	500fmol_03	Standard	▾	500
	500fmol_02	Standard	▾	500
	500fmol_01	Standard	▾	500
	1pmol_03	Standard	▾	1000
	1pmol_02	Standard	▾	1000
	1pmol_01	Standard	▾	1000
	10pmol_01	Standard	▾	10000
	10pmol_02	Standard	▾	10000
	10pmol_03	Standard	▾	10000

Fig. 13 Table format to generate standard curve in Skyline

Calibration Curve

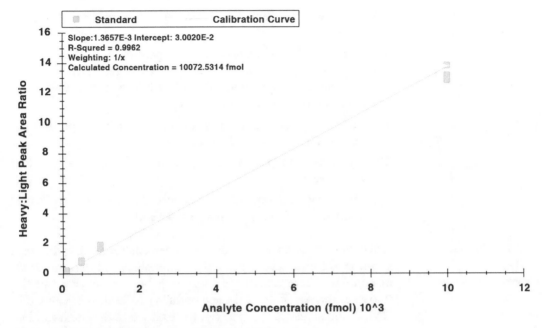

Fig. 14 Calibration curve for peptide AALEDTLAETEAR

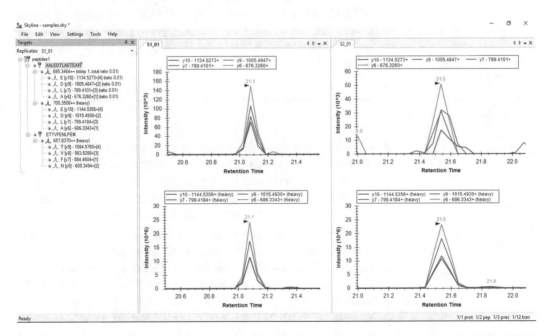

Fig. 15 Representative XIC of the endogenous peptide (light) and the heavy peptide (spiked) AALEDTLAETEAR in urine samples. The top panel represents the endogenous peptide, while bottom panel represents the heavy peptide

16. Go to View → Peak Areas → Replicate Comparison.

17. Go to view → Retention Time → Replicate comparison (*see* **Note 7**).

18. Select the heavy version of each peptide and check for its variation—(a) across all runs; (b) across technical replicates the coefficient of varation should be less than 20%. Check the retention time variation for each peptide.

19. Check the signal of the light peptide for each sample and the variation in RT across all runs.

20. Right click on peak area graph, click "Normalized to" and select heavy.

21. Check the variation of ratio across the technical replicates. rdotp value should be between 0.99 and 1.

3.2.5 Validation of Proteins Across Hundreds of Samples

Once the assay is set-up and data has been analyzed using the steps mentioned above, several proteins can be validated across hundreds of samples. There are several tools available for carrying out the statistical analysis (Table 2 list few of them). MSstats is integrated in Skyline and statistical analysis can be done using Skyline itself. For carrying out the statistical analysis using other software, the results can be exported from skyline and the analysis can be carried out. Here we show the representative example of peptide AALED-TLAETEAR of protein cytokeratin 19 (KRT19) which is found to be upregulated in urine of bladder cancer patients. In our data we have seen that the KRT19 is significantly upregulated with a fold change of 3.2 (Fig. 16).

4 Notes

1. Often it is not possible to avoid peptides with amino acids that are prone to modifications (or have posttranslational modification). If a peptide can be synthesized containing those modifications and the fragments ions are optimised for the analysis these peptides can be taken further for analysis.

2. For generating the response curve (or reverse response curve) the samples are spiked either with light or heavy synthetic peptides. If a matrix is available which does not have the peptide of interest, then the response curve can be generated by varying the light peptide, otherwise samples can be spiked with varying amounts of the heavy peptide (reverse response curve). In case of the reverse response curve, matrix needs not to be spiked with the light peptide if the matrix has a good signal of the endogenous peptide.

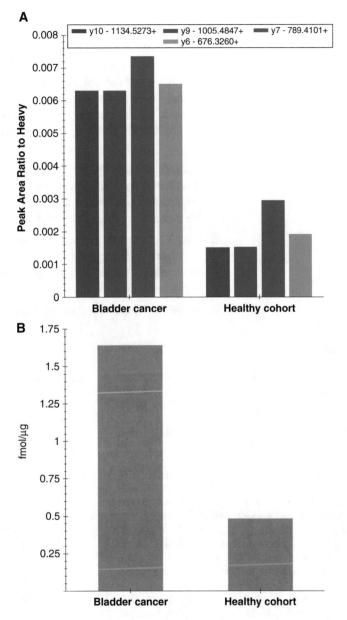

Fig. 16 Quantitation of (**a**) peptide AALEDTLAETEAR in urine samples (**b**) Protein cytokeratin 19 (KRT19) across bladder cancer and healthy individual urine samples

3. Depending on the sequence the peptide may have multiple charges. Some peptide may have +2/+3 or both. The use of a synthetic peptide defines the type of ions to be used for quantitation. If the assay is dependent upon the use of heavy tryptic peptides, then y ions are used for quantifying the peptides.

4. The signal intensity of peptides depends on the limit of detection and lower limit of quantitation. Peptide in a sample falling near to these limits may not have a good extracted ion chromatogram. In such cases check the XIC of heavy peptide used as internal standard to eliminate the possibility of technical errors of data acquisitions. Also, if the XIC of endogenous peptide is very poor try injecting the higher amount of protein to acquire PRM data (*see* **Note 6**).

5. While processing the standard curve the internal standard selected as the light peptide and was used to normalize the data while the heavy peptide was used in varying amounts. On the other hand, while analyzing the sample the heavy peptide becomes the internal standard.

6. The XIC of endogenous peptide (top right panel) is not good as others as the signal is near to lower limit of quantitation. In such cases check the XIC of heavy peptide to eliminate the technical errors of data acquisition.

7. Steps till 16 are the same in Subheading 3.2.3.

References

1. Zhao M, Li M, Yang Y, Guo Z, Sun Y, Shao C et al (2017) A comprehensive analysis and annotation of human normal urinary proteome. Sci Rep 7:3024

2. Aebersold R, Mann M (2003) Mass spectrometry-based proteomics. Nature 422:198–207

3. Picotti P, Bodenmiller B, Mueller LN, Domon B, Aebersold R (2009) Full dynamic range proteome analysis of S. cerevisiae by targeted proteomics. Cell 138:795–806

4. Picotti P, Rinner O, Stallmach R, Dautel F, Farrah T, Domon B et al (2010) High-throughput generation of selected reaction-monitoring assays for proteins and proteomes. Nat Methods 7:43–46

5. Peterson AC, Russell JD, Bailey DJ, Westphall MS, Coon JJ (2012) Parallel reaction monitoring for high resolution and high mass accuracy quantitative, targeted proteomics. Mol Cell Proteomics 11:1475–1488

6. Picotti P, Aebersold R (2012) Selected reaction monitoring-based proteomics: workflows, potential, pitfalls and future directions. Nat Methods 9:555–566

7. Sathe G, Na CH, Renuse S, Madugundu AK, Albert M, Moghekar A et al (2018) Quantitative proteomic profiling of cerebrospinal fluid to identify candidate biomarkers for Alzheimer's disease. Proteomics Clin Appl:e1800105

8. Bourmaud A, Gallien S, Domon B (2016) Parallel reaction monitoring using quadrupole-Orbitrap mass spectrometer: principle and applications. Proteomics 16:2146–2159

9. Hoofnagle AN, Whiteaker JR, Carr SA, Kuhn E, Liu T, Massoni SA et al (2016) Recommendations for the generation, quantification, storage, and handling of peptides used for mass spectrometry-based assays. Clin Chem 62:48–69

10. Lange V, Picotti P, Domon B, Aebersold R (2008) Selected reaction monitoring for quantitative proteomics: a tutorial. Mol Syst Biol 4:222

11. Vizcaino JA, Csordas A, Del-Toro N, Dianes JA, Griss J, Lavidas I et al (2016) 2016 update of the PRIDE database and its related tools. Nucleic Acids Res 44:D447–D456

12. Vizcaino JA, Cote RG, Csordas A, Dianes JA, Fabregat A, Foster JM et al (2013) The PRoteomics IDEntifications (PRIDE) database and associated tools: status in 2013. Nucleic Acids Res 41:D1063–D1069

13. Martens L, Hermjakob H, Jones P, Adamski M, Taylor C, States D et al (2005) PRIDE: the proteomics identifications database. Proteomics 5:3537–3545

14. Farrah T, Deutsch EW, Hoopmann MR, Hallows JL, Sun Z, Huang CY et al (2013) The state of the human proteome in 2012 as viewed

through PeptideAtlas. J Proteome Res 12:162–171

15. Desiere F, Deutsch EW, King NL, Nesvizhskii AI, Mallick P, Eng J et al (2006) The PeptideAtlas project. Nucleic Acids Res 34: D655–D658

16. Schmidt T, Samaras P, Frejno M, Gessulat S, Barnert M, Kienegger H et al (2018) ProteomicsDB. Nucleic Acids Res 46:D1271–d1281

17. Wilhelm M, Schlegl J, Hahne H, Gholami AM, Lieberenz M, Savitski MM et al (2014) Mass-spectrometry-based draft of the human proteome. Nature 509:582–587

18. Kim MS, Pinto SM, Getnet D, Nirujogi RS, Manda SS, Chaerkady R et al (2014) A draft map of the human proteome. Nature 509:575–581

19. Marimuthu A, O'meally RN, Chaerkady R, Subbannayya Y, Nanjappa V, Kumar P et al (2011) A comprehensive map of the human urinary proteome. J Proteome Res 10:2734–2743

20. Maclean B, Tomazela DM, Shulman N, Chambers M, Finney GL, Frewen B et al (2010) Skyline: an open source document editor for creating and analyzing targeted proteomics experiments. Bioinformatics 26:966–968

21. Tyanova S, Temu T, Sinitcyn P, Carlson A, Hein MY, Geiger T et al (2016) The Perseus computational platform for comprehensive analysis of (prote)omics data. Nat Methods 13:731–740

22. Cox J, Mann M (2012) 1D and 2D annotation enrichment: a statistical method integrating quantitative proteomics with complementary high-throughput data. BMC Bioinformatics 13(Suppl 16):S12

23. Choi M, Chang CY, Clough T, Broudy D, Killeen T, Maclean B et al (2014) MSstats: an R package for statistical analysis of quantitative mass spectrometry-based proteomic experiments. Bioinformatics 30:2524–2526

Chapter 13

Data Imputation in Merged Isobaric Labeling-Based Relative Quantification Datasets

Nicolai Bjødstrup Palstrøm, Rune Matthiesen, and Hans Christian Beck

Abstract

The data-dependent acquisition in mass spectrometry-based proteomics combined with quantitative analysis using isobaric labeling (iTRAQ and TMT) inevitably introduces missing values in proteomic experiments where a number of LC-runs are combined, especially in the growing field of shotgun clinical proteomics, where the protein profiles from the proteomics analysis of several hundred patient samples are compared and correlated to clinical traits such as a specific disease or disease treatment in order to link specific outcomes to one or more proteins. In the context of clinical research it is evident that missing values in such datasets reduce the power of the downstream statistical analysis therefore may hampers the linking of the expression of disease traits to the expression of specific proteins that may be useful for prognostic, diagnostic, or predictive purposes. In our study, we tested three data imputation approaches initially developed for microarray data for the imputation of missing values in datasets that are generated by several runs of shotgun proteomic experiments and where the data were relative protein abundances based on isobaric tags (iTRAQ and TMT). Our conclusion is that imputation methods based on *k Nearest Neighbors* successfully impute missing values in datasets with up to 50% missing values.

Key words Missing values, Clinical proteomics, Data imputation, Relative quantification, Isobaric tags

1 Introduction

Missing values are an unfortunate but common part of proteomics and research in general, and the proper handling of missing values can be crucial for the conclusions drawn from the analysis. A common strategy for handling missing values has been to omit them from the dataset, but this can be problematic as it decreases the power of the downstream statistical analysis and may also introduce unwanted bias in the investigated associations [1, 2]. Instead, utilizing the existing data to estimate plausible values to insert in place of the missing values, a process known as data imputation might therefore be a far better option.

Many different methods (algorithms) to impute data exist. They vary, however, greatly in the way they estimate missing values.

Rune Matthiesen (ed.), *Mass Spectrometry Data Analysis in Proteomics*, Methods in Molecular Biology, vol. 2051, https://doi.org/10.1007/978-1-4939-9744-2_13, © Springer Science+Business Media, LLC, part of Springer Nature 2020

Table 1
Different R functions and their type of method applied for data imputation

Name	R function	Type
MLE	norm::imp.norm	Global similarity
SVD (Bpca)	pcaMethods::pca	Global similarity
knn	impute::impute.knn	Local similarity
QRILC	imputeLCMD::impute.QRILC	Local similarity
MinDet	imputeLCMD::impute.MinDet	Local similarity
MinProb	imputeLCMD::impute.MinProb	Single value
min	Replaces the missing values by the smallest nonmissing value	Single value
zero	Replaces the missing values by 0	Single value
mixed	Msnbase::impute(x, "mixed")	NA
nbavg	Msnbase::impute(x, "nbavg")	Local similarity

The majority of methods currently applied to proteomic datasets, that generally provide quantitative information at the peptide level and relative quantification at the protein level or both, originate from imputation of missing values in microarray analysis. Generally, the different methods can be divided into three categories: single-value, local similarity, and global similarity (Table 1). The single value approach (Min.Prob) attempts to replace left-censored missing values by a random draw from a Gaussian distribution fixed around a minimal value [3]. The second method applied is a local similarity approach which finds the k-nearest neighbours (kNN) using a Euclidean metric for those columns where a value is not missing for the protein, in order to impute the missing values. Having found the kNN for each protein, the missing values are imputed by averaging the (nonmissing) elements of its neighbours. The third method is a global similarity approach termed singular value decomposition (SVD), which relies on some of the same principles as the kNN method, but differs in the estimation method. By design, SVD can only be applied to complete matrices, that is, no missing values. Therefore, all missing values are initially set to zero and an expectation maximization algorithm works iteratively until the change between estimations falls below a threshold. The SVD approach utilizes each step of the current estimates in order to compute a new estimate [4].

In the present work, we tested the Min.Prob, kNN, and the SVD methods (Table 1) for the imputation of missing data in a dataset that was generated by merging six shotgun proteomic datasets that relatively quantified 166 proteins in plasma samples from 48 patients using isobaric tagging by TMT. Hitherto, these imputation methods were previously applied to impute missing values in label-free proteomic datasets and aimed the imputation of missing peptide ion signals [4] but, to our knowledge, have not been applied to impute missing values in proteomic datasets based on relative quantification based on isobaric tagging (TMT or iTRAQ).

The dataset used for the present study was a subset of data from a previous study involving plasma samples from 500 patients admitted to the hospital with the suspicion of STEMI (ST segment elevated myocardial infarction) whereof 363 had received heparin as a part of the prehospital treatment procedure [5] before drawing of the blood sample used for the study. This data subset was comprised by six 10-plex TMT datasets, each containing ten LC-MSMS runs and quantified 8 patients relative to a pool of heparin-untreated patients, yielding at total of 48 patients, where 31 patients were treated with heparin prior to blood sampling and 17 patients who did not receive treatment. For detailed experimental details and clinical information, we refer to our recent published work [5]. After database search data were structured as a matrix with protein identifications in the rows and patients in the columns. A total of 387 proteins were identified whereof only 166 proteins contained a quantitative readout across all 48 patients (*see* Fig. 1) as a result of the nature of the data dependent acquisition mode and/or isolation interference. This significantly impaired the data set and thereby also the number of proteins available for the downstream statistics analysis. In order to increase the number proteins available for statistical analysis across all 48 patients, we aimed to

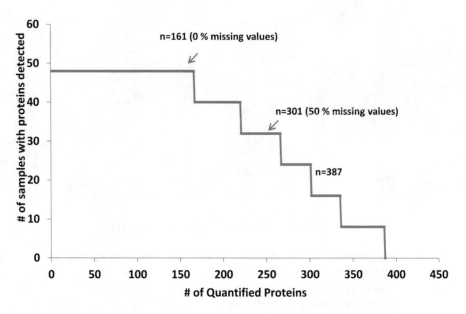

Fig. 1 Out of the 387 proteins detected in the 48 plasma samples, 166 proteins were found in all samples, and 301 was found in 50% of the samples

test various data imputation methods to pinpoint the most optimal ones for the imputation of missing values in a dataset based on isobaric tag-based quantitative proteomics. We tested three methods that are based on global (SVD), local (kNN) or single-value (Min.prob) imputation methods [4, 6]. For this testing we used the above described dataset with 166 quantified proteins with quantitative readouts across all 48 patients.

First we created test datasets with missing values by removing quantitative values, one at a time, TMT set-wise and in blocks of four identifications, as illustrated in Fig. 2, to create a test datasets which mimics the nature of missing data in traditional 10-plex TMT datasets. This process was repeated four times resulting in four different test datasets with increasing numbers/percentages of missing values for each of the 34 randomly chosen proteins ranging from 16.7% to 66.7% missing values as each TMT 10-plex dataset labels 8 patient samples and using one out of two control pools as

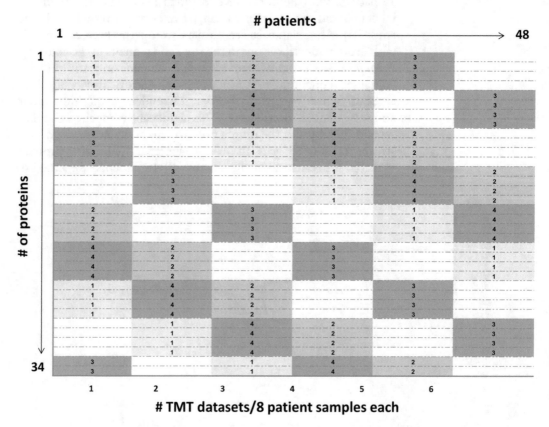

Fig. 2 Creation of the four datasets with increasing proportion of deleted relative abundances for 34 proteins. "1" denotes first round of deletion of relative abundances for four proteins for across eight patients from one TMT-dataset (16.67% missingness). "2" denotes the second round of deletions and to forth. White blocks indicate the remaining data after four rounds of deletion, leaving 32 missing values out of 48 data points (66.67% missingness) for imputation tests. Proteins to be deleted were randomly selected from the list of 166 proteins

reference sample. The three imputation methods were then applied to impute missing values in each of the four simulated test dataset. We then evaluated the accuracy of the imputations produced by each of the three methods by means of coefficient of variation (CV) calculations, and Student's t-test of differences in plasma levels of specific proteins between heparin-treated patients and patients that did not receive heparin. Finally, we chose the most optimal method to impute missing values of the original dataset described above.

As a first step of our evaluation of the three applied methods we compared the original values with the imputed values by calculating the coefficient of variation (CV) of the root-mean-square error (RMSE) for each of the 34 proteins in each of the four datasets with 16.7–67.7% missing values (Fig. 3). The CV (RMSE) is used as a measure of the difference between the original values and the imputed values, where RMSE is calculated as the square root of the average squared difference between the original (x) and the imputed (y) values for the total number of samples (n) for the ith observation for each individual protein. The CV (RMSE) normalizes the RMSE of each protein to the mean of the imputed values for that protein.

CV calculations have previously been applied to evaluated imputation methods when imputing missing values in proteomics datasets [7], and also as an overall data quantitative quality measure of LC-MS datasets [8]. Clearly, the kNN and SVD methods performs equally well with increasing CVs with increased number of missing values, whereas the Min.Prob method performed significantly worse with CVs that are several fold higher as compared to the kNN and SVD methods (Fig. 3). These results indicate that the

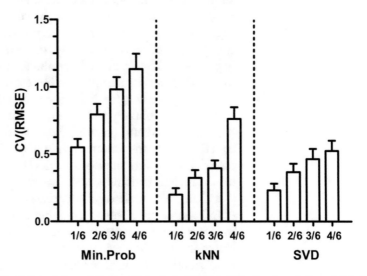

Fig. 3 Calculation of the coefficient of variation (CV) after imputation of 8 (1/6), 16 (2/6), 24 (3/6), or 32 (4/6) relative abundances out of 48 relative abundances per protein by the Min.Prob, kNN, and the SVD methods

kNN and SVD methods are equally well-suited for the imputation of missing values in TMT- and iTRAQ-based datasets. However, this evaluation does not provide any information on the upper limit of missing values these methods can provide consistent results.

Therefore, a more detailed investigation of the three imputation methods was done by randomly selecting four proteins (midkine, antithrombin II, C-C motif chemokine 21, and biglycan) which plasma concentration significantly increased upon heparin administration (Table 2), and then test how well the imputation methods could estimate the deleted values at increasing numbers of missing values. Four datasets that comprised of 16.7%, 33.3%, 50%, or 66.7% missing values were created as described above, that is, datasets where 8, 16, 24, or 32 out of 48 expression values for 34 affected proteins were removed and the missing values imputed using kNN, SVD, or Min.Prob imputation methods. After imputation data was grouped into nontreated and heparin-treated and the mean values and standard error of mean values for each group for the actual dataset with no missing values, and for the datasets with missing values for the selected proteins were calculated. P-values indicate significant expression differences between nontreated and heparin-treated groups. Clearly, for the four selected proteins the imputation by the Min.Prob method resulted in decreasing average values and increased P-values with increasing percentage of missing values. This indicates that the imputed values are lower than the original values, and the variance of the data also increases with the percentage of missing values. By contrast, this was only observed for the heparin-treated group for the dataset with 66.7% missing values when imputing the missing values using the kNN method. Imputation of missing values by the SVD-method seemed apparently to be close to the values for the original dataset except for biglycan. Here, the averages of the treated and heparin-treated groups seemed to be unaffected of increasing percentages of missing values. The increased P-values, however, indicate that the imputed values for this protein increased the CV of the data.

To explore this further a more detailed investigation was performed by grouping the imputed datasets into nontreated and heparin-treated and plotting the observed values (bright symbols) together with the imputed values (dark symbols) for the three tested imputation methods with 16.7%, 33.3%, 50%, or 66.7% missing values with midkine as an example (Fig. 4). From this figure it is clear, that the decreasing mean values with increasing number of missing values observed in Table 2 is due to the imputation of—in some cases—negative values. This tendency increases with increasing number of imputed values, clearly indicating this imputation method is not suited for the imputation of relative expression data based on isobaric tagging. By contrast, the SVD method impute equally well for the data obtained from the untreated group with up to 33.3% missing values and up to 50% missing

Table 2

Means ± SEM for the relative abundances of four randomly selected proteins in 48 patients grouped into untreated patients ($n = 16$) and patients treated with heparin ($n = 32$)

Protein name	Accession number	# Missing values	Observed (mean ± SEM) −	+	P-value	Min.Prob (mean ± SEM) −	+	P-value	kNN (mean ± SEM) −	+	P-value	SVD (mean ± SEM) −	+	P-value
Heparin														
Midkine	E9PLM6	0	1.080 ± 0.040	2.665 ± 0.172	<0.0001									
		8				0.839 ± 0.122	2.206 ± 0.192	<0.0001	1.098 ± 0.037	2.542 ± 0.145	<0.0001	1.214 ± 0.066	2.526 ± 0.142	<0.0001
		16				0.642 ± 0.172	1.868 ± 0.245	0.0012	1.108 ± 0.039	2.623 ± 0.141	<0.0001	1.289 ± 0.082	2.647 ± 0.133	<0.0001
		24				0.615 ± 0.115	1.438 ± 0.252	0.0241	1.169 ± 0.055	2.651 ± 0.151	<0.0001	1.406 ± 0.145	2.647 ± 0.137	<0.0001
		32				0.428 ± 0.155	0.834 ± 0.255	0.2692	1.218 ± 0.048	1.737 ± 0.149	0.0149	1.460 ± 0.156	2.657 ± 0.141	<0.0001
Antithrombin-II	P01008	0	1.22 ± 0.056	3.309 ± 0.244	<0.0001									
		8				0.980 ± 0.137	2.798 ± 0.326	0.0003	1.249 ± 0.067	3.241 ± 0.240	<0.0001	1.277 ± 0.112	3.319 ± 0.237	<0.0001
		16				0.737 ± 0.180	2.411 ± 0.357	0.0017	1.279 ± 0.063	3.302 ± 0.240	<0.0001	1.571 ± 0.161	3.550 ± 0.240	<0.0001
		24				0.604 ± 0.171	1.866 ± 0.358	0.0155	1.282 ± 0.066	3.187 ± 0.224	<0.0001	1.792 ± 0.203	3.618 ± 0.222	<0.0001
		32				0.420 ± 0.157	1.388 ± 0.352	0.0548	1.296 ± 0.038	2.124 ± 0.263	0.0255	1.867 ± 0.202	3.570 ± 0.223	<0.0001
C-C motif chemokine 21	O00585	0	1.219 ± 0.069	2.932 ± 0.246	<0.0001									
		8				1.048 ± 0.098	2.314 ± 0.302	0.0038	1.159 ± 0.052	2.777 ± 0.228	<0.0001	1.168 ± 0.064	2.871 ± 0.228	<0.0001
		16				0.855 ± 0.133	1.685 ± 0.287	0.0443	1.210 ± 0.062	2.515 ± 0.182	<0.0001	1.173 ± 0.058	2.624 ± 0.187	<0.0001
		24				0.565 ± 0.161	1.334 ± 0.267	0.0493	1.189 ± 0.062	2.341 ± 0.181	<0.0001	1.176 ± 0.054	2.370 ± 0.178	<0.0001
		32				0.487 ± 0.1140	0.784 ± 0.209	0.3286	1.202 ± 0.058	1.593 ± 0.118	0.0223	1.197 ± 0.058	2.077 ± 0.099	<0.0001
Biglycan (fragment)	C9JKG1	0	1.099 ± 0.068	1.550 ± 0.085	0.0008									
		8				0.943 ± 0.126	1.340 ± 0.137	0.0625	1.148 ± 0.064	1.527 ± 0.088	0.005	1.152 ± 0.064	1.564 ± 0.084	0.0016
		16				0.843 ± 0.145	1.097 ± 0.138	0.2455	1.173 ± 0.079	1.443 ± 0.080	0.0322	1.154 ± 0.063	1.508 ± 0.072	0.0021
		24				0.593 ± 0.130	0.763 ± 0.158	0.4717	1.091 ± 0.0472	1.280 ± 0.070	0.0682	1.099 ± 0.046	1.413 ± 0.063	0.0014
		32				0.526 ± 0.116	0.635 ± 0.149	0.6236	1.252 ± 0.044	1.381 ± 0.064	0.1676	1.251 ± 0.034	1.578 ± 0.061	0.0005

Data are shown for means of observed values and for means of relative abundances when increasing numbers of missing data are imputed by the Min.Prob, SVD, or the kNN methods. P-values indicate the significance between the mean of relative protein abundances in the treated group versus the heparin-treated group

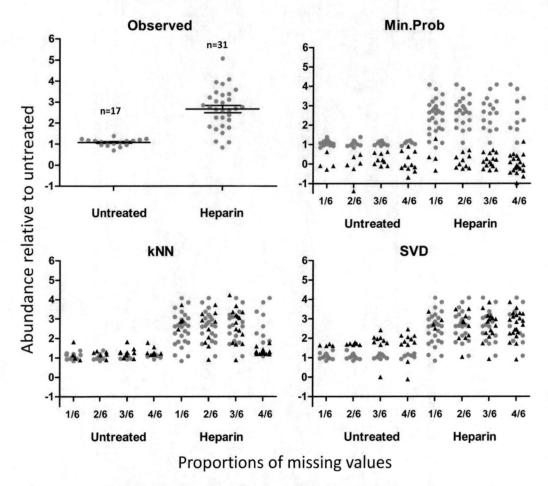

Fig. 4 Imputation of missing values for midkine at 1/6 (16.67%) to 4/6 (66.67%) missingness by the Min.Prob, kNN, and SVD methods. Light gray symbols indicate original values and black symbols indicate imputed values. Data are relative to an internal control (pool of samples from untreated patient)

values for the heparin-treated group with relatively unchanged mean values but with increasing standard error of the mean (SEM) (Fig. 3 and Table 2). The imputed values by the kNN method at highest number of missing values for the heparin-treated patients for midkine significantly dropped to values around one. This may, however, be explained the default settings of the method (missing values <50%, $k = 10$). In fact, changing the maximum allowed percentage of missing values to 67% (4/6) showed results similar to imputation results at 50% missing values, although with a slightly higher CV%.

In conclusion, we tested three methods for the imputation of missing values in datasets based on isobaric tag-based relative quantification of plasma proteins affected drug administration. One approach, the Min.Prob method that replace left-censored missing values by a random draw from a Gaussian distribution fixed around a minimal value, performed measurably worse than the other two tested methods. This method imputed negative values—even when imputing a low number of missing values (8 missing out of 48). As the Min.Prob method utilizes a Gaussian distribution centred at a minimal value, negative estimates are fairly common. The kNN and the SVD method, however, perform apparently equally well except for when the degree of missingness exceeds 50%. Here, the algorithm used by the kNN method then substitutes missing values with the overall mean of the sample, which is considerably less efficient than the SVD method. Nevertheless, the algorithm of the SVD method tends to impute negative values when the degree of missing values equals or exceeds 50%.

2 Materials and Software

The dataset used for imputation in the present study can be downloaded from https://drive.google.com/open?id=1JfqqG5XlYcbC 1CGugsTdcVYVtD-WNKDt

2.1 Noncommercial

1. R version 3.5.2 or newer (https://www.r-project.org/).

2. RStudio version 1.1.463 or newer (https://www.rstudio.com/products/RStudio/).

3. R-packages:
    (a) ImputeLCMD: Available via CRAN through RStudio.
    (b) Impute: Download from the Bioconductor homepage: http://www.bioconductor.org/packages/release/bioc/html/impute.html
    (c) pcaMethods: Can be downloaded from the Bioconductor homepage: http://www.bioconductor.org/packages/release/bioc/html/pcaMethods.html

3 Methods

3.1 Download and Installation of RStudio and R-Packages for Imputation of Missing Data

1. Download and install R version 3.5.2 or newer (https://www.r-project.org/) and RStudio version 1.1.463 or newer (https://www.rstudio.com/products/RStudio/).

2. Open RStudio and click on the "Packages" tab in the right lower window.

3. Click "Install" under the Packages tab and a new window will open. Select "Repository (CRAN)" in the upper field. Check "Install dependencies." Write "BiocManager" in the lower field and click "Install."

4. Load the BiocManager package by checking the box next to "BiocManager" now listed in the packages list under the Packages tab.

5. Install the "pcaMethods" package by writing "`install ("pcaMethods")`" in the console on the lower left side of the window and press "Enter."

6. Install the "impute" package by writing "`install ("impute")`" in the console on the lower left followed by pressing "Enter."

7. Install the "imputeLCMD" by clicking on the "Install" button under the "Package" tab, writing "imputeLCMD" in the text field under "Packages." Click "Install." Ensure that "CRAN" is selected and "Install dependencies" is checked.

3.2 Imputation of Missing Data by Using R-Packages in RStudio

1. Import datasets by clicking the "Environment" tab to the upper right, select "Import Dataset" and then select the relevant file format option from the dropdown menu. If more than one sheet in excel file then select the correct sheet in the drop-down menu on the lower left. Click "Import" on the lower right.

2. Check the ticbox next to the "impute," "imputeLCMD," and the "pcaMethods" packages under the "Packages" tab to the right.

3. To impute missing values using the Min.Prob (minimal probability method), name the imputed dataset by typing "file name"<-impute.MinProb("name of dataset") in the console on the lower left (*see* for example **Notes 1** and **2**).
IMPORTANT: The dataset to be imputed must not have columns containing any text (except for column titles)—only numbers are allowed (*see* **Note 3**). Ignore columns with text and columns with irrelevant data (e.g., patient data, clinical variables) and include only columns with relative quantitative data from TMT/iTRAQ experiments by adding "[c ("column"x":column"y")]" to "name of dataset" typed in the console: "file name"<-impute.MinProb("name of dataset [c("column"x":column"y")]"). "x" denotes the first column and "y" denotes the last column in the dataset to be included for the imputation of missing values.

4. To impute missing values using the kNN method do as described under **step 3** in Subheading 3.2, but replace impute.Min.Prob() with impute.knn() and add "as.matrix" to command written in the console:
"file name"<-impute.knn(as.matrix("name of dataset[c ("column"x":column"y")]")).

5. (*See* **Note 4**). The k-parameter is set to 10 by default and the percentage of missing values is set to 50% by default. The setting of the percentage of missing values should reflect the maximum percentage of missing values in the actual dataset.

6. To impute missing values using the SVD method, use the `pca` `()` function and define the SVD method by adding `meth-od="svdImpute"` to the function written in the console: "file name"<-pca(as.matrix("name of dataset[c("column"x": column"y")]"), method="svdImpute").

3.3 Export Datasets from RStudio to Excel

1. Export the imputed datasets to .xlsx format using the "writexl" package. Download the "writexl" package from CRAN as described for imputation with the "imputeLCMD" package in Subheading 3.4.

2. Set working directory: Click "Session" tab in RStudio→click "Set working directory→Choose directory→browse for directory". Exported files will be stored in the selected directory.

3. Use the `write_xlsx()` function to export datasets from R to the directory specified as default. Write a unique file name in front of the ".xlsx" to differentiate between files, as in the example given above. Depending on the imputation method used, slight variations to the code are necessary for the `impute.knn()` and `pca()` functions:
   Min.Prob:

```
write_xlsx(name of imputed dataset, path="name of exported
dataset.xlsx"))
```

   EXAMPLE: write_xlsx(dataset_A, path="dataset_B.xlsx"), name of imputed dataset: dataset_A, "name of exported dataset.xlsx": "dataset_B.xlsx"
   impute.knn:

```
write_xlsx(as.data.frame(name imputed of dataset$data),
path="name of exported dataset.xlsx")
```

   pca:

```
write_xlsx(as.data.frame(name of imputed dataset@comple-
teObs), path="name of exported dataset.xlsx")
```

4 Notes

1. If you have any doubts as to how a function should be used, then write "?" and the name of the function to open the R Documentation for that function (e.g., `?impute.MinProb`). Examples of proper use are also often given in the documentation.

2. To prevent the imputed results from being shown in the console, define the output to a variable using "<-" (e.g., `variable<-impute.MinProb(dataset)`).

3. If a dataset includes columns with text and a functions does not allow this, it can be solved by excluding the columns containing text from the function by stating which columns contains only numbers (e.g., "`impute.MinProb(dataset[c(from: to)]`").

4. Depending on the structure of the dataset, the `impute.knn()` function might return an error such as "`error in storage. mode(x)`". This can sometimes be alleviated by using the `as. matrix()` function (e.g., `impute.knn(as.matrix(dataset)))`). The same problem and solution applies to the `pca ()` function.

Acknowledgments

Odense University Hospital Research Fund (Grant R22-A1187-B615) is acknowledged for financial support.

References

1. Kang H (2013) The prevention and handling of the missing data. Korean J Anesthesiol 64:402–406

2. Beretta L, Santaniello A (2016) Nearest neighbor imputation algorithms: a critical evaluation. BMC Med Inform Decis Mak 16(Suppl 3):74

3. Lazar C, Gatto L, Ferro M, Bruley C, Burger T (2016) Accounting for the multiple natures of missing values in label-free quantitative proteomics data sets to compare imputation strategies. J Proteome Res 15:1116–1125

4. Troyanskaya O, Cantor M, Sherlock G, Brown P, Hastie T, Tibshirani R et al (2001) Missing value estimation methods for DNA microarrays. Bioinformatics 17:520–525

5. Beck HC, Jensen LO, Gils C, Ilondo AMM, Frydland M, Hassager C et al (2018) Proteomic discovery and validation of the confounding effect of heparin administration on the analysis of candidate cardiovascular biomarkers. Clin Chem 64:1474–1484

6. Chich JF, David O, Villers F, Schaeffer B, Lutomski D, Huet S (2007) Statistics for proteomics: experimental design and 2-DE differential analysis. J Chromatogr B Analyt Technol Biomed Life Sci 849:261–272

7. Webb-Robertson BJ, Wiberg HK, Matzke MM, Brown JN, Wang J, McDermott JE et al (2015) Review, evaluation, and discussion of the challenges of missing value imputation for mass spectrometry-based label-free global proteomics. J Proteome Res 14:1993–2001

8. Beck HC, Nielsen EC, Matthiesen R, Jensen LH, Sehested M, Finn P et al (2006) Quantitative proteomic analysis of post-translational modifications of human histones. Mol Cell Proteomics 5:1314–1325

Chapter 14

Clustering Clinical Data in R

Ana Pina, Maria Paula Macedo, and Roberto Henriques

Abstract

We are currently witnessing a paradigm shift from evidence-based medicine to precision medicine, which has been made possible by the enormous development of technology. The advances in data mining algorithms will allow us to integrate trans-omics with clinical data, contributing to our understanding of pathological mechanisms and massively impacting on the clinical sciences. Cluster analysis is one of the main data mining techniques and allows for the exploration of data patterns that the human mind cannot capture.

This chapter focuses on the cluster analysis of clinical data, using the statistical software, R. We outline the cluster analysis process, underlining some clinical data characteristics. Starting with the data preprocessing step, we then discuss the advantages and disadvantages of the most commonly used clustering algorithms and point to examples of their applications in clinical work. Finally, we briefly discuss how to perform validation of clusters. Throughout the chapter we highlight R packages suitable for each computational step of cluster analysis.

Key words Cluster stability, Cluster analysis, Clinical data, Stratification, Cluster validation, Cluster tendency, Cluster optimization

1 Introduction

Clinical data, the data related to the health status of a subject, include several types of data, from various sources, including electronic health records (EHR), complementary diagnostic tests, administrative data, health surveys, and clinical trials, among others. In recent years, much of these data have been registered on large and highly complex databases [1]. Additionally, the use of wearable devices (Internet of Things), currently allow us to register real-time data 24 h a day, providing information on the subjects' lifestyle, vital signs, glycemia, and various other parameters. Therefore, modern, clinical data has high volume, velocity, and variety, which are considered the 3 V's of Big Data [2]. Frequently, data Veracity, is often considered the fourth V of Big Data, is also an issue. However, storage of this data is useless without further treatment, given that human mind cannot easily tackle these highly complex datasets. Fortunately, increases in storage capacity have

Rune Matthiesen (ed.), *Mass Spectrometry Data Analysis in Proteomics*, Methods in Molecular Biology, vol. 2051, https://doi.org/10.1007/978-1-4939-9744-2_14, © Springer Science+Business Media, LLC, part of Springer Nature 2020

been accompanied by increased computing capacity, as well as the development of data mining algorithms that permit the exploitation and understanding of the stored data, with large advances in diverse clinical fields. These tools introduce the fifth V of Big Data, Value, into the equation. Although the integration of clinical data with all the omics data of each subject, which we call humanomics, still poses challenges [3], it will drive advancements in clinical knowledge. With the increasing sharing of data repositories, this will be a common future task. These technological developments are the base for the current paradigm shift, in which evidence-based medicine is being replaced by precision medicine, an approach more suited to major current clinical problems such as cancer and metabolic diseases.

Cluster analysis, one of the main tasks in data mining, uses algorithms to partition data, to form groups of observations, or clusters, maximizing the similarity of observations within the groups and the dissimilarity between the groups. Given that usually the group of interest to which each observation belongs is not known a priori, unlike classification problems, this kind of cluster analysis is said to be unsupervised. Cluster analysis can be used for different purposes [4], and its application in the clinical setting has been growing [5]. It can be used to deepen understanding of the data, being an excellent tool for data exploration and visualization, and for the reduction of dimensionality. The main aim of the process is revealing patterns hidden by the complexity of data and extracting knowledge. For these reasons it is one of the most powerful data mining tasks in the clinical field.

Cluster analysis should be methodical and follow a process. However, it is a dynamic, flexible and iterative process, in which, we frequently go back and forth between steps. In each step, several decisions must be made. Although these decisions are increasingly based on objective indicators, there still exist subjective decisions that will impact the final clusters. Therefore, clinical knowledge, as well as a profound understanding of the decision-making process, is crucial for analyzing and comprehending the results and drawing conclusions. Although clinical expertise is crucial when conducting this kind of work, it is important to ensure it does not bias our results, while remaining sufficiently open-minded to further analyze and validate surprising or unexpected results, that can give rise to disruptive knowledge.

Throughout this chapter we will point out some particularities of clinical data which should be considered, as they can be relevant in several steps of the cluster analysis process. Given that there is a wide variety of clustering algorithms and implementations, with different strategies, we will focus on the most commonly applied ones, in the clinical setting.

R is a programming language and a software environment for statistical computing and graphics, within which statistical and data mining techniques are implemented, mainly through packages [6]. R packages extend or add functionality, and usually include examples of code and documentation relating to their functions. Many useful packages can be freely downloaded from online repositories, such as CRAN (The Comprehensive R Archive Network, the official repository, maintained by the R community) (https://cran.r-project.org) and GitHub (https://github.com), for example. Frequently, multiple packages execute the same task or algorithm, with different implementations and functionality (Table 1). Therefore, it is very important to have a deep understanding of the packages used, and how they work.

Table 1
The most common clustering algorithms and their advantages and disadvantages

Advantages	Disadvantages	R implementations examples
Hierarchical clustering		

• Does not need pre-specification of number of clusters
• Accepts any kind of distance function
• Visualization
• Agglomerative good at identifying small clusters, divisive better identifying large clusters
	• High computational cost, it does not scale properly
	• Difficult to alter once the analysis starts
	• Different clusters forms according to linkage function
	• More prone to identify spherical and convex clusters
	• Need to define the cophenetic distance cut-of
	• Sensitive to outliers
		Agglomerative
		  • **stat : : hclust ()**
		  • **cluster: :agnes()**
		Divisive
		  • **cluster::diana()**

k-means		
• Simple to implement and understand • Fast and efficient for large datasets	• Require specification of number of clusters • Sensitive to the randomly chosen seeds • Some implementations use only Euclidean distance	• **stats: :kmeans()** Euclidean distance only Only numeric variables • **clustMixType** Accepts mix data type
k-medoids		

PAM
  • Simple to understand and implement
  • Less sensitive to noise and outliers than *k*-means
  • Allows using general dissimilarities of objects
	• Require specification of number of clusters
	• Sensitive to random initialization of medoids
	• Higher computational cost than *k*-means
	• More prone to identify spherical and convex clusters
	• Does not scale well for large datasets
		• **cluster::pam()**
		Euclidean and Manhattan distances
		Only numerical variables
		• **fpc::pamk()**
		Prints the suggested K based on silhouette, does not require the user to define the number of clusters

(continued)

Table 1
(continued)

Advantages	Disadvantages	R implementations examples

CLARA
  • The same as PAM
  • Deals with larger datasets than PAM
	• The same as PAM
	• The efficiency of performance depends upon the size of the dataset
		• **cluster::clara()**
		k-medoids extension that deals better with data containing large number of objects

Sell-organizing Maps		
• Easy to understand and interpret • Deals with large and complex data sets • Finds different clusters formats	• Requires many parameters to be set and optimized • It is computational expensive • When initialized randomly, it is sensitive to the initial seeds • The number of clusters must be previously defined	• **Kohunen::supersont()** One or several grids Mixed variable types Different distance measures
Density-based		

DBSCAN
  • Identifies clusters of any shape
  • Does not need do pre-specify the number of clusters
	• High computational cost
	• Some algorithms require parameters optimization
		DBSCAN
		  • **fpc::dbscan()**
		  • **dbscan::dbscan()**
		OPTICS
		  • **dbscan::()**

Model-based		

EM
  • Simple implementation
  • Fits a model to data
  • Does not require the pre-specification of number of clusters
  • Deals with missing values
	• Slow rate of convergence
	• Not suitable for clusters with low number of observations and with collinearity issues
		EM
		  • **mclust**

Grid-based		
• Different shape clusters • Different density clusters • Low computational cost • Scalability • Does not require the pre-specification of number of clusters • Well suited for spatial data	• Require parameter setting and optimization • Some algorithms allows numerical data only	• **subspace**

Although some limitations of R have been pointed out, regarding its use in mining big data, where it has been considered less efficient than other languages, some solutions have been developed to overcome these obstacles. In particular, when working with big data, we can use packages which boost processing, or turn to cloud computing [7].

While we refer to some useful packages, it is out of the scope of this chapter to comprehensively list them all. To locate required R packages, the CRAN (https://cran.r-project.org) *tasks view* tool allows browsing of the packages by categories. Searching on RDocumentation (https://www.rdocumentation.org), a documentation aggregator of R, can also be very helpful. R offers a wide array of packages to accomplish a cluster analysis process. They are the result of the great effort of many contributors, and they should be acknowledged. In this chapter the following notation is used: package names are in bold (**package**), the packages' functions names are in bold followed by "()" (**function()**, or **package::function()**).

In this chapter, for simplicity, each step of the cluster analysis process is described in a different section (Fig. 1). To give some simple examples, we will explore an artificial clinical data set, comprising 2010 observations of six variables (categorical and numerical), which we refer to as "*Our_data*". As a development environment, we will use RStudio (https://www.rstudio.com) [8]. The contents of each section are as follows: Subheading 2 is devoted to data preprocessing; Subheading 3 focuses on feature extraction and selection; Subheding 4 deals with the choice of a distance measure; in Subheding 5 summarizes some of the most well-known and clinically applied clustering algorithms,

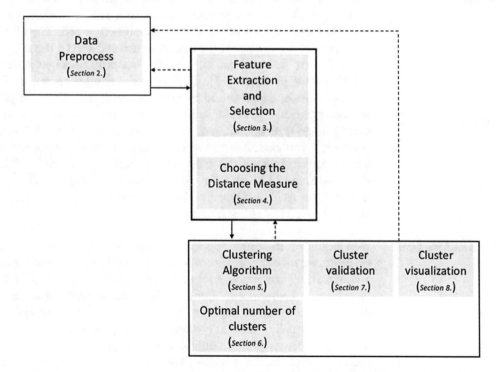

Fig. 1 Cluster analysis methodology scheme. The cluster analysis methodology is a dynamic, flexible, and iterative process, going forth (arrow) and back (arrow with dotted line) when needed

highlighting some advantages and drawbacks; Subheading 6 focuses on cluster tendency and determination of the optimal number of clusters; Subheading 7 deals with how to validate the resulting clusters; Subheading 8 discusses the choice of the clustering algorithm; and Subheading 9 briefly summarizes the cluster profiling.

2   Data Preprocessing

Data preprocessing is one of the most important and time-consuming steps when applying any data-mining tool. This step aims to render the data in an appropriate format and with a good quality for applying a given algorithm. Clinical data, in particular, has some characteristics, which make this a crucial step.

R has several packages devoted to the data-mining process, which also address preprocessing. Although some of these packages, such as **caret** [9], are oriented to prediction models, their preprocessing functions can be applied to preprocessing datasets for cluster analysis.

2.1 Structuring and Formatting the Data

Due to the variety of sources, clinical data can be in highly varied formats (sometimes, even in an unstructured format). Many algorithms in R use tidy datasets, where each variable corresponds to a column, each observation corresponds to a row, and thus, each variable value of an observation corresponds to a cell. Therefore, it is important ensure that the data is structured and in the appropriate format. This task can be accomplished with packages such as the **tidyverse package** [10]. This package includes several useful packages for data analysis and the preprocessing step. For example, the **tidyr package** [11] allows several columns to be gathered into one (**gather()**) or spreading of one column into multiple columns (**spread()**) and also allows us to deal with columns that might not be formatted as required (**separate()**, **extract()**, **unite()**). The **dyplr package** [12] allows to select columns (**select()**), filtering of observations (**filter()**) and addition of new variables (**mutate()**), alongside a host of other functions.

2.2 Identifying Errors

Depending on its source, clinical data can be prone to errors. For example, we may expect fewer errors when working with data from a clinical trial, compared with data from EHR. Errors should be identified and eliminated, after excluding the existence of an error pattern. Data visualization is an excellent tool to aid in this task.

Errors usually appear as outliers. They may be identifiable on a univariate analysis plot, or only become apparent when analyzing multivariate plots. For example, Fig. 2a plots the waist circumference distribution of a population (*Our_data*). In this plot, we can easily identify one error, given that the waist circumference cannot

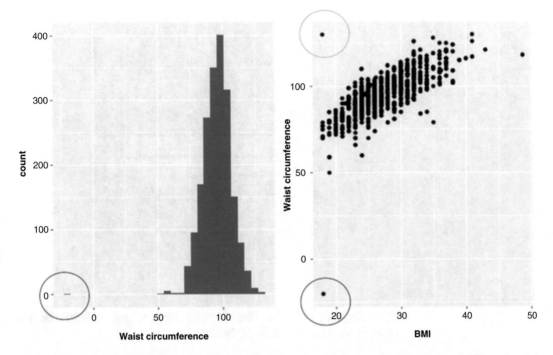

Fig. 2 *Our_data* plots: (**a**) Histogram of waist circumference distribution. (**b**) Scatterplot of BMI and waist circumference. The red circle points out an error, in which the waist circumference has a negative value and can be identified in both plots. The orange circle identifies an odd value, in which the minimum BMI has the maximum waist circumference

have a negative value. However, if we analyze the distribution of waist circumference plotted with body mass index (BMI) (Fig. 2b), we detect an odd value besides the previous error. The subject that has the maximum of the waist circumference and the minimum of BMI, might represent an error. The errors should be discarded, provided there is not enough information to correct them.

2.3 Dealing with Missing Values

Missing values are frequently present in clinical datasets, and require careful consideration, given that the majority of available algorithms do not deal with missing values [4].

If the missing values belong to the same column, we may consider to eliminate it. However there is not a preestablished cutoff to base this decision, which will depend on the information contained on the variable, despite the presence of missing values, as well as on the missing pattern present in the data.

If missing values are scattered throughout the data set, we must choose whether to (1) eliminate those observations, or (2) keep them and imputing those values. There are different possible methods to treat missing values and the best one depends on the missing pattern present in the data [13], and on the problem to be solved.

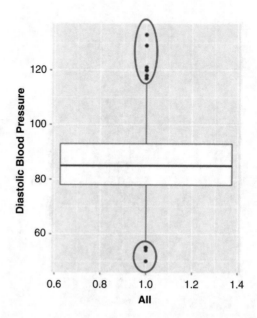

Fig. 3 *Our_data*. Box plot of Diastolic blood pressure, (**ggplot2** [85]). The blue circle shows outliers considering the IQR

In the first approach, we discard the observations with the missing values, after excluding the existence of a missing value pattern, provided we have a sufficient number of observations. Therefore, we trade the bias of imputing the missing values for the number of observations. This should be the preferred strategy whenever possible. The second option might be useful, if we have a smaller number of observations and can accept the bias of imputing missing values. In this case, we can employ alternative imputation methods (*see* **Note 1**). Still, in R, there are functions that implement clustering algorithms which can deal with missing values, like **kohonen::supersom()** [14], and therefore, in this case we do not have to eliminate or replace them at this stage. However, this parameter must be set, and these algorithms eventually exclude these observations. Whatever option has been taken, it must be considered in the analysis of the results.

2.4 Outliers

Outliers are odd observations that do not result from error. They are usually defined by establishing a dispersion measure cutoff value. Visualization methods can be very helpful in identifying outliers. For example, a histogram or a boxplot (Fig. 3) of a numeric variable can detect the presence of outliers.

When exploring a clinical data set with outliers we should first understand their meaning. We can choose to include them throughout the analysis, or exclude them from the remaining dataset, and explore them separately.

If we decide to include these observations, we should be aware of their impact on distance measures (Subheading 4) and clustering algorithms (Subheading 5) which have different behavior with outliers. Given that the data used in most clustering algorithms are scaled, the presence of outliers might increase the difficulty of finding clusters.

Outliers can be identified using several different strategies. First, by using a dispersion measure cutoff, above or below which the values are considered outliers and are excluded. This can be accomplished with **dplyr::filter()** [12], and with R basic functions for standard deviation, interquartile range, or median absolute deviation (**sd()**, **iqr()**, **mad()**), among others, however, defining the cutoff value is not straightforward. Another approach is using a cluster algorithm (*see* **Note 2**) to identify the outliers based on the considered variables. Like other preprocessing steps, this decision should be considered when analyzing the results and drawing conclusions.

2.5 Data Transformation

In this section we will cover several aspects of preprocessing: data transformation to achieve a normal distribution, binning numerical variables, encoding categorical variables, and scaling the data.

For a cluster analysis it is not mandatory that data must follow a normal distribution. However, this may be important for the performance of some algorithms [4]. The transformation can be achieved through the usual methods (logarithm and square root transformations, for example). This strategy may unearth patterns that would not be otherwise revealed, but the analysis of the results might not be as straightforward as when using the original data. Thus, backward transformations may be required to understand the results.

The majority of cluster analysis algorithms work better with numerical variables than categorical variables (Subheading 4). However, in clinical problems, it is common to be more interested in the behavior of several classes of a variable, than in its numerical distribution. In this case we must transform a numerical variable into a categorical variable. For example, binning the BMI variable into the different classes (lean, overweight, etc.), or the estimated glomerular filtration rate (eGFR) into kidney dysfunction classes, etc. If the classes ranges are already established, as in the previous examples we can use **dylpr::mutate()** [12], together with **ifelse()**, to define them. However, if the classes are not established, we can use, for example, **RcmdrMisc::binVariable()** [15], with which we can define the number of bins as well as the binning method to be used.

The majority of clinical datasets will have mixed data (categorical and numerical variables). Unfortunately, some clustering algorithms accept only numerical variables, and for those, it becomes necessary to address the categorical variables. One way to tackle them is to flag the variables, or transform them into dummy variables, in a process called encoding. When we encode a categorical variable, we get $k - 1$ new variables (being k equal to the number of classes present in the variable). That might not be a relevant issue when encoding binary variables, like gender for example. However, in variables with more classes, we might be negatively contributing to the sparsity of the data. There are R packages that accept categorical variables, in which this transformation is encapsulated in the package and executed before running the clustering algorithm. In these cases, there is no requirement to encode the variables. Another approach is to transform categorical variables into numerical variables. However, this is only possible when the classes can be ordered in some way, and the differences between the classes can be considered equal. For example, BMI classes can be considered ordered classes. However, in the case of a categorical variable in which each class represents different eating patterns for example normal, volume eater and sweet eater, transforming it into a numerical variable by ordering the classes, would be a serious mistake.

Many clustering algorithms use distance measures as dissimilarity measures. In these cases, data should be scaled, which can be done with several methods [4], namely the min-max normalization, the Z-score standardization, and decimal scaling (*see* **Note 3**). As many variables have different scales, if we do not scale the data, we are giving more importance to the variables with higher values, based on an apparent higher distance. For example, in a data set with the two nonscaled variables, blood creatinine, and mean blood pressure, the algorithm will find clusters based more on blood pressure than on creatinine, given that the blood pressure values are much higher than the creatinine values. By scaling them, they will have comparable ranges and similar impact on the algorithm. Although these methods can be applied by writing simple functions in R, there are some packages that can be used to accomplish this task. To apply the Z-score we can use **scale()** from the R base functions [6]. With **rescaler()** from the **reshape package** [16], we can scale the data using one of three methods, the Z-score, the Min-Max normalization or the Median/Mad. Lastly, in some cases we may need to scale a variable, depending on another, usually categorical, to find some patterns of interest. For example, it is already known that there is gender linked differences for the risk cutoff values of some metabolic disease risk factors. Thus, in exploring risk factors such as waist circumference for example, we would scale them by gender. This task is easily done using the **dplyr package** [12], as well as other previously mentioned R functions.

3 Feature Selection and Feature Extraction

In high dimensional data sets, the high number of variables translates into an important sparsity of the data in the hyperspace. This phenomenon, known as curse of dimensionality [17], can occur when tackling a clinical clustering problem with a high number of features. Some strategies have been developed to minimize it, by reducing the dimensions, namely feature selection and feature extraction.

3.1 Feature Selection   When describing a phenomenon, some variables are more important than others. The feature selection step tries to identify the most important variables for solving the clustering problem in hand, reducing the dimensionality. This step will contribute to improve learning performance, lowering storage and computation complexity and therefore diminishing the required time to run the models and produce better results. Because feature selection maintains the original data, it is more readable and prone to interpretation, which is a really important issue in the clinical field [18, 19]. However, although in classification and regression analysis we have an outcome of interest that can aid to identify and select the relevant features for the analysis, in cluster analysis this is not as straightforward. Also we should keep in mind that different subsets of selected features will differently lead to different cluster solutions [20].

Still, there are some helpful strategies [21]. Differently from a supervised problem, in which the aim is to find the more relevant features to the prediction, in the unsupervised problems the most relevant features will be the ones that capture most information of the dataset. Therefore, besides improving the cluster performance, we simultaneously gain insight about the most relevant features.

We can readily discard the variables with a very low variability, those that have a high value of missing values or the ones that are highly correlated. For that, a cutoff must be established, based on the dataset, and on the problem to solve [4].

Clinical knowledge is fundamental in this step and should always complement any other strategy. Furthermore, in many clinical clustering problems we might do a knowledge-based feature selection, thus, not using any kind of data-driven approach. Moreover, if we support our choices only in the data mining algorithms, we might choose some features that can be good in defining the clusters, however might be clinically irrelevant. However, cluster analysis is first and foremost an exploratory tool, and as in other steps of this kind of analysis, it is important that we keep a critical open mind, and do not let the knowledge limit the pattern discovery, for we might miss some relevant new knowledge. Furthermore, with the increasing analysis of conjoined datasets comprehending clinical data, as well as omics data, for example, the data driven approach can turn out crucial for our analysis.

Another approach is selecting the features based on a third one. For example, we may select clinical features, based on genomic data of the subjects. In the latter case, we would treat it as a feature selection for a supervised problem.

When using an algorithm to select the variables, four things can happen. First, we can find meaningful variables for the cluster that are clinical relevant, so we keep them. Second, we find meaningful variables for the cluster that are considered to be clinical irrelevant. In this case, we should be very careful before discarding them, for there might be hidden knowledge. Third, we can find unmeaningful variables for the cluster that are clinical irrelevant and thus, we can discard them. Lastly, if we unexpectedly find unmeaningful variables for the cluster that are thought to be clinical relevant, it can also represent knew knowledge. When a subset of features is selected using a pure data driven strategy, we must validate its stability by adding perturbation to the data.

The data driven strategies for feature selection include feature ranking, and subset selection techniques. For more details on this subject the reader is redirected to Alelyani et al. and Pacheco [19, 22].

In the feature ranking a scoring is given to each variable (e.g., **FSelector package** [23]). It is independent of the cluster technique, but each variable performance is analyzed individually, and the user must set a cutoff value. It can be used to support a knowledge-based selection.

In the subset selection techniques, a search of the variables space is done, with a heuristic approach and considering a quality measure. Besides selecting the features, they also determine the size of the subset, with a higher computational cost. In these methods, the variables are analyzed all together, unlike the ranking method. Also, they can be independent (filter selection) or dependent of the clustering algorithm (wrapper, embedded). The filter selection method uses a quality measure independent of the clustering algorithm to be used. For example, **findcorrelation()** of the **caret package** uses correlation as quality measure for filter selection [9]. The wrapper algorithms evaluate the quality of feature selection, based on the clustering results quality. As such, they are dependent on the clustering algorithm that is used, and thus might have some drawbacks. If we are not using the best clustering algorithm they might not be the real relevant features. The **clustvarsel package** [24], is an example of an implementation of this method for model based clustering algorithms (Subheading 5). Some of these algorithms are already a part of the clustering algorithm and are called embedded methods. It is the case of the **wskm package** [25], that uses the k-means clustering algorithm (Subheading 5).

3.2 Deriving New Variables and Feature Extraction

In the clinical setting is very common the usage of indexes, resuming several variables, for which a clinical meaning was asserted. Using these indexes can serve two purposes. First, it can help in dimensionality reduction. For example, instead of the *Weight* and the *Height* variables, we might use the *BMI* variable, thus eliminating one feature of our analysis. Likewise, we can use the *Mean Blood Pressure* instead of the *Systolic Blood Pressure* and *Diastolic Blood Pressure* variables. The second purpose is to reveal patterns that are not revealed when using the variables separately. For example, we might not find any interesting pattern when analyzing the *Blood Creatinine*, but this pattern can be obvious when analyzing the *Estimated Glomerular Filtration Rate* (eGFR), a very well-known estimated index of renal function. For deriving new variables, the functions **mutate()** and **ifelse()**, from the **dplyr package** [12] can be used (*see* **Note 4**).

Feature extraction is a strategy of dimensionality reduction based on transformations of the original dataset. Given that data is transformed, it may turn out more difficult to understand and interpret the results. Principal component analysis (PCA) [26], is probably the most widely used and known method of feature extraction, however it only deals with numeric variables. PCA computes new variables (eigenvectors), that are a linear combination of the original ones, ordered by the explained variance of the data (eigenvalues). It can be done using two different strategies: the spectral decomposition approach, focusing on the covariances or correlation matrix between the variables, and the value decomposition approach, which uses the covariances or correlations between the individuals [21]. For the first approach we can use, for example, **princomp()** from the **stats package**. **procomp()**, also from the **stats package** [6] and **PCA()** form the **FactoMineR package** [27], uses the latter approach.

4 Dissimilarity Measures

The aim of any cluster algorithm is to group together similar instances. Thus, a dissimilarity measure must be defined, which depends on the variable type. Despite some algorithms implemented in R require users to provide a dissimilarity matrix, many allows the setting of an optional proximity measure to calculate it. However, the latter, might not admit any kind of categorical dissimilarity measures, which we should be aware of, when choosing the dissimilarity measure as well as the algorithm to be used (Subheading 5).

Defining the best similarity measure for the variables is crucial for the goodness of the results. For numeric variables, the similarity is usually represented by a distance or correlation measure (Euclidean, Manhattan, Pearson correlation, and others) [4]. However, for categorical variables this is not straightforward, and this can be handled differently according to the variable in question. One way to deal with it is to use a similarity measure that does not have a distance connotation (Hamming distance, matching coefficient, Tanimoto distance, and Jaccard distance). Another strategy, if the variable is binary or with some ordinal variables, is to treat them like a numeric variable. When dealing with a categorical variable with more than two classes, which do not have a specific order, or in which we cannot assume the same distance between the classes, we cannot treat it this way, for we will be introducing serious errors in the analysis. In this case, it can be encoded to a dummy variable. This can be problematic, even impossible, if the variable has many classes, which can easily happen specially in cases integrating clinical with omics data.

If the algorithm we want to use requires for a dissimilarity matrix, there are some useful packages in R. **daisy()** from the **cluster package** [28], returns a dissimilarity matrix allowing to choose between different distance measures, and can handle mixed type data. Also, **get_dist()** from the **factoextra package** [29] can be used to accomplish this task. Although the later package can handle only numeric and binary categorical data, we can compute correlation-based distance. When choosing a correlation-based dissimilarity measure, the resulting clusters will have observations that are well correlated, although they might be far apart in the multidimensional space.

5   Types of Clustering Algorithms

Clustering algorithms are usually classified in partition, hierarchical, density based, grid-based, and model-based algorithms. However, given the increasing variety of existent clustering algorithms, they are already being classified in nine categories: based on partition, based on hierarchy, based on fuzzy theory, based on distribution, based on density, based on graph theory, based on grid, based on fractal theory, and based on model [30]. Given the variety and number of existent algorithms, it would be impossible to discuss all. Also, the description of these algorithms have been done in several excellent works, for which we refer the reader [31–33]. Therefore, our aim in this section it is to summarize the advantages and drawbacks of the most well-known and used algorithms, referring some R implementations (Table 1), and clinical work that have used them. We do not intend to make a literature survey, but illustrate the algorithms used to solve some clinical problems.

5.1   Hierarchical Clustering

Hierarchical clustering is probably the most used in the clinical field. These algorithms build a hierarchy of nested clusters by both a bottom-up (agglomerative) approach and a top-down (divisive) approach. The nested clusters are usually represented as a tree-graph, the dendrogram (Fig. 4), a very powerful visualization tool because it is very intuitive and easily understandable.

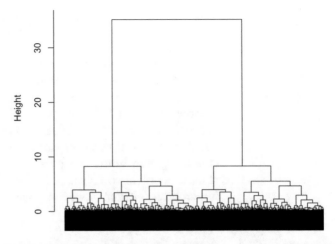

Fig. 4 Dendrogram. The **hclust()** was applied to *Our_data*. The Euclidean distance measure and the Ward's linkage method were used

The hierarchical clustering algorithm builds a cluster solution, without requiring the prespecification of the number of clusters. Still, it requires a cutoff for the cophenetic distance to determine the number of clusters, which can be determined by statistical measurement of large jumps in the similarity measures. Moreover, the hierarchical cluster algorithm implementations in R, usually require a dissimilarity matrix, and therefore can use any kind of distance function. If the data set contains mixed type variables, we can use a function that deals with them in calculating a dissimilarity matrix (e.g., **daisy()** from the **cluster package** [28]). Whereas the agglomerative algorithm is better for identifying small clusters, the divisive approach performs better at identifying larger clusters. The hierarchical clustering has also some important drawbacks. They have a high computational cost, and therefore they do not scale properly. The clustering solutions depend on the chosen linkage method (*see* **Note 5**). They also are sensitive to the presence of outliers and are not suitable to identify nonconvex clusters [33].

The hierarchical algorithm [34] is very well-known and frequently used in bioinformatics, in the omics field. In the clinical setting it has been used to tackle problems in very different settings, such as pneumological diseases [35], rheumatologic diseases [36], and metabolic diseases [37]. Also, it is widely used in combination with other algorithms, aiding in visualization and in determining the number of clusters [38]. Despite an increasing number of cluster analysis being done to phenotype and stratify the diseased population, some of them present a limited number of subjects. Additionally, for the same clinical problem, the used methodology can be very heterogeneous, regarding the used algorithm and their parameters, the selected variables, etc. [39]. This might turn more difficult to compare and summarize the results, and thus to draw conclusions. Therefore, whenever possible, we should aim at homogenizing the cluster analysis regarding specific clinical problems.

hclust() from the **stats package** [6] is one of the most used R hierarchical agglomerative algorithm (AGNES) implementations. Another example of the implementation of AGNES is **agnes()** from **cluster package** [28]. This package also has an implementation of the divisive algorithm DIANA (**diana()**). They all require a dissimilarity matrix, and the user can set the linkage method. Different clustering algorithms can perform differently, depending on the clinical problem to solve. For exploring and comparing different hierarchical cluster solutions we suggest using **Dendextend package** [40].

5.2 k-means

k-means is one of the most well-known and used partition algorithms. It is a distance-based algorithm, in which, each cluster is represented by its centroid (the mean of the observations in the group). It is very simple to implement and understand, and it is fast and efficient for large datasets [41]. However, given that it is a distance-based algorithm (usually uses the Euclidean distance), it does not deal with categorical variables, without further transformation, and is sensitive to outliers. Furthermore, final clustering solution it is highly impacted by the random initialization of the seeds [42]. Also, besides requiring the user to previously specify the number of clusters, it is more prone to find spherical clusters, even if this is not their natural shapes [43]. Not being able to find non-convex clusters, it can be limited in solving some complex clustering problems.

Together with the hierarchical cluster algorithm, *k*-means has been one of the most used clustering algorithms to solve clinical problems. Ahlqvist et al. used an R *k*-means implementation, as well as a hierarchical cluster to stratify diabetic subjects, based on five variables [37]. Also, it has been used to investigate Parkinson's disease heterogeneity [44], or identifying patterns of general practitioner service utilization in people with diabetes [45]. Regarding heart failure patients stratification, and using a *k*-means clustering algorithm, Ahmad et al. identified four clinical distinct phenotypes, that differed significantly not only on the outcomes but also in response to therapeutics [46].

kmeans(), from the **stats package** [6], implements the *k*-means algorithm and deals only with numeric variables. Moreover, this package uses only the Euclidean distance. **kmeans()** from the **amap package** [47], like the one from the **stats package**, also uses only numerical variables, but differently, allows to choose a distance measure other than the Euclidean. There are some specific extensions of the *k*-means algorithm. For example, the **clustMixType package** [48], is an extension of *k*-means that can deal with both numerical and categorical variables.

5.3 *k-medoids*

k-medoids is an algorithm related to *k*-means. However, in *k*-medoids, each cluster is represented by the observation that minimizes the average dissimilarity between all the other members within that group (medoid). Although *k*-medoids is considered a more robust algorithm than *k*-means, it does not necessarily perform better on real clinical datasets [49].

   k-medoids, as *k*-means, is easy to understand but less sensitive to noise or outliers. Moreover, this algorithm implementation usually allows other distances measures, besides Euclidean distance. However, it also requires the prespecification of number of clusters, it is sensitive to the random initialization of medoids and it is not suitable to find nonconvex clusters [42]. Additionally, the higher computational cost, compared to *k*-means, is one of the reasons it is not widely used. In R it is implemented by **pam()** of the **cluster package** [28]. The CLARA algorithm, which is implemented in R by **clara()** of the **cluster package** [28], is an extension of the *k*-medoids that is more efficient dealing with a data with high number of observations.

5.4 *Self-Organizing Maps*

Self-organizing Maps is a neural network-based algorithm [50], that maps its observation to a neuron, which will be the centroid of a given cluster (Fig. 5). At the end of the training, the input space

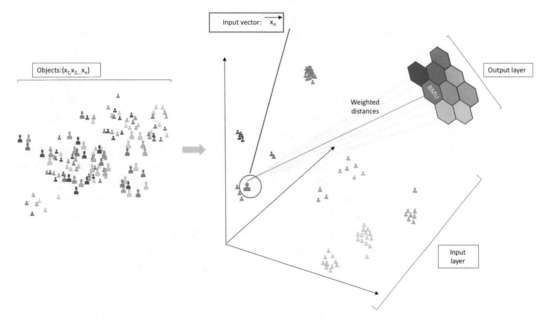

Fig. 5 Self-organizing map scheme: The input vectors correspond to the population objects in the input layer (multidimensional space). In each iteration, the weighted distances of each vector ($\vec{x}_n$) to every grid unit are calculated, and the unit with the lowest distance (Best Matching Unit—BMU) moves toward the input vector. In the output layer (a topological 2D grid), each unit corresponds to the centroid of the objects that it represents

(observations in the d-space) is represented on a 2D grid of neurons that preserves the topological relations between the data. It uses two very intuitive visualization tools: the U-matrix and the component planes. In the U-matrix we can visualize the distance between the neurons, and subsequently the distances between the clusters. The component planes are a heatmap, one for each parameter, representing the variables distribution throughout the neurons or clusters [51].

SOMs have been widely used as a data dimensionality reduction and visualization tool, given its characteristics. It is easy to understand and interpret. It deals with large and complex data sets [50], and finds different clusters formats [52]. However, SOMs also have some disadvantages. It requires many parameters to be set and optimized and is computational expensive. When initialized randomly, it is sensitive to the initial seeds. The number of clusters must be defined previously [53, 54]. To surpass the latter issue, we can use two different strategies: (1) by calculating the optimal number of clusters based on a quality measure (Subheading 6.2) and (2) by using an emergent SOM [55, 56]. In this algorithm we use very big grids, which allow us to better understand the data structure and get the possible number of clusters. Then applying another clustering algorithm on the top of a SOM (typically a hierarchical clustering algorithm) can guide us in defining the optimal number of clusters.

supersom() of the **kohonen package** [57], can deal with mixed type variables, and allows to run a simple SOM algorithm, or to join different kinds of variables in simultaneous grids. Additionally, it permits to use different dissimilarity measures. This function is an implementation of an extension of the original algorithm, using several parallel grids to solve a clustering problem. This can be relevant in the clinical field, for we can analyze different kinds of information separately.

SOMs have been used in several clinical clustering problems [58, 59]. Additionally, supersom extension in R [57], which allows grouping variables in different grids, also shows advantages in clustering clinical data. Toppila could find different phenotypes, regarding T1D patients complications, using a supersom, followed by a hierarchical algorithm [38]. Additionally, Pina et al. using a supersom to stratify a population based on metabolic parameters, found different phenotypes regarding metabolic disease [60].

5.5 Density-Based Clustering

The density-based algorithms are developed, based on the notion that clusters are dense regions in the data space, separated by lower density regions. There are many different algorithms with advantages and limitations (e.g., DBSCAN [61], OPTICS [62]). Overall, they have the advantages of identifying clusters of any shape and do not need the prespecification of the number of clusters. However, they have a relatively high computation cost, and most of them

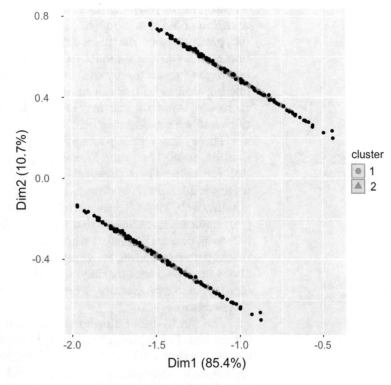

Fig. 6 Density based clustering. The **dbscan::dbscan()** was applied to *Our_data*, and is plotted with the **factoextra:fviz_cluster()**

require the optimization of some parameters (e.g., cluster radius and minimum number of objects for DBSCAN, the density threshold for OPTICS).

For the DBSCAN implementation in R, **fpc::dbscan()** [63] and **dbscan::dbscan()** [64] can be used. The **dbscan package** has also an implementation of the OPTICS algorithm that, unlike the DBSCAN, can be applied to data of diverse densities. Additionally, it includes **knndist()**, to optimize the cluster radius. We refer the reader to the packages documentation. The solutions derived from the mentioned packages can be visualized with **factoextra:: fviz_cluster()** [29] (Fig. 6).

Density-based algorithms have been less used, compared with the previous mentioned methods. There are more reports of their use in the fundamental sciences than in the clinical field. However, given their nature and characteristics it can be worthwhile to know them, for they might be convenient in solving some clinical clustering problems, particularly in combination with other algorithms, for example in radiomics [65].

5.6   Model-Based Clustering

In the model-based clustering, the data is a mixture of the probabilities distribution of two or more clusters. It uses a soft assignment of the objects to the clusters, as the probability of belonging to each cluster [66]. For this methodology the expectation-maximization (EM) algorithm [67] is the most widely used to estimate the parameters of the finite mixture probability. It is simple to implement and fits a model to the data [68]. Moreover, it does not require to prespecify the number of clusters, and it deals with missing values. Still, it can have a slow rate of convergence. Furthermore, is not suitable for clusters with low number of observations and with collinearity issues [69].

There are several R packages to apply a model-based clustering method, using Maximum Likelihood, Bayesian, and other estimation methods. These packages which offer different initialization methods are suited for different types of data (https://cran.r-proj ect.org/web/views/Cluster.html). For example, **mclust package** [70] implements functions that uses the EM algorithm for parameter estimation for normal mixture models, with different covariance matrix parametrization, for a range of k components. To learn more about these packages we suggest reading "A quick tour of mclust" (https://cran.r-project.org/web/packages/mclust/vignettes/mclust.html).

The model-based clustering has been successfully applied to solve some clinical problems, namely in the medical image field. For example, using an EM clustering algorithm, Hwang proposed a technique to automatically detect bowel bleeding regions, in Wireless Capsule Endoscopy videos. The authors claim that the proposed bleeding detection method achieves 92% sensitivity and 98% specificity [71].

5.7 Grid-Based Clusters

In grid-based clustering methods, a set of grid cells is defined in the object space. Then, each object is assigned to a cell of the grid, and the densities of each cell are computed. The clustering is done on the grid, not on the data. STING, Wave Cluster, and CLIQUE are examples of this kind of algorithms. The grid-based algorithms have several advantages. They have fast run-times, are scalable, and are able to detect clusters with different shapes and densities [72–74]. Besides, they are efficient when applied to solve large multidimensional data clustering problems. However, they still require parameter setting and optimization. Furthermore, they can perform worst if there are highly irregular data distributions, with local important variations. Also, some algorithms only allows numerical data [74].

The **subspace package** is an interface to "OpenSubspace," an open source framework for evaluation and exploration of subspace clustering algorithms in WEKA. To the best of our knowledge, at this time, there is no pure grid-based clustering algorithm fully implemented in R.

The grid-based algorithms, given its characteristics have been used in genomics, and also on medical image processing, in conjunction with other algorithms [75], and spatial epidemiological problems [76].

6 Cluster Tendency and Optimal Number of Clusters

One important question in cluster analysis is if the data has natural clusters. As discussed, every cluster analysis will provide a clustering solution, which will vary according to the selected variables, the dissimilarity measure, and the clustering algorithm used and its parameters. Thus, we should first aim to know if there are natural meaningful clusters. Also, for some algorithms we will have to know the number of clusters. Last, we want to assure that they are clinically relevant, regarding for example the diagnosis, the disease progression, the prognosis, or therapeutic outcomes.

6.1 Cluster Tendency

To evaluate the cluster tendency, we can use a mixed approach of data visualization and PCA analysis. If the data have some natural clusters, based on the used variables, the PCA plot might reveal them. Also, Hopkins statistics, and visual assessment of cluster tendency (VAT) (*see* **Note 6**), implemented in R, might be useful in this task. To deepen this subject we suggest reading "Assessing Cluster Tendency" in Kassambara [77].

6.2 Optimal Number of Clusters

The number of clusters is one of the parameters that have to be set in some algorithms, such as *k*-means and SOM, among others. Considering the same variables in the same dataset, it would be expectable to find the same number of clusters through the execution of all the algorithms. However, given that the performance of the algorithms is very different, considering the complexity of the problem, this is very often untrue.

The optimal number of clusters might be determined based on the field knowledge. However, there are some data-driven strategies that help us with this task.

One very well-known method, for defining the number of clusters, is the elbow method (*see* **Note 7**). Originally is computed with the within cluster sum of squares (WSS), but several other internal validity indexes (Subheading 7) can also be used. Additionally, the average silhouette method, used as a cluster validation method, can equally be used to assess the optimal number of clusters. **fviz_nblusct()** of the **factoextra package** [29] implements both these methods in R (Fig. 7). Although this function allows to compute and to visualize the plots for the elbow, silhouette and gap statistics methods, it only accepts to use *k*-means, PAM and the hierarchical algorithms.

There are several internal quality indexes that can be used to assess the optimal number of clusters. Similarly, to the elbow method, these indexes can be plotted with the *k* number of cluster,

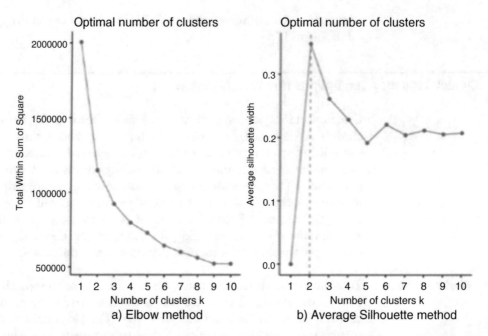

Fig. 7 Defining the optimal number of cluster in *Our_data,* with the **factoextra::fviz_nblcust**. The *k*-means algorithm was used, defining *k* with the (**a**) Elbow method (method = "wss") and (**b**) The average silhouette method (method = "silhouette")

giving a visual idea of their optimal number. **NbClust()** of the **NbClust package** [78] uses 30 indexes to evaluate the optimal number of clusters and gives several individual and summarized information as the output. The user must set the distance measure, and the clustering method. Although it accepts the most used algorithms, like *k*-means and hierarchical cluster, it does not permit algorithms like SOM for example.

7   Cluster Validation and Cluster Stability

Besides assuring that the data has natural clusters, it is important to assess the goodness of those clusters, assuring that the cluster structure is appropriate to the data. Plus, comparing the clustering solutions throughout different populations, it can bring consistency to the results, a very important issue in the clinical setting. Additionally, we should assess the cluster stability, assuring that our clustering structure is not derived from randomness.

7.1 Cluster Validation

We can consider three types of cluster validation analyses [79], for which we will use proper cluster validation criteria (*see* **Note 8**). In the external cluster validation, we compare the clustering solution structure with a known structure, using external validation criteria. In the internal cluster validation, we assess how well the clustering solution fits the data. Finally, the relative validation compares the execution of the same clustering algorithm with different parameters. For the internal and relative validation, internal criteria are used. Given that clustering analysis are unsupervised problems, the internal validity indexes are the most used in cluster validation, and therefore, in the remaining of this section we will focus on these criteria. Nevertheless, the external validation, of the results, with different data sets is critical in the clinical setting, for the consolidation of the extracted knowledge, and it should be done, whenever possible, particularly with different geographic populations.

The internal validation criteria are commonly based on the compactness and separateness of the clusters. They can also assist in the choice of the clustering algorithm, of the cluster number, and in the algorithm parameters optimization. As already mentioned in Subheading 6.2, **NbClust()** of the **NbClust package** [78] provides 30 indexes to internal validate the cluster solutions. **fpc** [63], **clv** [80], and **clValid packages** [81], further discussed in the next section, can also be used for this purpose. **factoextra::fviz_silhouette()** [29] allow us to easily plot silhouette coefficients of observations [77], which we want to maximize (Fig. 8) Additionally, some algorithms implemented in R, have an embedded internal validation index, that is used to choose the best result from the several iterations. That is the case of **kohonen::supersom()** [57], for example.

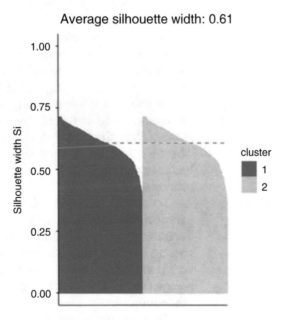

Fig. 8 Clusters silhouette plot. A *k*-means solution of *Our_data* with *k* = 2 was used (**factoextra::fviz_silhouette()**)

7.2 *Cluster Stability*   Cluster stability evaluates to what extent the defined clustering structure is present in the data. Thus, if the cluster solution is stable, it must be steady when challenged with small perturbations in the data.

There are several R packages that tackle the cluster stability subject, as **clv** and **clValid**.

cls.stab.sim.ind() and **cls.stab.opt.assign()**, from the **clv package** [80], evaluates the cluster stability. These functions can use hierarchical, k-means, and k-medoids algorithms, applied to a subset of the data.

clValid package [81] is directed at validating cluster analysis results. It allows to choose between nine different cluster algorithms (hierarchical, k-means, k-medoids, fuzzy clustering, SOM, model-based clustering, and others). It has different validation measures available, including the stability measures: average proportion of nonoverlap (APN); the average distance (AD); the average distance between means (ADM) and; the figure of merit (FOM). The stability is evaluated by comparing these measures between our solution and the solutions derived by removing one column at a time.

8 Choosing a Cluster Algorithm

Choosing the best clustering algorithm is an important decision that must consider different factors. From the problem to solve, the dataset characteristics, as well as the clustering algorithm implementation, computation time, etc.

In simple cluster analysis problems, with well-separated spherical clusters, most of the algorithms can find their structure, and independently of the algorithm used, we probably get a similar cluster solution. Unfortunately, often clinical problems are high-dimensional and complex, with noisy data, and irregular cluster shapes. Thus, the cluster algorithm selection is not straightforward. The "No free lunch theorem" [82] states that no algorithm is better in solving all the problems, so one should explore different solutions.

To choose between different clustering algorithms, we compare internal validation indexes between the different solutions. Fortunately, there are some R packages that assist us in this difficult task. In fact, the previous mentioned **clValid package** [81] can be very useful. Still, it does not include all the algorithms and their parameters. In those cases, we might have to directly compute the validation indexes.

Another more recent strategy is the ensemble clustering [83]. The notion that some complex algorithms structures cannot be captured by a single clustering algorithm, led some authors to develop a methodology, that similarly to the ensemble methods for classification and regression, take advantage of combining several clustering algorithms to get a better final solution. As we aim to increasingly integrate several types of complex data, namely integrating clinical and various omics data, these methods might show to be relevant in these cases. The **diceR package** [84] is an implementation of the ensemble cluster in R, that the reader might want to explore.

9   Cluster Profiling

The clustering profiling is the most thrilling step of the process because it is at this stage that new knowledge might be disclosed. It aims at deeply characterize and understand each cluster. Additionally, we compare the clusters, and test clinical hypothesis with standard statistical methods and tools.

To understand a clustering solution is important to know which variables are the most relevant to explain each cluster. Cluster visualization is one of the most important tools in this step. However, given that at this stage we already know the cluster to which each observation belongs, this question can be solved in a supervised manner.

Besides classical statistical methods, that allow us to test clusters characteristics, and compare them, cluster visualization is one of the most essential tools to cluster profiling.

Some algorithms are usually evaluated by some specific visualization tools that aid us in clustering profiling. This is the case of the hierarchical clustering, which dendrograms can be customized, and analyzed together with heatmaps, for example. Also, the SOM provides the U matrix and the component planes, both very intuitive visualization tools (Subheading 5). However, to deepen the visual analysis of the results of these and other clustering algorithms, we can use several of a wide pallet of visualization tools.

In R, there are a lot of packages aiming at visualization. The **ggplot2 package** [85], also a part of the **tidyverse**, is a widely used package for data visualization, based on the Grammar of Graphs [86]. It gives a variety of options of visualization, allowing the user to build different kinds of plots, adding different layers to the same plot and customizing them. Given that data visualization is a very important task, throughout all the cluster analysis process, the reader might want to learn more deeply about this package, for which we suggest reading *ggplot2:Elegant Graphics for Data Analysis* by Hadley Wickam [85].

10   Conclusion

Cluster analysis applied to clinical problems, is one of the most powerful data mining techniques, allowing data exploration, visualization, and the extraction of patterns that the human mind could not otherwise understand. It has already given important insights in several clinical fields. Clinical heterogeneity is increasingly recognized in several conditions and diseases and leads the need to stratify the population. In these scenarios, the "one size fits all" clinical approach strategy has been showing poor outcomes, whether in pathophysiology understanding, in the diseases prevention, diagnosis, and even in therapeutic outcomes.

The cluster analysis process demands several decisions that should be kept in mind along the way and must be accounted for when analyzing the results and in the moment of drawing conclusions. Multiple strategies have been developed to assist in all the decisions to be made, however, clinical cluster analysis still depends much on the clinical knowledge, as well on data mining expertise.

There is a wide array of clustering strategies and methods. When there are many and different solutions, it is usually because none of them is an optimal solution. Every clustering algorithm has advantages and drawbacks. If all the algorithms performed equally solving clinical problems, we should always choose the most simple and straightforward. Unfortunately, this is not the case. Moreover, clinical problems often have characteristics that make it difficult to find an optimal algorithm. Further, increasingly we aim at solving cluster problems that join different kinds of data. Specially in the clinical setting, this is the current reality.

The biggest challenges are coming as more data is collected joining several kinds of clinical and all the omics data. Until a "universal" algorithm is developed, which alone can tackle these kinds of problems, we are left to use several algorithms, simultaneously or in a stepwise manner, or with an ensemble approach. The potential of knowledge discovery in this kind of analysis is probably as great as its complexity and we should be joining all the knowledge and capacities of a multidisciplinary team. This is the base of this new and exciting era of precision medicine.

11   Notes

1. Prior to computing the missing value, we need to first create a flag column, indicating if the value is missing in the original data set. This will aid in the later results analysis and can be easily accomplished, for example, using **mutate()** of the **dplyr**

A)
```
> summary(Our_data$BMI)
   Min. 1st Qu.  Median    Mean 3rd Qu.    Max.    NA's
   18.0    24.0    27.0    27.4    30.0    48.7      71
```

B)
```
Our_data<-Our_data%>%
  mutate(BMI_NA=if_else(is.na(BMI),"1","0"))
```

C)
```
> summary(Our_data$BMI_NA)
    0     1
 1939    71
```

D)
```
Our_data[is.na(Our_data$BMI),"BMI"]<-mean(Our_data$BMI, na.rm = TRUE)
```

E)
```
> summary(Our_data$BMI)
   Min. 1st Qu.  Median    Mean 3rd Qu.    Max.
   18.0    25.0    27.0    27.4    30.0    48.7
```

F)
```
Our_data$BMI<-impute(Our_data$BMI, mean)
```

Fig. 9 Rstudio screenshots for **Note 1**

package [12], also part of the **tidyverse**. In R, the missing values of both numeric and categorical variables appear as **NA**. They can be identified with **is.na()**. Also, using **which()** (which(is.na)), we can identify the observations indexes in which the values are missing.

Example:

In *Our_data* the BMI variable has 71 missing values (Fig. 9a).

Prior to impute them we create a flag, and using the "%>%" operator we can add additional flag columns (Fig. 9b).

Now, there is a new column, BMI_NA, that takes the value "1" if the value is missing on the original dataset, and "0" otherwise (Fig. 9c). This column can be used to detect missed data patterns, and analyze the results, with missing value imputation.

To compute the missing values we can use the simplest strategy that fits the data, and the problem to be solved, namely, substitute them by a constant value, a global value (e.g., the mean or the median), or a value predicted using the other variables in the data set [4].

The missing value substitution for the mean, the median of the variable or a constant value are easily done by attributing the value (**mean()**, **median()**, or a constant) using the R base functions (Fig. 9d, e).

However, there are also some more specific functions that can be used, like **impute()** from the **Hmisc package** [87], which can impute the missing values in the same way (Fig. 9f), and also with a random value from the variable distribution, and can be more simply applied to a data set.

A prediction algorithm can also be used to impute the missing value, based on the other observations. Although this strategy is more interesting, it is also the more complex. For the latter, for example, **mice** [88], **Amelia** [89], **missForest** [90], and **Hmisc::argImpute()** [87] can be used. For a deeper approach on this subject we suggest reading the package documentation.

2. Detecting outliers can be a task of a clustering algorithm [91]. This approach allows the outlier detection based on all the variables used in the algorithm, whereas the more classical methods only identifies outliers based on each variable at a time. Therefore, we apply a clustering algorithm on the scaled data (Subheading 2.5), then we identify and separate the clusters that represent outliers, rescale the data, and continue with the cluster analysis. The outliers are also analyzed.

3. Several methods can be used to scale the variables before running the clustering algorithm [4]. The most used are the Z-score standardization (Eq. 1), and the Min-Max Normalization (Eq. 2). However, the standardization based on the median and on the median absolute deviation (Eq. 3) as well as the decimal scaling (Eq. 4) can also be used. The method should be selected based on the data and problem to be solved, as well as on the algorithm to be used (Subheading 2.4) and can importantly impact the results.

Z-score

$$Z = X - \mu/\sigma \qquad (1)$$

Min-Max normalization

$$Z = X_i - \min(x)/\max(x) - \min(x) \qquad (2)$$

Median/MAD

$$Z = X_i - Me/MAD \qquad (3)$$

A)

```
Our_data_scaled<-rescaler.data.frame(Our_data,type="range")
```

B)

```
> summary(Our_data_scaled$BMI)
   Min. 1st Qu.  Median    Mean 3rd Qu.    Max.
 0.0000  0.2280  0.2932  0.3061  0.3909  1.0000
```

Fig. 10 Rstudio screenshots for **Note 3**

```
Our_data<-Our_data%>%
  mutate(BMI=Weight/((Height)^2))%>%
  mutate(MAP=(2*Diastolic_AP+Systolic_AP)/3)
```

Fig. 11 Rstudio screenshots for **Note 4d**

Decimal scaling

$$X_{\text{decimal}} = X/10^d \tag{4}$$

d = number of digits of the largest absolute value in the data

We will use **reshape::rescale()**, to rescale our data. Given that our dataset has a binary variable, and nonnormal distributed numeric variables, we will use the Min-Max normalization method (Fig. 10a, b).

4. Using **dplyr::mutate()** we can derive new variables, as for example BMI or Mean Arterial Pressure (MAP), thus reducing the data dimensionality (Fig. 11).

5. The linkage method (Fig. 12) refers to the way the distance between clusters is calculated, given that a cluster comprehends several observations. In hierarchical clustering algorithms, this parameter needs to be set to compute the intercluster distance. The selected linkage method will impact the results [92]. In the single linkage method, the distance between the clusters is the minimum distance between observations of both clusters. In the complete linkage, the intercluster distance is the maximum distance between observations of both clusters. In average linkage, the distance between the clusters is the mean distance between observations of both clusters. The centroid linkage, measures the distance between the clusters centroids. In the Ward linkage, the distance between the clusters is the sum of the squared deviations from points to centroids.

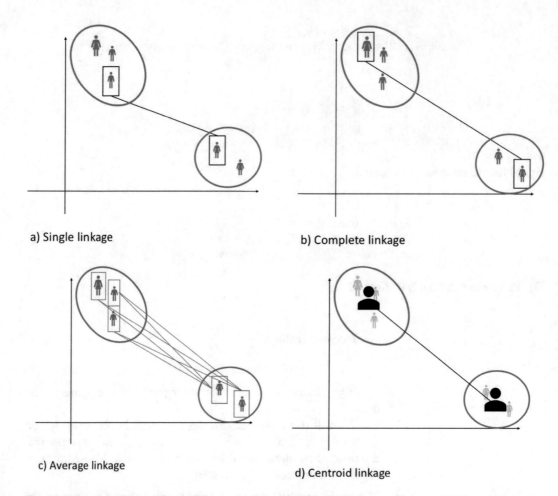

a) Single linkage

b) Complete linkage

c) Average linkage

d) Centroid linkage

Fig. 12 Linkage method. (**a**) Single linkage. (**b**) Complete linkage. (**c**) Average linkage. (**d**) Centroid linkage

6. To evaluate the cluster tendency in *Our_data*, in this example, we use **factoextra::fviz_dist()** [29], applied to the scaled data set, to display the dissimilarity matrix (Fig. 13), suggested by Kassambara [77]. As we can see in Fig. 10, *Our_data* seems to have a natural cluster structure, since two groups of subjects present lower dissimilarity values.

7. The elbow method is a visual way to identify the optimal number of clusters. Originally, plots the explained variance (within cluster sum of squares (WSS)), and the k number of clusters. The point in which there is a steep bend in the curve is usually assumed to correspond to the best number of clusters. Several other quality measures, besides the explained variance, can also be used. In these cases, we choose the number of clusters that maximize or minimize the measure value, depending on the chosen quality index.

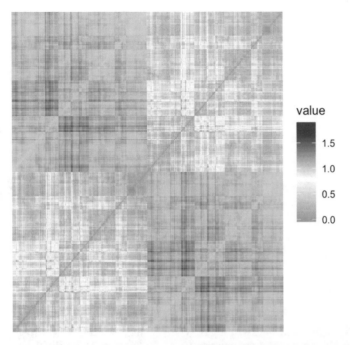

Fig. 13 Visual inspection of cluster tendency, using **factoextra::fviz_dist()**. Dissimilarity matrix of *Our_data* scaled. The purple decodes the dissimilarity of observations: purple is the maximum of dissimilarity, and red the minimum of dissimilarity

8. Cluster validity criteria measure the goodness of the cluster results. Whereas the external validation criteria assume previous knowledge about the data, the internal validation criteria is based only on the information intrinsic to the data [93]. Thus, the latter are the criteria usually applied to clustering solutions validation.

From the wide array of existent internal validity indexes [93], Davies–Bouldin index (Eq. 5) [94], Silhouette index (Eq. 6) [95], and Dunn index (Eq. 7) [96], are only three examples of commonly used internal validity indexes.

Davies–Bouldin index

$$\text{BD} = \frac{1}{c} \sum_{i=i}^{c} \text{Max}_{i \neq j} \left\{ \frac{d(X_i) + d(X_j)}{d(c_i, c_j)} \right\} \quad (5)$$

Where,

c = number of clusters

i, j = clusters labels

$d(X_i)$ and $d(X_j)$ = cluster samples to cluster centroids distances

$d(c_i, c_j)$ = distance between centroids

Silhouette index

$$s(i) = \frac{(b(i) - a(i))}{\text{Max}\{a(i), b(i)\}} \tag{6}$$

Where,

$a(i)$ = average distance between the ith sample and all X_j samples

$b(i)$ = minimum average distance between the ith sample and samples clustered in X_k, $(k = 1, \ldots c, k! = j)$

Dunn Index

$$\text{Dunn} = \min_{1 \ll i \ll c} \left\{ \min \left\{ \frac{d(c_i, c_j)}{\max_{1 \ll i \ll c}(d(X_k))} \right\} \right\} \tag{7}$$

Where,

$d(c_i, c_j) = X_i$ and X_j intercluster distance

$d(X_k) = X_k$ intracluster distance

c = number of clusters

References

1. Raghupathi W, Raghupathi V (2014) Big data analytics in healthcare: promise and potential. Health Inf Sci Syst 2:1–10. https://doi.org/10.1186/2047-2501-2-3

2. Laney D (2001) 3D data management: controlling data volume, velocity, and variety. Appl Deliv Strat 949:4. https://blogs.gartner.com/doug-laney/files/2012/01/ad949-3D-Data-Management-Controlling-Data-Volume-Velocity-and-Variety.pdf. Accessed 21 Jan 2019

3. Huang S, Chaudhary K, Garmire LX (2017) More is better: recent progress in multi-omics data integration methods. Front Genet 8:1–12. https://doi.org/10.3389/fgene.2017.00084

4. Larose DT, Larose CD (2015) Clustering. In: Data mining and predictive analytics, 2nd edn. Wiley, Chichester, UK

5. Islam S, Hasan M, Wang X et al (2018) A systematic review on healthcare analytics: application and theoretical perspective of data mining. Healthcare 6(54):1–43. https://doi.org/10.3390/healthcare6020054

6. R Core Team (2018) R: a language and environment for statistical computing. R Foundation for Statistical Computing, Vienna, Austria. https://www.R-project.org/. Accessed 21 Jan 2019

7. Walkowiak S (2016) Big data analytics with R: utilize R to uncover hidden patterns in your big data. Packt Publishing Limited, Birmingham, UK

8. RStudio Team (2016) RStudio: integrated development environment for R. RStudio, Inc, Boston, MA. http://www.rstudio.com/. Accessed 21 Jan 2019

9. Kuhn M et al (2018) caret: classification and regression training. R package version 6.0-80. https://CRAN.R-project.org/package=caret. Accessed 21 Jan 2019

10. Wickham H (2017) tidyverse: easily install and load the "tidyverse". R package version 1.2.1. https://CRAN.R-project.org/package=tidyverse. Accessed 21 Jan 2019

11. Wickham H, Henry L (2018) tidyr: easily tidy data with "spread()" and "gather()" functions. R package version 0.8.1. https://CRAN.R-project.org/package=tidyr. Accessed 21 Jan 2019

12. Wickham H, François R, Henry L, Müller K (2018). dplyr: a grammar of data manipulation. R package version 0.7.6. https://CRAN.R-project.org/package=dplyr. Accessed 21 Jan 2019

13. Bennett DA (2001) How can I deal with missing data in my study? Aust N Z J Public Health 25:464–469

14. Wehrens R, Buydens L (2007) Self- and super-organizing maps in R: the kohonen package. J Stat Softw 21(5):1–19

15. Fox J (2018) RcmdrMisc: R commander miscellaneous functions. R package version 2.5-1. https://CRAN.R-project.org/package=RcmdrMisc. Accessed 21 Jan 2019

16. Wickham H (2007) Reshaping data with the reshape package. J Stat Softw 21(12):1–20

17. Bellman R (1957) Dynamic programming. Princeton University Press, Princeton, NJ

18. Saeys Y, Inza I, Larrañaga P (2007) A review of feature selection techniques in bioinformatics. Bioinformatics 23:2507–2517. https://doi.org/10.1093/bioinformatics/btm344

19. Alelyani S, Tang J, Liu H (2013) Feature selection for clustering: a review. In: Data clustering: algorithms and applications. CRC Press, Boca Raton, FL, pp 110–121

20. Dy J, Brodley C (2004) Feature selection for unsupervised learning. J Mach Learn Res 5:845–889

21. Anukrishna PR, Paul V (2017) A review on feature selection for high dimensional data. In: I2017 International conference on inventive systems and control (ICISC), pp 1–4

22. Pacheco E (2015) Unsupervised learning with R: work with over 40 packages to draw inferences from complex datasets and find hidden patterns in raw unstructured data. Packt Publishing, Birmingham, UK

23. Romanski P, Kotthoff L (2018) FSelector: selecting attributes. R package version 0.31. https://CRAN.R-project.org/package=FSelector. Accessed 21 Jan 2019

24. Raftery LS, Raftery AE (2018) clustvarsel: a package implementing variable selection for Gaussian model-based clustering in R. J Stat Softw 84:1–28. https://doi.org/10.18637/jss.v084.i01

25. Williams G, Huang J, Chen X, Wang Q, Xiao L (2015) wskm: weighted k-means clustering. R package version 1.4.28. http://CRAN.R-project.org/package=wskm. Accessed 21 Jan 2019

26. Jolliffe IT (2010) Principal component analysis. Springer, New York

27. Le S, Josse J, Husson F (2008) FactoMineR: an R package for multivariate analysis. J Stat Softw 25:1–18. https://doi.org/10.18637/jss.v025.i01

28. Maechler M et al (2018) cluster: cluster analysis basics and extensions. R package version 2.0.7-1. https://cran.r-project.org/web/packages/cluster/cluster.pdf. Accessed 21 Jan 2019

29. Kassambara A, Mundt F (2017) factoextra: extract and visualize the results of multivariate data analyses. R package version 1.0.5. https://CRAN.R-project.org/package=factoextra. Accessed 21 Jan 2019

30. Xu D, Tian Y (2015) A comprehensive survey of clustering algorithms. Ann Data Sci 2:165–193. https://doi.org/10.1007/s40745-015-0040-1

31. Kaufman L, Rousseeuw PJ (2005) Finding groups in data: an introduction to cluster analysis. Wiley, New York

32. Han J, Kamber M, Pei J (2012) Data mining: concepts and techniques. Elsevier, Amsterdam

33. Xu R, Wunsch DC (2010) Clustering algorithms in biomedical research: a review. IEEE Rev Biomed Eng 3:120–154. https://doi.org/10.1109/RBME.2010.2083647

34. Abdullah Z, Hamdan AR (2015) Hierarchical clustering algorithms in data mining. Int J Comput Elect Autom Control Inf Eng 9(10)

35. Williams E, Colasanti R, Wolffs K et al (2018) Classification of tidal breathing airflow profiles using statistical hierarchal cluster analysis in idiopathic pulmonary fibrosis. Med Sci 6:75. https://doi.org/10.3390/medsci6030075

36. Vincent A, Hoskin TL, Whipple MO et al (2014) OMERACT-based fibromyalgia symptom subgroups: an exploratory cluster analysis. Arthritis Res Ther 16:1–11. https://doi.org/10.1186/s13075-014-0463-7

37. Ahlqvist E, Storm P, Karajamaki A et al (2018) Novel subgroups of adult-onset diabetes and their association with outcomes: a data-driven cluster analysis of six variables. Lancet Diabetes Endocrinol 6:361–369. https://doi.org/10.1016/S2213-8587(18)30051-2

38. Toppila I (2016) Identifying novel phenotype profiles of diabetic complications and their genetic components using machine learning approaches. Aalto University, Helsinki, Finland

39. Burgel P-R, Paillasseur J-L, Roche N (2014) Identification of clinical phenotypes using cluster analyses in COPD patients with multiple comorbidities. BioMed Res Int 2014:420134, 9 pages. https://doi.org/10.1155/2014/420134

40. Galili T (2015) dendextend: an R package for visualizing, adjusting, and comparing trees of hierarchical clustering. Bioinformatics. https://doi.org/10.1093/bioinformatics/btv428

41. Swarndeep Saket J, Pandya S (2016) An overview of partitioning algorithms in clustering techniques. Int J Adv Res Comput Eng Technol 5:2278–1323

42. Berkin P (2006) Grouping multidimensional data. In: Grouping multidimensional data. Springer, Berlin, pp 25–71

43. Boomija MD (2008) Comparison of partition based clustering algorithms. J Comput Appl 1:18–21

44. Lewis SJG, Foltynie T, Blackwell AD et al (2005) Heterogeneity of Parkinson's disease in the early clinical stages using a data driven approach. J Neurol Neurosurg Psychiatry 76:343–348. https://doi.org/10.1136/jnnp.2003.033530

45. Ha NT, Harris M, Preen D et al (2018) Identifying patterns of general practitioner service utilisation and their relationship with

potentially preventable hospitalisations in people with diabetes: the utility of a cluster analysis approach. Diabetes Res Clin Pract 138:201–210. https://doi.org/10.1016/j.diabres.2018.01.027

46. Ahmad T, Lund LH, Rao P et al (2018) Machine learning methods improve prognostication, identify clinically distinct phenotypes, and detect heterogeneity in response to therapy in a large cohort of heart failure patients. J Am Heart Assoc 7:1–15. https://doi.org/10.1161/JAHA.117.008081

47. Lucas A (2018) amap: another multidimensional analysis package. R package version 0.8-16. https://CRAN.R-project.org/package=amap. Accessed 21 Jan 2019

48. Szepannek G (2018) clustMixType: k-prototypes clustering for mixed variable-type data. R package version 0.1-36. https://CRAN.R-project.org/package=clustMixType. Accessed 21 Jan 2019

49. Velmurugan T (2015) Clustering lung cancer data by k-means and k-medoids algorithms. In: International conference on information and convergence technology for smart society, pp 17–21

50. Kohonen T (1990) The self-organizing map. Proc IEEE 78:1464–1480. https://doi.org/10.1109/5.58325

51. Hajek P, Henriques R, Hajkova V (2014) Visualising components of regional innovation systems using self-organizing maps-evidence from European regions. Technol Forecast Soc Change 84:197–214. https://doi.org/10.1016/j.techfore.2013.07.013

52. Paul M, Shaw CK, David W (1996) A comparison of SOM neural network and hierarchical clustering methods. Eur J Oper Res 93:402–417

53. Cabanes G, Bennani Y (2010) Learning the number of clusters in self organizing map. In: Self-organizing maps. Intech, Croatia. https://doi.org/10.5772/9164

54. Kohonen T (2001) Self-organizing maps. Springer, Berlin

55. Henriques R, Bacao F, Lobo V (2012) Exploratory geospatial data analysis using the GeoSOM suite. Comput Environ Urban Syst 36:218–232. https://doi.org/10.1016/j.compenvurbsys.2011.11.003

56. Ultsch A (2007) Emergence in self organizing feature maps. WSOM 2007 - 6th Int work self-organizing maps

57. Wehrens M, Kruisselbrink J (2018) Self- and super-organising maps in R: the kohonen package. J Stat Softw 21(5)

58. Markey MK, Lo JY, Tourassi GD, Floyd CE (2003) Self-organizing map for cluster analysis of a breast cancer database. Artif Intell Med 27:113–127. https://doi.org/10.1016/S0933-3657(03)00003-4

59. Vanfleteren LEGW, Spruit MA, Groenen M et al (2013) Clusters of comorbidities based on validated objective measurements and systemic inflammation in patients with chronic obstructive pulmonary disease. Am J Respir Crit Care Med 187:728–735. https://doi.org/10.1164/rccm.201209-1665OC

60. Pina AF, Patarrão RS, Ribeiro RT, Penha-Gonçalves C, Raposo JF, de Oliveira RM, Gardete-Correia L, Duarte R, Boavida JM, Andrade R, Correia I, Medina JL, Henriques R, Macedo MP (2018) Are the normal glucose tolerance individuals totally outside of the diabetes spectrum? Diabetologia 61:S143

61. Bhuyan R, Borah S (2013) A survey of some density based clustering techniques. In: Conf. advancements in information, computer and communication

62. Ankerst M, Breunig M, Kriegel H-P, Sander J (1999) OPTICS: ordering points to identify the clustering structure. In: ACM SIGMOD international conference on Management of data. ACM Press, New York, pp 49–60

63. Hennig C (2018) fpc: flexible procedures for clustering. R package version 2.1-11.1. https://CRAN.R-project.org/package=fpc. Accessed 21 Jan 2019

64. Hahsler M, Piekenbrock M (2018) dbscan: density based clustering of applications with noise (DBSCAN) and related algorithms. R package version 1.1-3. https://CRAN.R-project.org/package=dbscan. Accessed 21 Jan 2019

65. Celebi ME, Aslandogan YA, Bergstresser PR (2005) Mining biomedical images with density-based clustering. In: International conference on information technology: coding and computing (ITCC'05) - Volume II. IEEE, Washington, DC, pp 163–168

66. Bouveyron C, Brunet C (2013) Model-based clustering of high-dimensional data: a review. Comput Stat Data Anal 71:1–47. https://doi.org/10.1016/j.csda.2012.12.008

67. Rubin DB, Dempster AP, Laird N (1977) Maximum likelihood from incomplete data via the EM algorithm. J R Stat Soc 39:1–38

68. Couvreur C (1997) The EM algorithm: a guided tour. Comput Intens Methods Control Signal Process 1997:209–222. https://doi.org/10.1007/978-1-4612-1996-5

69. Fraley C, Raftery AE (1998) How many clusters? Which clustering method? Answers via model-based cluster analysis. Comput J 41:578–588. https://doi.org/10.1093/comjnl/41.8.578

70. Scrucca L, Fop M, Murphy TB, Raftery A (2017) mclust 5: clustering, classification and density estimation using Gaussian finite mixture models. R J 8(1):205–233

71. Hwang S, Oh J, Cox J et al (2006) Blood detection in Wireless Capsule Endoscopy using expectation maximization clustering. In: Progress in Biomedical Optics and Imaging - Proceedings of SPIE

72. Saini S, Rani P (2017) A survey on STING and CLIQUE grid based clustering methods. Int J Adv Res Comput Sci 8:2015–2017

73. Mann AK, Kaur N (2013) Grid density based clustering algorithm. Int J Adv Res Comput Eng Technol 2:2143–2147

74. Ilango M, Mohan V (2010) A survey of grid based clustering algorithms. Int J Eng Sci Technol 2:3441–3446

75. Yue S, Shi T, Wang J, Wang P (2012) Application of grid-based K-means clustering algorithm for optimal image processing. Comput Sci Inf Syst 9:1679–1696. https://doi.org/10.2298/CSIS120126052S

76. Waller L, Gotway C (2004) Applied spatial statistics for public health data. Wiley, Hoboken, NJ

77. Kassambara A (2017) Practical guide to cluster analysis in R: unsupervised machine learning. STHDA

78. Charrad M, Ghazzali N, Boiteau V, Niknafs A (2014) NbClust: an R Package for determining the relevant number of clusters in a data set. J Stat Softw 61:1–36

79. Halkidi M, Batistakis Y, Vazirgiannis M (2001) On clustering validation techniques. J Intell Inf Syst 17:107–145. https://doi.org/10.1023/A:1012801612483

80. Nieweglowski L (2013) clv: cluster validation techniques. R package version 0.3-2.1. https://CRAN.R-project.org/package=clv. Accessed 21 Jan 2019

81. Brock G, Pihur V, Datta S (2008) clValid: an R package for cluster validation. J Stat Softw 25:1–22

82. Wolpert D, Macready G (1997) No free lunch theorems for optimization. IEEE Trans Evol Comput. https://doi.org/10.1109/4235.585893

83. Alqurashi T, Wang W (2018) Clustering ensemble method. Int J Mach Learn Cyber. https://doi.org/10.1007/s13042-017-0756-7

84. Chiu DS, Talhouk A (2018) DiceR: an R package for class discovery using an ensemble driven approach. BMC Bioinformatics 19:17–20. https://doi.org/10.1186/s12859-017-1996-y

85. Wickham H (2016) ggplot2: elegant graphics for data analysis. Springer, New York

86. Wilkinson L (2005) The grammar or grammar of graphics. Springer, New York

87. Harrell FE Jr, with contributions from CD and many others (2018) Hmisc: Harrell miscellaneous. R package version 4.1-1. https://CRAN.R-project.org/package=Hmisc. Accessed 22 Jan 2019

88. van Buuren S, Groothuis-Oudshoorn K (2011) mice: multivariate imputation by chained equations in R. J Stat Softw 45:1–67

89. Honaker J, King G, Blackwell M (2011) Amelia II: a program for missing data. J Stat Softw 45:1–47

90. Stekhoven DJ, Bühlmann P (2012) Missforest-non-parametric missing value imputation for mixed-type data. Bioinformatics 28:112–118. https://doi.org/10.1093/bioinformatics/btr597

91. Ilango V, Subramanian R, Vasudevan V (2012) A five step procedure for outlier analysis in data mining. Eur J Sci Res 75:327–339

92. Steinbach M, Ertöz L, Kumar V (2004) New directions in statistical physics. In: The challenges of clustering high dimensional data. Springer, Berlin, pp 273–309

93. Rendón E, Abundez I, Arizmendi A, Quiroz E (2011) Internal versus external cluster validation indexes. Int J Comput Commun 5(1):27–34

94. Davis DL, Bouldin DW (1998) A cluster separation measure. IEEE Trans Pattern Anal MachIntel PAMI 1(2):224–227

95. Rousseeuw PJ (1987) Silhouettes: a graphical aid to the interpretation and validation of cluster analysis. J Comp Appl Math 20:53–65

96. Dunn JC (1973) A Fuzzy relative of the ISODATA process and its use in detecting compact well-separated clusters. J Cyber 3:32–57

Review of Issues and Solutions to Data Analysis Reproducibility and Data Quality in Clinical Proteomics

Mathias Walzer and Juan Antonio Vizcaíno

Abstract

In any analytical discipline, data analysis reproducibility is closely interlinked with data quality. In this book chapter focused on mass spectrometry-based proteomics approaches, we introduce how both data analysis reproducibility and data quality can influence each other and how data quality and data analysis designs can be used to increase robustness and improve reproducibility. We first introduce methods and concepts to design and maintain robust data analysis pipelines such that reproducibility can be increased in parallel. The technical aspects related to data analysis reproducibility are challenging, and current ways to increase the overall robustness are multifaceted. Software containerization and cloud infrastructures play an important part.

We will also show how quality control (QC) and quality assessment (QA) approaches can be used to spot analytical issues, reduce the experimental variability, and increase confidence in the analytical results of (clinical) proteomics studies, since experimental variability plays a substantial role in analysis reproducibility. Therefore, we give an overview on existing solutions for QC/QA, including different quality metrics, and methods for longitudinal monitoring. The efficient use of both types of approaches undoubtedly provides a way to improve the experimental reliability, reproducibility, and level of consistency in proteomics analytical measurements.

Key words Computational mass spectrometry, Quality control approaches, Large scale data analysis, Cloud technology, Reproducible analysis pipelines

Abbreviations

BSA	Bovine serum albumin
CWL	Common workflow language
DAC	Data Access Compliance
DACO	Data Access Compliance Office
DDA	Data-dependent acquisition
DIA	Data-independent acquisition
FDR	False discovery rate
GUI	Graphical user interface
HPC	High-performance computing
HUPO	Human Proteome Organization
LC	Liquid chromatography

Rune Matthiesen (ed.), *Mass Spectrometry Data Analysis in Proteomics*, Methods in Molecular Biology, vol. 2051, https://doi.org/10.1007/978-1-4939-9744-2_15, © Springer Science+Business Media, LLC, part of Springer Nature 2020

LCL	Lower control level
MS	Mass spectrometry
PCAWG	Pan-cancer analysis of whole genomes
PSI	Proteomics Standards Initiative
QA	Quality assessment
QC	Quality control
SOP	Standard operating procedure
SPC	Statistical process control
SRM	Selected reaction monitoring
UCL	Upper control level
WMS	Workflow management system

1   Introduction

Mass spectrometry (MS) is currently the main workhorse of high-throughput proteomics approaches [1]. Its unrivalled sensitivity and specificity are particularly suited to detect, identify, and compare the abundances of different analytes, including some active components contained in samples, such as peptides, proteins and also metabolites. In fact, MS offers a high throughput solution for the analysis of clinically relevant samples, for both day-to-day operations and other applications such as disease biomarker discovery [2]. Whereas MS has been in use for scientific discovery in proteomics for more than 15 years, still, instrumentation and related analysis algorithms undergo rapid developments. Thanks to the improvements in instrumentation, the impact of technical variability in the analytical measurements has decreased. One example would be the lessened impact of masking ions due to higher scan frequencies and sensitivity, and in turn to achieve a more robust list of identified peptides (for instance, the use of a higher number of possible Top N and wider exclusion window settings in data-dependent acquisition (DDA) approaches).

However, in order to leverage the most of the analytical power for data acquisition in present-day instruments, a number of additional analytical steps have to be undertaken prior to the MS analysis. Each of these steps introduces their own sources of variability, adding up and influencing each other directly, so it is essential that extra care and expertise are applied to perform them in a reliable manner [3]. Logically, this becomes most crucial in clinical applications. Examples of these sources of variability are, most prominently, protein digestion (used to improve sensitivity by increasing the amount of ionizable and eventually detectable peptides), and the liquid chromatography (LC) step (used to separate the analytes across the elution time). Like in other analytical fields, QA and QC can help to monitor the variability of the measurements and the instrument health. They can provide a valuable insight into the performance of different instrument setups and guide the selection of fitting parameters for the analysis.

According to Apweiler et al. [4], we can define "clinical proteomics" as any proteomics approach applied to a clinically relevant sample, on a spectrum of situations ranging from preclinical discovery to applied diagnostics. But this definition also includes the selection, validation, and assessment of standard operating procedures (SOPs) to deliver robust methods for the clinical daily use, which must encompass data analysis reproducibility and data quality. For proteomics MS approaches in particular, reproducibility is a key feature for its routine applicability in a clinical environment. Certainly, once the technical variability of the measurements has decreased to an acceptable level through QC/QA approaches, data analysis reproducibility is the next issue.

In fact, the fast-paced increase in data volumes generated from modern clinical proteomics studies highlights some issues related to data analysis reproducibility. Mainly, there is the need to handle the increased throughput, in terms of computational resources, but also the existence of more complex experimental designs (e.g., in the case of QC approaches). Growing volumes of data also trigger the need for more robust analysis workflows since the availability of more data usually means that a higher coverage of the value range distribution is reached. This represents an increased chance of extreme values occurring on the edges of such distribution. Therefore, robust analysis workflows need to be ready to deal with a wider data spectrum as an input.

Robust protocols and SOPs help to minimize biological variability in the analysis. Computational analysis methods are in principle deterministic and replicable by their algorithmic nature. However, in practice, data analysis in proteomics is usually performed in several steps [e.g., peptide and protein identification, FDR (False Discovery Rate) control, QC, peptide and protein quantification, and statistical analysis]. Each step can be usually performed by different algorithms and tools, each of them with their own particularities and supported use cases. The high number of software tools available (taking also into account the availability of different versions of the same tool) makes challenging the selection of the appropriate ones when building analysis workflows. Additionally, the input parameters needed to run the software, which are dependent on different extrinsic and intrinsic factors, represent another source of variability for the analysis results. Furthermore, input parameters can vary between different versions of the same software. Therefore, in all cases, the concrete software choices (including their versions) and the relevant input parameters have to be carefully documented.

However, the complexity of the protocols applied, the sensitivity of routinely used LC-MS workflows to the changes in the analytical environment, and the increasingly complex nature of clinical samples often demand the adjustment of the input parameters to achieve the best results. Therefore, both a comprehensive documentation of the input parameters and the explanation of the given experimental variability by either QC and/or QA approaches help to achieve experimental reproducibility. The availability of QC samples [5, 6], which are frequently employed in related fields of clinical research, and the existence of a well-planned experimental design [7, 8] can help to increase accuracy and confidence in the results [9]. This also helps with the decision to include the appropriate analysis steps (e.g., normalization and calibration), where necessary. Another key aspect behind the design and development of data analysis pipelines is the choice of the data formats used to exchange data between the different steps of the analysis. This can influence the availability and level of detail of various aspects of the intermediate and final analysis results, including among others, confidence scores and evidence traces.

The choice of software is influenced not only by personal preference, but mainly by constraints in the compute infrastructure, such as the operating system or the ability to cope with the high-throughput of increased data volumes. In order to build the data analysis into a routine operation and to be able to cope with the increased volumes of data, different workflow systems have been designed for bioinformatics data analysis. Recent developments have favored the integration of these systems into cloud computing environments, a fact resulting from the existence of active development communities, a competitive pricing for computation and their increased technological flexibility. A data analysis pipeline well documented in the abovementioned aspects is then ready to be included in a SOP, and can then be used to strengthen the data analysis reproducibility. Through this book chapter, we aim to achieve the following:

- Highlight existing challenges and solutions in the construction of robust data analysis workflows.

- Introduce concepts that can help to achieve data analysis reproducibility and improve data reuse.

- Explain how the use of QC and/or QA approaches can improve the data analysis and explain experimental variability.

- Detail how data analysis reproducibility and QC can go hand-in-hand.

2 Robust and Reproducible Data Analysis Pipelines

Data analysis reproducibility is important for many aspects of science. Additionally, data reanalysis plays a pivotal role for achieving an independent confirmation of the analysis results. Therefore, data analysis needs to be transparent enough so that it can be run in different sites, even with a different IT infrastructure and using comparable input parameters. Similarly, data analysis must be applicable for the replication of studies performed to compare results. Furthermore, for successful data reuse by third parties, analysis reproducibility is key. With increasing data volumes and more complex studies the dependence on the software used for the data analysis grows in parallel, and therefore, the need to produce reproducible data analysis becomes essential.

In fact, some scientific journals have made mandatory to make the source code available to improve the transparency and openness of the peer review process. However, code openness represents only one aspect of the analysis. Using that code to produce software running appropriately, its correct configuration, and its compatibility to other software used are indeed equally important factors for achieving data analysis reproducibility. With the goal of reproducibility in mind, funding agencies and journals are becoming increasingly more interested in this area [10].

The construction of robust and reproducible data analysis involves from a software standpoint, both the definition of the analysis steps and the creation of the analysis workflow. The first requirement can be summarized as the ability to reproduce the same analysis results when using the same input data, even in different compute environments. The recording of input parameters, as the second requirement, represents the next key component. The production of auditable documentation of the analysis results through appropriate provenance tracking can be used to strengthen reproducibility [11]. The third requirement is to deliver a sensible and comparable output result for a wide range of input data.

The capability to successfully move between different compute environments becomes essential if scalability is the issue. In fact, moving from a development environment capable of handling small datasets in a reasonable amount of time, to a more powerful computational environment, ensuring in parallel that all components work as expected, is very far from trivial. The main underlying reason is that different compute environments are indeed different and both scalability and parallelization are not minor issues to solve (e.g., due to job submission queuing or algorithmic limitations). We will discuss the related issues in a later section. It should be noted that development environments are readily accessible and usually come with graphical user interfaces (GUI) that can help in the design process. However, more powerful computational environments (e.g., computer clusters or cloud infrastructures) are only remotely accessible and therefore it can be potentially very costly to debug software in them. Optimally, the workflow used should be independent of the compute environment, and should be able to work automatically, involving the execution of the different steps in the right order. Next, we will go through the main factors to consider one by one.

2.1 Versioning, Automation and Source Code Repositories

A significant proportion of the software used to date for proteomics data analysis is free and often open source with very permissive licenses. For the latter, the source code is usually organized and made available in source code repositories, which make it easy for developers to distribute and track new releases and versions. Commonly, versioning follows the semantic versioning convention "MAJOR"."MINOR"."PATCH", where each position is enumerated but incremented independently in agreement to the semantics behind the release. In general, whereas major version changes may indicate incompatible releases, minor version changes involve the addition of new or updated features. Finally, a patch version can reflect a change performed to fix a bug or incompatibility. It is then essential to keep track of the software version used in the construction and execution of any analysis workflow step.

GitHub, Bitbucket, and SourceForge are, among others, the widely used common source code repositories which provide simple ways for the developers to distribute new releases, and for the users to inspect for each software tool the following issues:

- The version tree, including the version history.
- The "changelog," summarizing all the feature changes implemented between different versions, highlighting probable compatibility issues between newer and older versions.

2.2 Software Containers

Another crucial point for software to work in a reproducible manner is the availability of a stable computing environment. Most software tools are depending on underlying software libraries (with different versions) which can have certain functionality that is developed independently. As an example, the dependency on the . NET library in the various versions of the Windows operating system. To make software run reliably, one has to make stable the installed software libraries. A crude alternative option is to simply freeze the system after the installation of all necessary software. This has obvious drawbacks like the lack of support for software updates and the concurrent library dependencies, or the requirement to be able to support several parallel data analysis processes. A different option increasingly popular is to put software into "silos," the

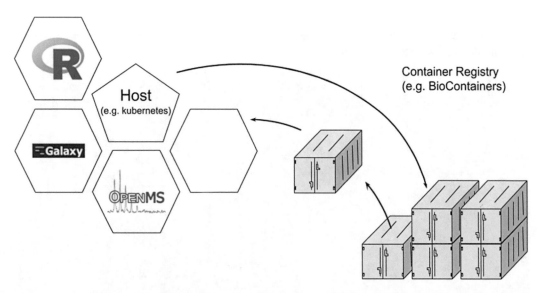

Fig. 1 Schema of the described software architecture. Software containers can be pulled from a registry to run in isolation on the host system without the need to install further software. Kubernetes allows to run software containers in multiple hosts and connect them via virtual networks

so-called software containers, like those provided through Singularity (https://www.sylabs.io/singularity/ or https://www.sylabs.io/docs/) or Docker (https://www.docker.com). These containers contain the software tools together with all its dependencies, and can run on top of the operating system without the need to install extra software or dependencies (Fig. 1). Compared to virtual machines, with which the same can be accomplished, containers are more lightweight and more portable. They bring the important software but leave the operating system level control and virtualization to the container engine. Also, their construction is layered and codified which results in more transparency and reproducibility, especially with the uses of versioning, automation and source code repositories (see previous section).

Fortunately, many bioinformatics (and proteomics in particular) software tools are already available as software containers through online registries like DockerHub (http://hub.docker.com), Singularity Hub (http://singularity-hub.org), and BioContainers [12]. *See* Table 1 for an example of computational proteomics software that is currently available in containers. Other resources that act as registries for tools, such as bio.tools [21], can provide instructions about how to build such containers.

2.3 Open Data Standard Formats

In a multistep data analysis process, data has to be transferred between the different software tools responsible of the different steps of the analysis. For both primary inputs as well as intermediate result files, the use of open data standards provides great benefits.

Table 1
Selection of popular and highly used proteomics MS software and frameworks, their current versions (by November 2018), and code availability

Software	GUI	Workflow system integration	Version (October 2018)	Open code repository	Container availability
SearchGUI [13]	✓	✗	v3.3.3	✓	✗
MaxQuant [14]	✓	?	1.6.2.3	✗	✗
MZmine2 [15]	✓	?	v2.33	✓	✗
ProteoWizard [16]	✓	✓	3.0.18240	✓	✓
mspire [17]	✓	?	v0.10.8.0	✓	✗
mass-up [18, 19]	✓	?	v1.0.13	✓	✗
OpenMS [20]	✓	✓	Release 2.3.0	✓	✓
Percolator [19]	✗	✓	rel-3-02-01	✓	✓

It should be noted that all software developments do not follow the same versioning scheme

Using open standard formats at every step has the added benefit of providing an audit trail for each analysis step. In the proteomics field, the HUPO (Human Proteome Organisation) PSI (Proteomics Standards Initiative) [22–24] has defined over the years different open community standards for data representation, in order to facilitate data comparison, exchange and verification. The primary data input to any proteomics analysis will be the acquired mass spectra. The widely adopted data standard mzML [25], can be used to represent the MS data in an open and vendor-independent fashion. The data can therefore be read by a multitude of software, especially when compared to the use of vendor specific software libraries, which make achieving reproducibility more difficult [26]. The mzML format can indeed be readily used with many popular peptide/protein identification algorithms [27–30]. Furthermore, the same interoperability benefits come when using the other PSI standard formats, namely, mzIdentML, mzTab, and mzQuantML, for representing peptide/protein identification and quantification results [31–34], and qcML (PSI standard format under development), for representing QC and QA related information [35]. Tables 2 and 3 include a list of the most prominent software implementing the standard formats mzidentML and mzTab, respectively. QC approaches will be the main focus of the second part of this book chapter.

2.4 Workflow Management Systems (WMS)

The actual construction of multistep data analysis workflows has to follow the actual sequence of the steps of the analysis. There are several Workflow Management Systems (WMS) available that can be used to encode the actual analysis steps and their order. Most of

Table 2
List of tools that implement an up-to-date export to the mzIdentML format (version 1.1), by November 2018

Tool	Formats	URL
IP2 [36]	Native support	http://www.integratedproteomics.com/
Mascot (Matrix Science) [27]	Native support	http://www.matrixscience.com/
MS-GF+ [29]	Native support	https://omics.pnl.gov/software/ms-gf
OpenMS [20]	Native support	https://www.openms.de/
Peaks (Bioinformatics Solutions Inc.) [37]	Native support	http://www.bioinfor.com/
PeptideShaker [38]	Native support	http://compomics.github.io/projects/peptide-shaker.html
Scaffold (Proteome Software) [39]	Native support	http://www.proteomesoftware.com/products/scaffold/

More information is available at http://www.psidev.info/tools-implementing-mzIdentML#

Table 3
List of tools that implement an up-to-date export to the mzTab format (version 1.1), by November 2018

Tool	Formats	URL
Mascot (Matrix Science) [27]	Native support	http://www.matrixscience.com/
MaxQuant [14]	Native support	http://maxquant.org/
OpenMS [20]	Native support	https://www.openms.de/

them have a GUI (Fig. 2), to build and execute such analysis workflows for a wide range of bioinformatics tools. Some of the most popular WMS are:

- CWL (Common Workflow Language) [40]
- Galaxy [41]
- KNIME [42]

WMS can integrate different software tools, input and output formats, and can describe the input parameters in detail. Software integration in WMS must be specific to each software tool version and should usually include documentation detailing all the input parameters available and how they should be handled. It is important to highlight that most input parameters have default settings,

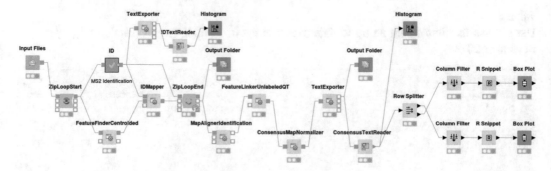

Fig. 2 An exemplary proteomics workflow constructed and visualized using KNIME

but others (e.g., the protein sequence database) have to be specified. The impact of the used parameters on the data analysis reproducibility will be discussed in a later section. Workflow engines can then sequentially and automatically execute all analysis steps. Management of compute resource availability is usually met by the WMS interfacing with scheduling systems, for achieving an efficient parallelization. Therefore, usually a far greater throughput and reproducibility can be achieved while in parallel decreasing the amount of manual work required from bioinformaticians and/or computational scientists.

Finally, an added benefit of organizing the data analysis pipelines as workflows is that they can be versioned as well, and then stored in repositories for archival purposes. The availability of both analysis pipelines as workflows and containerized software reduce the chance that a particular workflow is not available or cannot be executed due to missing or incompatible software.

2.5 Handling the Increase in Data Volumes

One increasingly used analysis approach in (clinical) proteomics is exemplifying why it is important to be able to handle the increase in data volumes appropriately. It is increasingly accepted that, in terms of reliability and accuracy, data-independent acquisition (DIA) approaches (e.g., SWATH-MS) are becoming comparable with selected reaction monitoring (SRM) techniques, which are still considered the gold standard proteomic quantification methods [43]. In fact, DIA approaches can combine the high throughput of DDA approaches with the high reproducibility of SRM [44, 45]. However, that improvement comes at the cost of a steep increase in data volumes, reaching routinely hundreds of GB for an average experiment. Data analysis workflows must therefore be able to handle such very high data volumes. In parallel, more complex experimental designs are becoming the norm, involving higher numbers of technical and/or biological replicates. These requirements involve handling the data in parallel and scaling out the computation, either in computer clusters (HPC, high-performance computing) or using cloud infrastructures. The heterogeneity of scheduling systems a WMS has to interface on different compute clusters makes cloud infrastructures a compelling alternative choice. Compute clouds provide infrastructure-as-a-service, making it easy to deploy the computational platform for a specific task on-demand (e.g., for a given workflow).

2.6 Handling Data Availability and Data Security

An increasingly used method to provide data available in a scalable fashion is to use cloud storage. One common requirement is that data access needs to be made secure as well. One project exemplifying such capabilities in the genomics field is the pan-cancer analysis of whole genomes (PCAWG) study. The cloud-based computing infrastructure developed for this project has the capability to handle data uploads and updates very efficiently. Additionally, it can provide raw read data, genome alignments and associated metadata (to and from Amazon S3 systems), where read-level data is controlled and is made accessible and decryptable only with the approval of Data Access Compliance Office (DACO) (https:// dcc.icgc.org/icgc-in-the-cloud/aws, accessed 01.09.2018).

When designing and developing such infrastructures, there can often be "local" legislative issues related to data localization that need to be considered. A technical solution is to rely on local clouds, restricting the data and the analysis to data centers within a given area of jurisdiction (e.g., a specific country). Even tighter access restrictions can be met by using private or the so-called "on-premises" clouds, where compute infrastructure is confined to a specific institution or data center. This has, strictly speaking, few characteristics in common with cloud computing principles, except for the existence of a compatible technical implementation and the use of cloud management software, usually OpenStack. Generally, the creation of data silos makes data even more secure than more commonly used solutions, like shared filesystems. In these cases, if data security is compromised for one given project, additional projects could also be exposed. Data Access Compliance (DAC) policies can then be implemented in a more flexible and granular manner without much technical overhead, by creating virtual or physical data silos with controlled shared access, without putting other silos at risk. A great advantage of this form of cloud computing is the potential to make data available for analysis without actually transferring it.

2.7 Cloud Technologies: Kubernetes

Due to the reasons explained in previous sections, it is required that data analysis workflows are implemented seamlessly in different compute environments. The most efficient way to achieve this migration is by using systems that can be abstracted from the actual compute infrastructure. A software system called Kubernetes (https://kubernetes.io/) represents one of the most widely used approaches to achieve this goal since it combines such abstraction with the use of containerized software. Kubernetes is a container orchestration system that provides automated container deployment, scaling, and software management capabilities (Fig. 1). A Kubernetes cluster can be constructed in an HPC environment as well as in different cloud environments.

Albeit being primarily designed to run service-centric computations, Kubernetes can also be used efficient to run data-centric compute jobs, which are much more appropriate for cloud environments. Service-centric computation efforts require the provision of a constant level of service. One example of translating this concept to proteomics data analysis would be the availability of an analysis service for MS/MS identification, and another one for FDR calculation and filtering of the results. Due to the many different stages of a proteomics analysis workflow, many different "services" would be needed in parallel. In contrast, data-centric compute job scheduling does not need to have available a constant service. Instead, the scheduling of the computation as the data comes in represents the main requirement. This second option scales much better and is more resource conservative, a fact that is essential in cloud computing, where computation could be billed.

However, a benefit from the service-centered origins of Kubernetes is the availability of recovering mechanisms for failed software containers. As a concrete example, if a job is not running due to instance failing, the job execution will be retried a selectable number of times before "admitting defeat." This functionality increases the robustness of a workflow, especially in cloud environments. In these cases, instances might be affected by different points of failure that originate from the highly virtualized infrastructure (network congestions, host oversubscription, hard disk failures, etc.).

It is important to highlight that, in addition to interfacing with other compute clusters, WMS can also interface with a Kubernetes cluster, given an integration solution. Leveraging Galaxy to work in Kubernetes has been exemplified with the Galaxy/Kubernetes integration [46] produced within the metabolomics-centric PhenoMeNal project [47] (http://portal.phenomenal-h2020.eu).

3   Quality Control Approaches

Putting data analysis pipelines into robust and reproducible workflows represents just one aspect of making data analysis reproducible. As mentioned before, every analysis pipeline, however robust, needs a number of input parameters that could change the analysis results, when not documented properly. As an example, comparative studies using common samples have shown low reproducibility [48]. Therefore, providing access to the contributing factors of variation gives a valuable insight and alleviates the task of performing combined analysis [49]. Next, we will explore, how metadata—input parameters and data that can influence input parameters, such as thresholds and tolerances—can be employed to provide the means to reproduce and explain occurring experimental variability. We will also show how QC can help to explain variability. This eventually enables the original data to exceed its initial value by being reused in a more accurate manner by the community.

One of the greatest impacts for any proteomics analysis is the protein sequence database used for matching the mass spectra. Inclusion of the right protein sequences in the database is crucial for the identification of proteins present in a sample. This may include the accession numbers of protein sequences in, for example, UniProt [50] or the complete FASTA file generated in the case of sequences coming from tailored samples like those increasingly used in proteogenomics experiments. The worse the sequence database matches the true contents of the analyzed sample, the higher the chances for random matches are, which will inadvertently introduce variability in the final results.

As shown with the search database input parameters, data analysis has the potential to introduce variability. It is well accepted that the more parameters are involved and the greater the number to combine, the higher the potential for variability in the analysis results is. This characteristic is not necessarily tied to a specific software or algorithm. A sensible adoption of parameters and their adjustment (e.g., sequence database size, protein modifications used for the search and length/cleavage sites in different experimental settings) can be supported by also carefully reporting the experimental protocols used. Together with a proper versioning of software and analysis workflows, these can provide the basis for the integration of data analysis approaches in SOPs.

Additionally, the availability of comprehensive metadata, both biological and technical, alongside the core experimental data can help to find and evaluate the right data to be reused. Recording of appropriate metadata will greatly improve the chance of the data actually being reused and that way, to increase the recognition for the original data producers [51]. In fact, virtually everyone in the community, including for example collaborations spanning multiple laboratories, can greatly benefit from the structured collection and assessment of data and metadata. Supplying detailed and readily accessible metadata (e.g., in the corresponding scientific publication or in a public data repository) will help to contextualize the experimental data and thereby will enable a better-informed data reuse. Data users from connecting or overlapping fields of research, where implied knowledge of common protocols is not ensured, will profit even more [52]. As a key point, the use of open data formats, as those introduced earlier, can already help with the documentation and improve transparency. In fact, many of the described formats are designed to include the analysis input parameters. Even more, in the case of qcML, this format is designed to keep QC data, a crucial concept for reproducibility.

In addition to the potential variability in the analysis results resulting from differing input parameters, the variability of the data itself can also be reflected in the analysis results. Any type of variability needs to be addressed, particularly in clinical settings. For enabling data analysis reproducibility, it is essential that the experimental variability is well documented and put in the same context than the input parameters and the actual data. In the next section, we will explore in which ways data reproducibility can be affected by the experimental variability and what QC approaches can do to increase the robustness in data analysis.

3.1 Performance Metrics

QC approaches can be used to control the experimental variability, providing a more stable (and sustainable) level of consistency in the analytical measurements, by monitoring the experimental instrumentation and the analytical performance. Related fields of research to proteomics such as metabolomics, separation sciences, and chemistry, have advanced SOPs available for many clinical applications. In fact, much of the "language" in clinical proteomics is dominated by clinical chemistry [4], and many QC procedures can be adapted for clinical proteomics approaches as well. It is generally accepted that the experimental variability increases in parallel with the number of preparation steps. The use of Integrated laboratory information management systems (LIMS) that can track information represents a definite advantage for documentation purposes.

One example of experimental variability that can hinder the experimental reproducibility in proteomics is related to the number of missed cleavages and the peptide size distribution. Both of them are protocol dependent but could also point to sample preparation issues. The expected peptide size will influence the number of peptide/protein identifications through the lengths of the peptide sequences matched against the spectra. Therefore, longer peptides could be undetected if a lower number of missed cleavages were expected and chosen as the input parameter for the analysis. Tracking the used digestion enzyme batch and to perform a further investigation into autolysis products would be two of the possible follow-up actions in case any issues are detected. A general loss of peptide/protein identifications and a higher variability in the sequence of the identified peptides could also point to a higher adsorption, which could be derived from changed lab materials or updated protocols.

Another example of a source of experimental variability that can be detected using QC is when a broad distribution of observed mass errors is observed for identification results. These which might indicate that the mass tolerance settings used as input parameters might be incorrect. Additionally, several peaks in the mass error distribution (Fig. 3) or a broad distribution in general may hint at the frequent identification of spectra with neutral losses, modifications within the given mass tolerance, or loss of the lock-mass,

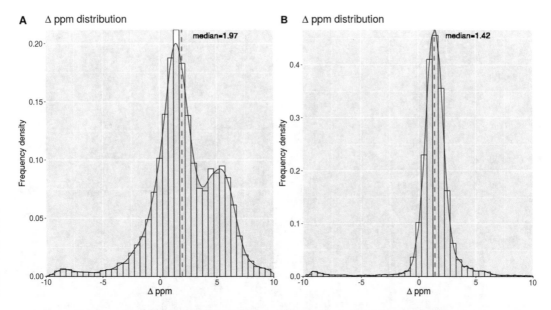

Fig. 3 Distribution of the peptide identification mass error in an *E. coli* lysate standard MS run. (**a**) Contains a second peak to the right indicating a concentration of off-mass identifications within the given mass tolerances. This effect can lead to missed identifications in the identification analysis. (**b**) Distribution of the peptide identification mass error after correcting the modification and precursor mass input parameters

hence requesting an adjustment of the protein modification, precursor mass, and/or mass tolerance parameters.

Controlling simple quality metrics such as the number of identified peptides or the observed mass error in the identification results, and combining them with a proper recording of experimental metadata, will enable a better experimental reproducibility. However, low peptide/protein identification numbers can also point to issues related to the instrumentation. If this low rate for identification results were observed together with stable MS/MS spectra numbers, the overall issues could be related to noise levels or, as explained above, to the choice of protein sequence database. It is therefore also important to control quality metrics before the analysis is performed.

Tabb et al. [53] coined the term of "IDfree" metrics for those that can be calculated before the data analysis is performed, which usually starts with the identification process. These include metrics such as:

- Charge state distributions

- Ion injection times

- Noise levels

- Dynamic sampling (number of successfully identified MS2 spectra per cycle)

- Characteristics of the ion current and the chromatogram (Fig. 4)

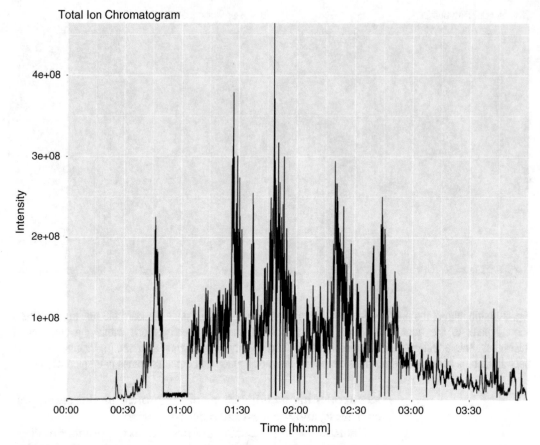

Fig. 4 Total ion chromatogram coming from a MS run with an electrospray ionization failure in the early retention time area, just after 50 min. The result is a sharp decrease in the total ion current and observed intensity. This problem could provide an explanation to the reduced number of MS2 spectra suitable for successful identification

- The abundance of common signals
- Evaluation of MS1 or MS2 intensities over the chromatographic gradient

The existence of variability in these selected metrics will undoubtedly have a direct impact in the data analysis results. Figure 4 shows a total ion chromatogram coming from a MS run with a failure in the electrospray ionization in the early retention time area, just after 50 min. The result is a sharp drop in the total ion current and observed intensity. Such problems can explain the variable performance of the subsequent analysis, for example a reduced number of MS2 spectra suitable for successful identification and/or a low identification rate.

The characteristics of the analytical environment can also be used as metrics, such as column temperatures, pump pressures, or other instrument parameters [54]. They will influence the data analysis results as well, albeit in a more indirect way than the "IDfree" metrics mentioned above. Nonetheless, these metrics are especially important in the longitudinal monitoring of the health of the instrumentation, which we will consider later in Subheading 3.3.

It is also important to highlight that there are several software packages that be used to perform a wide variety of QC metric calculations on any given sample [35, 55, 56], to capture the instrument performance, and to compile QC reports. For more information, QC software and the related metrics have been recently reviewed by Laukens et al. [57].

3.2 Use of QC Samples

Using QC in day-to-day operations can indeed reduce the experimental variability or mitigate the impact of mechanisms introducing technical variability. The overall goal is to achieve a sustainable quality of data acquisition. Reporting the observed variability together with quality metrics is most informative to everyone involved in the data analysis process including reviewers, reproducers, and of course the initial data analysts. A well-established method of controlling the analytical setup involves the use of QC samples, which are samples of known content and characteristics. They can provide confidence that the system is performing as expected or give indicators of developing issues when run before and after an analytical sample, and regularly interspersed. Another advantage of the QC samples is that they can be used to calibrate, evaluate the data analysis performance, and give further insight to the experimental variability considering metrics that are not "IDfree." Many different metrics (some of them have already been mentioned above) have indeed been proposed to measure the performance of an instrument, data analysis or in broader terms, experimental setups [54, 55, 58, 59].

As mentioned, running QC samples between regular samples can greatly increase the confidence in the analytical system by proving that the instrumentation is performing as expected, and also by comparing and assessing the analytical variability between different batches, and by eliminating different sources of variability during troubleshooting. In addition to the usefulness of QC samples for normalization purposes, their most obvious application is the assessment of the analytical reproducibility. Depending on the QC sample content and the metrics applicable, operators can be pointed at a variety of incidences (e.g., mass calibration problems) that are causing loss of performance so the situation can be corrected.

The application of using QC samples has been shown in the literature. Sabido et al. [60] provide an accessible way to monitor and compare intralab and interlab instrument performance and provide a quick evaluation using a software called QCloud. They use two different, commercially available standards, called QC1 and QC2, to assess a list of useful quality metrics. QC1 is a trypsin-digested BSA (bovine serum albumin) MS standard (CAM modified, meaning that was reduced and alkylated with iodoacetamide) that is recommended for quick and frequent checks. QC2 is a more complex HeLa protein digest standard, that can be used less frequently but can provide a more in-depth QC related information. The metrics assessed with QCloud include a range of "IDfree" metrics like median injection time, total ion current, chromatographic resolution, number of spectra, number of PSMs (Peptide Spectrum Matches), uniquely identified peptides, and mass accuracy for selected peptides, among other informative metrics. The software QCloud can correlate them with the acquisition time and related instrument maintenance events, which provides a better insight into the instrument performance. In a second publication, Mechtler et al. [61] also recommend two types of QC standards for the evaluation of instrument performance. First, they describe a sensitivity standard consisting of a mixture of digested BSA peptides, cytochrome c, and phosphopeptides, containing varying concentrations going from a low to a high *fmol* range. QC samples must be run twice per day to assess basic LC performance and the capability to consistently identify both low and highly abundant peptides.

Overall, a variety of predigested standards comprising different complex ranges of proteomics mixtures are available from several of the leading MS vendors and reagent companies. Options for QC samples might range from single peptides, mixtures of peptides, protein digests, and complex lysates. There are different options which can be used to customize QC practices. The type of standard to use should be driven by the actual metrics that are most suitable for the actual experimental set-ups.

It should also be considered that while the interspersion of real samples with QC samples results in a higher sensitivity for the detection of suboptimal conditions and issues, the process decreases the analysis throughput. However, when using highly valuable samples in the case of clinical settings, this represents only a minor disadvantage when compared with the benefits of having a robust QC process in place. The availability of such vetted system also simplifies the decision process of whether a dataset is acceptable for publication or not.

3.3 Longitudinal Monitoring

The highest level of control by monitoring QC samples is leveraged by performing a longitudinal monitoring along the instrumentation's operation time. Performance can be tracked over time, keeping issues met and the occurring experimental variability in acceptable value ranges. The actual question what is considered acceptable can be calculated using statistical process control (SPC) methods [62], established via reference sets, or imposed by predefined industry standards. Limits and measurements are compiled in different control charts along with a time axis, and are ideally annotated with manual intervention events, such as column changes, cleaning, or general maintenance operations. There are several software tools that have been developed which can be used to perform longitudinal QC monitoring (semi)automatically:

- QCloud [60]
- SimPatiQCo [63]
- iMonDB [54]
- MSStatsQC [6]
- SProCoP (SRM) [64]

For longitudinal monitoring, it is worthwhile noting that many of the "IDfree" metrics can be used in addition to parameters of the analytical environment, employing both QC samples and/or regular sample runs. In order to follow SPC methods, however, different types of samples should be quality controlled in separate groups, since different value ranges will be expected for at least some of the metrics. Also, the different groups should have either an adequate size for one to be able to calculate robust statistics or a reference set must be available for the type of sample to enable the establishment of appropriate control limits. With separate groups, however, the longitudinal monitoring might not be able to cover the instruments' up-time as regularly as scheduled dedicated QC samples would.

3.4 Use of Control Limits

A set of predefined acceptance criteria for the set of considered QC metrics must be fulfilled after each QC sample is run. These predefined control limits are commonly used in QC of related analytical fields [65] and in many clinical settings. Of course, the derived limits can be calculated empirically or by using statistical methods. A more modern way to choose the limits is to consider the value distribution of a given metric from a reference set, that is, observing the normal range of operation, and to model the limits around the so-called "three-sigma" rule. For a range of values that are approximately normally distributed, 68% of the values are within (+/−) one sigma, or standard deviation of the mathematical mean. Additionally, 95% of the values are within two "sigmas," and 99.7% within three "sigmas." Since observations above these limits will be increasingly rare, increased observations could indicate a process that is potentially out of control. However, it is good to remember that even a process that satisfies the predefined criteria could still perform poorly, especially when only a single metric is considered instead of a set.

3.5 Control Charts

Control charts [66] make SPC visually interpretable since metric values are plotted along the instrument time axis. The calculated mean acts as a center line and the calculated "sigma" (standard deviation) limits are indicated as lines above (Upper Control Level, UCL) and below (Lower Control Level, LCL) the mean value (Fig. 5). They show if a given metric's values are stable over time

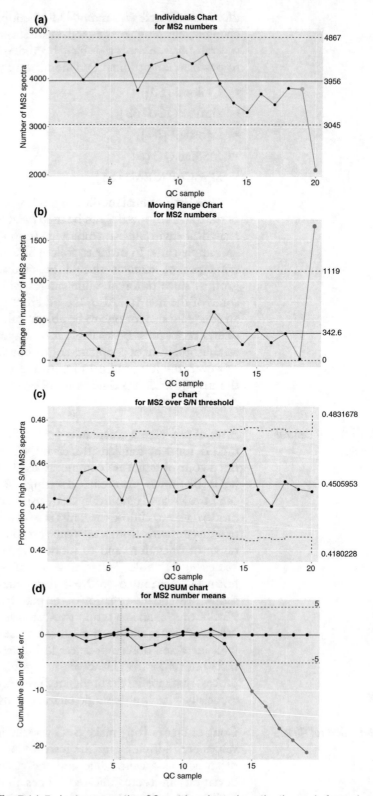

Fig. 5 (a) $\bar{x}$-chart representing QC metric values along the time axis for a given QC sample, highlighting the control limits that are used to show analytical measurements that could be out of statistical control (example highlighted in

around a target value or range of values. Different types of control charts have been developed with this purpose. They can be specialized representations used to visualize certain aspects of the data in a better way, like the detection of value shifts of different sizes. Control charts can involve different statistics, for example for performing sample subgrouping in applications where sampling each measurement is impractical. Next, we will briefly describe the main types of control charts:

- A $\bar{x}$-chart (read x bar chart) can be used to monitor the mean of a given quality parameter within the actual measurement (as circumstances require for different possible subgroups), at regular intervals.

- A mR-chart (moving range) can be used to monitor the variance of a given quality parameter in a moving range for consecutive individual measurements.

- A p-chart enables the visualization of binary measurement types of tests, as fractions of "pass" or "fail" values per time point, applying binomial distribution derived limits.

Fig. 5 (continued) red). Persistent runs containing values below or above average are marked as a violation of the Western Electric rules for control charts [67] (highlighted in yellow). (**b**) The mR-chart can be used to determine if a given process is stable and predictable. It can plot variation, calculated from the ranges of successive observations, showing variability between one data point and the next. Any lack of control in the individuals chart may be due to unstable variation, not due to actual changes in the process's core. If the mR-chart is under control, an out-of-control individual chart can point to changes in the process. However, the moving ranges are serially correlated so cycles may appear that do not indicate real problems in the underlying process. In some cases, it may be advisable to use the median of the moving range rather than the mean, especially when a few larger values inflate the estimate of the standard deviation. (**c**) A p-chart is a control chart used for data coming from subgroups with varying size. Attributes are controlled by a "pass/fail"-type inspection, effectively applying the specification limits to the data before they are plotted. Here, the S/N attribute of the acquired MS2 spectra is chosen. As the number of MS2 spectra can vary, proportions of "pass/fail" from a targeted minimum S/N ratio for successful identification are plotted. Overall, p-charts can show how the process changes over time for subgroups of varying size. (**d**) CUSUM-chart plotting the negative and positive cumulative standard error of the sample group (here just one sample). This metric is usually more sensitive to small changes and can detect potential problems earlier. However, the chart relies on a target value (usually the mean) and on an estimation of the standard deviation coming from calibration data (left of the dotted horizontal line). The decision interval can then be chosen from the calibration data (as the standard deviation and the mean can be calculated from that), and the result can be used to control the new incoming data process

- CUSUM charts are time-weighted control charts that can be used to detect small changes by drifts in the mean or variation of a given metric.

The "Shewhart individuals control" usually combines the $\bar{x}$-chart for individuals and the mR-chart to detect changes that are outside of the natural variability. More information on the use of control charts can be found in Vitek et al. [6] and Bramwell [62].

The use of regularly interspersed QC sample runs, as a source input for the control charts, provides a useful measure of control over the metrics used. This also helps to maintain a stable analytical system under statistical control. Additionally, it will also give insight to the development of the expected data variability over time, and provide vital information for the explanation of the variability in particular cases.

The mean and limit lines are used to divide the chart into different zones. In addition to the rule related to exceeding the three "sigmas" limit, additional ones have been devised in manufacturing [67] and clinical chemistry approaches [68] to detect aberrant processes. These rules aim to detect unnatural patterns within the charts, like certain frequencies within different zones, that could indicate the need for intervention. Basically, they represent a way to detect outliers. Like this, we can use control charts to detect trends and intervene early. Overall, by choosing the right control limits and interspersing regular QC samples in the analysis, a level of confidence in the performance of our analytical setup for valuable clinical sample can be ensured.

4   Conclusions

We have shown the importance of having stable data analysis environments for enabling reproducible science. Designing common, robust and portable data analysis workflows are important goals for a future of increasingly larger datasets. We detailed WMS and software containerization as possible solutions for enabling robust data analysis. Next, we explained the role of input parameters in the data analysis and stressed the importance of performing a careful documentation. Last, we described how QC workflows can help to keep analytical measurements under statistical control. There are several software solutions to facilitate QC approaches, as referenced in the previous paragraphs. *See* Table 4 for an overview of available QC metrics from different software solutions. QC can minimize the experimental variability, which gets reflected by achieving more robust and reproducible results. The consistent application of QC greatly benefits the confidence to one's own data and ultimately benefits the data reuse and increases the long-term value of the datasets.

Table 4
List of QC metrics from referenced QC tools as stated in the corresponding publications

---

Quameter [55]
C-1A: Fraction of repeat peptide IDs with divergent ID (−4 min)
C-1B: Fraction of repeat peptide IDs with divergent ID (+4 min)
C-2A: Interquartile retention time period (period [min])
C-2B: Interquartile retention time period (Pep ID rate)
C-3A: Peak width at half-height for IDs (median value)
C-3B: Peak width at half-height for IDs (interquartile distance)
C-4A: Peak widths at half-max over RT deciles for IDs (first decile)
C-4B: Peak widths at half-max over RT deciles for IDs (last decile)
C-4C: Peak widths at half-max over RT deciles for IDs (median decile)
DS-1A: Ratios of peptide ions IDed by different numbers of spectra (once/twice ratio)
DS-1B: Ratios of peptide ions IDed by different numbers of spectra (twice/thrice ratio)
DS-2A: Spectrum counts (MS1 scans/full)
DS-2B: Spectrum counts (MS2 scans)
DS-3A: MS1 max/MS1 sampled abundance ratio IDs (median all IDs)
DS-3B: MS1 max/MS1 sampled abundance ratio IDs (Med bottom 1/2)
IS-1A: MS1 during middle (and early) peptide retention period (MS1 jumps >10×)
IS-1B: MS1 during middle (and early) peptide retention period (MS1 falls >10×)
IS-2: Precursor m/z for IDs (median)
IS-3A: IDs by charge state (relative to 2+, charge 1+)
IS-3B: IDs by charge state (relative to 2+, charge 3+)
IS-3C: IDs by charge state (relative to 2+, charge 4+)
MS1-1: Ion injection times for IDs: MS1 (median)
MS1-2A: MS1 during middle (and early) peptide retention period (S/N median)
MS1-2B: MS1 during middle (and early) peptide retention period (TIC median)
MS1-3A: MS1 ID max (95/5 percentile)
MS1-3B: MS1 ID max (median)
MS1-4A: MS1 intensity variation for peptides (within series)
MS1-4B: MS1 intensity variation for peptides (between/in)
MS1-5A: Precursor m/z—peptide ion m/z (median)
MS1-5B: Precursor m/z—peptide ion m/z (mean absolute)
MS1-5C: Precursor m/z—peptide ion m/z (ppm median)
MS1-5D: Precursor m/z—peptide ion m/z (ppm interQ)
MS2-1: Ion injection times for IDs: MS2 (median)
MS2-2: MS2 ID S/N (median)
MS2-3: MS2 ID peaks (median)
MS2-4A: Fraction of MS2 identified at different MS1 max quartiles (ID fract Q1)
MS2-4B: Fraction of MS2 identified at different MS1 max quartiles (ID fract Q2)
MS2-4C: Fraction of MS2 identified at different MS1 max quartiles (ID fract Q3)
MS2-4D: Fraction of MS2 identified at different MS1 max quartiles (ID fract Q4)
P-1: MS2 ID score (median)
P-2A: Tryptic peptide counts (identifications)
P-2B: Tryptic peptide counts (ions)
P-2C: Tryptic peptide counts (peptide)
P-3: Peptides counts (semi/tryp)

(continued)

Table 4
(continued)

SIMPATIQCO [63]
Number of MS1 and MS/MS scans
MS1 and MS/MS scans where lock-mass was detected
MS1 and MS/MS scans where maximum injection time was reached
Average MS1 lock mass deviation
Average MS1 and MS/MS ion injection time and total scan time
Total number of identified PSMs and proteins
Sequence coverage of proteins that match an entry in the "proteins of interest" list
For definable "peptide lists" (e.g., for QC 1 sample: cytochrome c peptides and synthetic phosphopeptides): peptide peak elution time, apex intensity, peak width and area
MS1 and MS/MS ion injection time and scan time
Lock mass detection and deviation
MS1 and MS/MS total ion current
MS1 and MS/MS "(%) of target value"
Number of PSMs identified per minute m/z of triggered precursors

QCloud [60]
Peak area
Mass accuracy
Retention time drift
Median injection time MS1
Median injection time MS2
Chromatographic resolution peak capacity
Total ion current
MS1 spectra count
MS2 spectra count
Chromatogram count TIC slump
Total number of missed cleavages
Total number of identified proteins
Total number of uniquely identified proteins

MSStatsQC [6]
LC peak retention times
LC peak intensities
LC peak widths
Total peak areas of peptides
MS1 and MS/MS ion injection times
Number of acquired MS1 and MS/MS scans
Mass accuracy

iMonDB [54]
Turbopump power
Turbopump speed
Turbopump temperature
ESI capillary temperature
Vacuum pressure
Syringe pump flow rate
Total number of PSMs
Total number of identified peptides
Total number of uniquely identified peptides
Mean delta ppm median delta ppm
ID ratio
MS quantification results details
Number of features

Acknowledgments

The authors would wish to acknowledge funding from ELIXIR Implementation Studies, BBSRC [grant number BB/P024599/1], Wellcome Trust [grant number 208391/Z/17/Z], and EMBL core funding.

References

1. Meo AD et al (2014) What is wrong with clinical proteomics? Clin Chem 60:1258–1266

2. Foster JM et al (2011) A posteriori quality control for the curation and reuse of public proteomics data. Proteomics 11 (11):2182–2194

3. Klont F et al (2018) Assessment of sample preparation bias in mass spectrometry-based proteomics. Anal Chem 90:5405–5413

4. Apweiler R et al (2009) Approaching clinical proteomics: current state and future fields of application in fluid proteomics. Clin Chem Lab Med 47:724–744

5. Cairns DA et al (2008) Integrated multi-level quality control for proteomic profiling studies using mass spectrometry. BMC Bioinformatics 9:519

6. Dogu E et al (2017) MSstatsQC: longitudinal system suitability monitoring and quality control for targeted proteomic experiments. Mol Cell Proteomics 16:1335–1347

7. Clough T et al (2012) Statistical protein quantification and significance analysis in label-free LC-MS experiments with complex designs. BMC Bioinformatics 13(Suppl 1):S6

8. Piehowski PD et al (2013) Sources of technical variability in quantitative LC–MS proteomics: human brain tissue sample analysis. J Proteome Res 12(5):2128–2137

9. Villanueva J, Carrascal M, Abian J (2014) Isotope dilution mass spectrometry for absolute quantification in proteomics: concepts and strategies. J Proteome 96:184–199

10. Easing the burden of code review (2018) Nat Methods 15(9):641

11. Kanwal S et al (2017) Investigating reproducibility and tracking provenance - a genomic workflow case study. BMC Bioinformatics 18:1–14

12. Leprevost FD et al (2017) BioContainers: an open-source and community-driven framework for software standardization. Bioinformatics 33(16):2580–2582

13. Barsnes H, Vaudel M (2018) SearchGUI: a highly adaptable common interface for proteomics search and de novo engines. J Proteome Res 17(7):2552–2555

14. Cox J, Mann M (2008) MaxQuant enables high peptide identification rates, individualized p.p.b.-range mass accuracies and proteome-wide protein quantification. Nat Biotechnol 26:1367–1372

15. Pluskal T et al (2010) MZmine 2: modular framework for processing, visualizing, and analyzing mass spectrometry-based molecular profile data. BMC Bioinformatics 11:395

16. Kessner D et al (2008) ProteoWizard: open source software for rapid proteomics tools development. Bioinformatics 24 (21):2534–2536

17. Prince JT, Marcotte EM (2008) mspire: mass spectrometry proteomics in ruby. Bioinformatics 24(23):2796–2797

18. Lopez-Fernandez H et al (2015) Mass-Up: an all-in-one open software application for MALDI-TOF mass spectrometry knowledge discovery. BMC Bioinformatics 16:318

19. Käll L, Canterbury J, Weston J (2007) Semi-supervised learning for peptide identification from shotgun proteomics datasets. Nature 4:923–925

20. Röst HL et al (2016) OpenMS: a flexible open-source software platform for mass spectrometry data analysis. Nat Methods 13:741–748

21. Ison J et al (2016) Tools and data services registry: a community effort to document bioinformatics resources. Nucleic Acids Res 44: D38–D47

22. Deutsch EW et al (2015) Development of data representation standards by the human proteome organization proteomics standards initiative. J Am Med Inform Assoc 22(3):495–506

23. Deutsch EW et al (2017) Proteomics standards initiative: fifteen years of progress and future work. J Proteome Res 16:4288–4298

24. Orchard S, Hermjakob H, Apweiler R (2003) The proteomics standards initiative. Proteomics 3:1374–1376

25. Martens L et al (2011) mzML—a community standard for mass spectrometry data. Mol Cell Proteomics 10:R110.000133

26. Martens L, Vizcaíno JA, Banks R (2011) Quality control in proteomics. Proteomics 11:1015–1016

27. Perkins DN et al (1999) Probability-based protein identification by searching sequence databases using mass spectrometry data. Electrophoresis 20:3551–3567

28. Eng JK, Jahan TA, Hoopmann MR (2013) Comet: an open-source MS/MS sequence database search tool. Proteomics 13:22–24

29. Kim S, Pevzner PA (2014) MS-GF+ makes progress towards a universal database search tool for proteomics. Nat Commun 5:5277

30. Fenyö D, Beavis RC (2003) A method for assessing the statistical significance of mass spectrometry-based protein identifications using general scoring schemes. Anal Chem 75:768–774

31. Jones AR et al (2012) The mzIdentML data standard for mass spectrometry-based proteomics results. Mol Cell Proteomics 11: M111.014381

32. Vizcaíno JA et al (2017) The mzIdentML data standard version 1.2, supporting advances in proteome informatics. Mol Cell Proteomics 16:1275–1285

33. Griss J et al (2014) The mzTab data exchange format: communicating mass-spectrometry-based proteomics and metabolomics experimental results to a wider audience. Mol Cell Proteomics 13:2765

34. Walzer M et al (2013) The mzQuantML data standard for mass spectrometry-based quantitative studies in proteomics. Mol Cell Proteomics 12:2332–2340

35. Walzer M et al (2014) qcML: an exchange format for quality control metrics from mass spectrometry experiments. Mol Cell Proteomics 13:1905–1913

36. Xu T et al (2015) ProLuCID: an improved SEQUEST-like algorithm with enhanced sensitivity and specificity. J Proteome 129:16–24

37. Zhang J et al (2012) PEAKS DB: de novo sequencing assisted database search for sensitive and accurate peptide identification. Mol Cell Proteomics 11:M111.010587

38. Vaudel M et al (2015) PeptideShaker enables reanalysis of MS-derived proteomics data sets. Nat Biotechnol 33:22–24

39. Searle BC (2010) Scaffold: a bioinformatic tool for validating MS/MS-based proteomic studies. Proteomics 10(6):1265–1269

40. Amstutz P et al (2016) Common workflow language, v1.0

41. Afgan E et al (2018) The Galaxy platform for accessible, reproducible and collaborative biomedical analyses: 2018 update. Nucleic Acids Res 46:W537–W544

42. Berthold MR et al (2009) KNIME - the Konstanz information miner. ACM SIGKDD Explor Newsl 11:26

43. Gillet LCL et al (2012) Targeted data extraction of the MS/MS spectra generated by data independent acquisition: a new concept for consistent and accurate proteome analysis. Mol Cell Proteomics 11:1–45

44. Röst HL et al (2014) OpenSWATH enables automated, targeted analysis of data-independent acquisition MS data. Nat Biotechnol 32:219–223

45. Collins BC et al (2017) Multi-laboratory assessment of reproducibility, qualitative and quantitative performance of SWATH-mass spectrometry. Nat Commun 8:1–11

46. Moreno P et al (2018) Galaxy-Kubernetes integration: scaling bioinformatics workflows in the cloud. bioRxiv. Preprint

47. Peters K et al (2018) PhenoMeNal: processing and analysis of Metabolomics data in the Cloud. bioRxiv. Preprint

48. Albar JP, Canals F (2013) Standardization and quality control in proteomics. J Proteome 95:1–2

49. Tabb DDL et al (2010) Repeatability and reproducibility in proteomic identifications by liquid chromatography-tandem mass spectrometry. J Proteome Res 9:761–776

50. Bateman A et al (2017) UniProt: the universal protein knowledgebase. Nucleic Acids Res 45: D158–D169

51. Tabb DL (2013) Quality assessment for clinical proteomics. Clin Biochem 46:411–420

52. Rodriguez H, Pennington SR (2018) Revolutionizing precision oncology through collaborative proteogenomics and data sharing. Cell 173:535–539

53. Wang X et al (2014) QC metrics from CPTAC raw LC-MS/MS data interpreted through multivariate statistics. Anal Chem 86:2497–2509

54. Bittremieux W et al (2015) iMonDB: mass spectrometry quality control through instrument monitoring. J Proteome Res 2015:150323163122004

55. Ma ZQ et al (2012) QuaMeter: multivendor performance metrics for LC-MS/MS proteomics instrumentation. Anal Chem 84:5845–5850

56. Gatto L, Wen B (2018) proteoQC: an R package for proteomics data quality control. R package version 1.16.0. https://github.com/wenbostar/proteoQC

57. Bittremieux W et al (2017) Computational quality control tools for mass spectrometry proteomics. Proteomics 17:3–4

58. Rudnick PA et al (2010) Performance metrics for liquid chromatography-tandem mass spectrometry systems in proteomics analyses. Mol Cell Proteomics 9:225–241

59. Bielow C, Mastrobuoni G, Kempa S (2016) Proteomics quality control – a quality control software for MaxQuant results. J Proteome Res 15(3):777–787

60. Chiva C et al (2018) QCloud: a cloud-based quality control system for mass spectrometry-based proteomics laboratories. PLoS One 13:1–14

61. Köcher T et al (2011) Quality control in LC-MS/MS. Proteomics 11:1026–1030

62. Bramwell D (2013) An introduction to statistical process control in research proteomics. J Proteome 95:3–21

63. Pichler P et al (2012) SIMPATIQCO: a server-based software suite which facilitates monitoring the time course of LC-MS performance metrics on orbitrap instruments. J Proteome Res 11:5540

64. Bereman M et al (2014) Implementation of statistical process control for proteomic experiments via LC MS/MS. J Am Soc Mass Spectrom 25:581–587

65. Dong M, Paul R, Gershanov L (2001) Getting the perfect peaks: system suitability for HPLC. Todays Chemist At Work 10(9):38–42

66. Shewhart WA (1939) Statistical method from the viewpoint of quality control. Department of Agriculture, Washington, DC, pp 1–7

67. Western Electric (1958) Statistical quality control handbook. Western Electric, Indianapolis

68. Westgard JO, Barry PL, Hunt MR (1981) A multi-rule Shewart chart for quality control in clinical chemistry. Clin Chem 27:493–501

Chapter 16

Review of Batch Effects Prevention, Diagnostics, and Correction Approaches

Jelena Čuklina, Patrick G. A. Pedrioli, and Ruedi Aebersold

Abstract

Systematic technical variation in high-throughput studies consisting of the serial measurement of large sample cohorts is known as batch effects. Batch effects reduce the sensitivity of biological signal extraction and can cause significant artifacts. The systematic bias in the data caused by batch effects is more common in studies in which logistical considerations restrict the number of samples that can be prepared or profiled in a single experiment, thus necessitating the arrangement of subsets of study samples in batches. To mitigate the negative impact of batch effects, statistical approaches for batch correction are used at the stage of experimental design and data processing. Whereas in genomics batch effects and possible remedies have been extensively discussed, they are a relatively new challenge in proteomics because methods with sufficient throughput to systematically measure through large sample cohorts have only recently become available. Here we provide general recommendations to mitigate batch effects: we discuss the design of large-scale proteomic studies, review the most commonly used tools for batch effect correction and overview their application in proteomics.

Key words Quantitative proteomics, Batch effects, Statistical analysis, Experimental design

1 Introduction

Proteins are the main effectors of biochemical function inside living organisms and often represent the most direct cause for a specific phenotype. Consequently, their study is of key importance in understanding the molecular mechanisms of health and disease. Not surprisingly, proteins present a huge market as targets of pharmaceutical intervention or diagnostic biomarkers.

Over the past decades, mass spectrometry-based proteomics has emerged as a key methodology for capturing the complexity of proteomes and to quantify their response to perturbations [1]. Traditionally, proteomics focused on the deep, high-coverage characterization of a limited number of samples. However, in order to capture the often subtle connection between proteotype (i.e., the acute quantitative state of a proteome) and phenotype, high-

Rune Matthiesen (ed.), *Mass Spectrometry Data Analysis in Proteomics*, Methods in Molecular Biology, vol. 2051,
https://doi.org/10.1007/978-1-4939-9744-2_16, © Springer Science+Business Media, LLC, part of Springer Nature 2020

throughput studies consisting of hundreds, if not thousands, of samples are required. Until recently, proteomic analyses of cohorts consisting of hundreds of samples were prohibitively expensive and time-consuming, as profiling of such cohort required several months of machine time [2, 3]. Recent technological advances, particularly DIA/SWATH-type mass spectrometry and related approaches, have allowed for a major increase in sample throughput, without sacrifice in quantitative robustness, sensitivity and reproducibility [1]. Now studies, measuring hundreds of samples with high quantitative accuracy are becoming a new routine [4–6]. This shift has opened the door to new and exciting research possibilities, but it has also focused the spotlight on novel data analysis challenges, characteristic of such high-throughput proteomic studies.

The number of samples, that can be prepared in one batch and measured by mass spectrometry, is limited by the complex logistics of the experiment. Importantly, changes in sample preparation and measurement conditions introduce systematic nonbiological biases known as batch effects.

In bottom-up proteomics batch effects can be introduced at three steps: (1) the biomaterial acquisition step, for example, by unequal retrieval or storage of clinical specimens, (2) the preparation of peptide mixtures from samples, or (3) the mass spectrometric measurement itself (*see* Fig. 1).

Consider a proteomic experiment designed to identify proteins, differentially abundant in tissue biopsies of cancer patients and corresponding healthy controls. First, the tissue samples are collected. Second, the samples are prepared for MS analysis. At this stage, proteins are extracted, digested, and possibly enriched for post-translational modifications of interest. Finally, the peptide mixture is measured by mass spectrometry. Depending on the scale of the study, the tissues might be collected by one or more study nurses/pathologists at different hospitals. Furthermore, sample preparation might require multiple days/weeks and involve different scientists and reagent batches. In large-scale cohort studies, it is also conceivable that samples are prepared and measured in different locations, or several mass spectrometry instruments are used, each one with slightly different performance characteristics. All these factors—reagents, personnel, instrumentation, and location—might introduce systematic nonbiological variation that needs to be taken into consideration.

If not properly accounted for, batch effects can bias, or, in severe cases, invalidate the conclusions of the experiment and make the results irreproducible [7]. Stringent experimental protocols improve reproducibility, but, unfortunately, they cannot completely prevent batch effects. Batch effects can, however, be mitigated by experimental design and batch effects correction during data processing. Here we will provide recommendations for the

associated batch effects

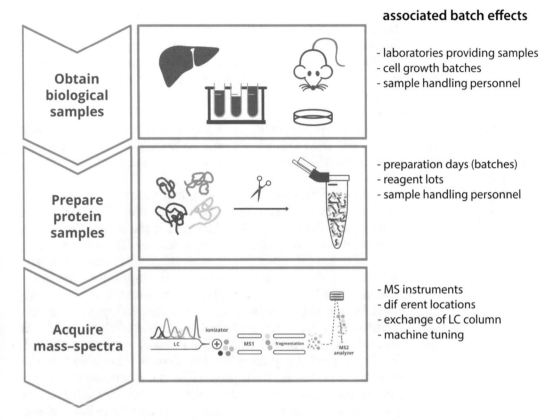

- laboratories providing samples
- cell growth batches
- sample handling personnel

- preparation days (batches)
- reagent lots
- sample handling personnel

- MS instruments
- diferent locations
- exchange of LC column
- machine tuning

Fig. 1 Sources of batch effects in proteomics. Batch effects can be introduced in the data at various stages of the experimental workflow. Proteomics shares common sources of batch effects with other biology related workflows (i.e., sample collection and processing) but also has unique sources of bias that require dedicated correction methodologies (i.e., MS acquisition)

design of large-scale proteomic experiments and describe the methods that can be used for batch effects analysis and correction.

2   Experimental Design and Batch Effects

Batch effects need to be considered from the earliest stages of planning a project. The choice of experimental design will, in the end, dictate if one can adjust for batch effects or not.

Experimental design can be defined as "a protocol that defines the population of interest, selects the individuals for the study and/or allocates them to treatment groups, and arranges the experimental material in space and time" [8]. Experimental design determines whether the data can be used to answer the questions of the study. Good designs mitigate the experimental error and the impact of technical artifacts [9]; conversely, bad designs obscure the signal and can make the data biased beyond repair [10, 11]. Results, obtained from such data potentially lead to erroneous conclusions and findings that cannot be replicated [7]. Good design also maximizes the cost–effect ratio, as selecting the right population and number of subjects to be profiled also determines whether the study will have sufficient statistical power to detect the effect of interest.

2.1 Biological and Technical Factors in Experimental Design

To ensure reproducibility of measurements, experimental design should take into account not only *biological factors* of interest but also *the technical factors* affecting measurement error and introducing bias to the data.

Technical factors, sometimes called *nuisance variables* [9], introduce unwanted association between samples. There are three types of such association. The first type is the *predictable factors*, such as hospitals or protein digestion batches. The second type is the association that arises due to *stochastic factors*, such as humidity in the room, instrumental downtime, or unplanned changes in reagent batches. These are difficult to incorporate in the experimental design explicitly, but one can note such factors in the sample annotation to use later at the correction step. When tracking is problematic (e.g., precise values of room humidity or temperature), the day of profiling/preparation can serve as reasonable proxy of such technical factor. Both predictable and stochastic factors are *discrete*, that is, they affect the samples as a group, or form a batch/grouping factor, that is the minimal statistical unit in batch effects correction models.

Finally, the sample association can be *continuous*. In mass spectrometry samples are profiled one after the other and the LC/MS system performance can change over time, introducing a continuous drift associated with sample running order.

All types of batch effects—discrete and continuous, predictable and stochastic—need to be evaluated at the batch effects correction step to choose an appropriate batch correction strategy and ensure biological signal extraction. At the stage of experimental design, knowledge of factors that will or might affect the samples is crucial to prevent confounding of biological and technical factors.

2.2 Confounding, Randomization and Blocking

The biological signal can be obscured entirely if the biological conditions are *confounded* with technical factors. This can happen, for instance, when distinct groups of samples, like disease and matched normal samples, are handled by different people or even processed in different centers. Confounding of the result by technical covariates can be prevented by sample randomization and blocking (*see* Fig. 2).

Randomization of samples prevents bias by breaking the connection between technical and biological factors. Failure to randomize samples among groups, defined by technical covariates, makes the correction of batch effects impossible and jeopardizes the results [12].

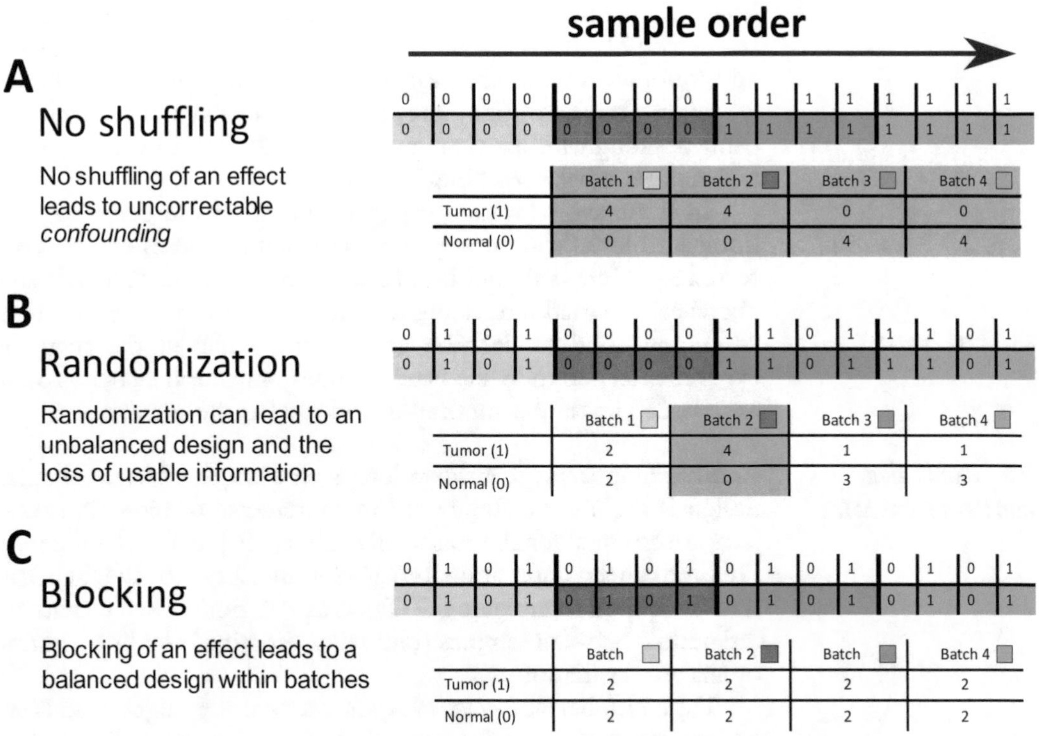

Fig. 2 Designing to minimize batch effects. It is very important to take batch effects into account from the very first stages of an experimental workflow. Here, we illustrate common experimental design strategies: (**a**) No shuffling (**b**) Randomization and (**c**) Blocking. Cases, when tumor/normal status is confounded with batch are shaded in pink in corresponding summary tables

A possible negative outcome of sample randomization is the generation of *unbalanced experimental designs*. For example, for studies comparing tumor vs. normal tissues, a certain digestion batch could end up containing mostly tumors, and thus not being sufficiently balanced by normal tissue samples (*see* Fig. 2b). Unbalanced batches add to the *nominal sample size*, but not to the *effective sample size* [13]. Nominal sample size is a total number of biological subjects, or samples profiled. Effective sample size is the number of samples, that would achieve the same power for biological signal detection if no batch effects were present. It is not uncommon that higher effective sample size could be achieved when batches with unbalanced sample grouping would be excluded from analysis. It is especially important to consider the ratio of nominal to effective sample size as nominal sample size determines the experimental costs, but the ratio of nominal to effective sample size indicates the degree of bias in group difference estimation with statistical tests [13].

The problem of unbalanced design can be mitigated by *blocking*. Blocks are predefined experimental units that contain a balanced number of samples from each group [8] (*see* Fig. 2c). Note, however, that the order of samples within the block normally is not restricted [8]. In large-scale studies the chance of stochastic biases and continuous signal drift is increased. The majority of such biases are order-associated; therefore, proper balancing can be achieved using a sliding window, that ensures every 20–30 consecutive samples represent balanced "blocks."

In summary, allocating experimental groups in balanced randomized blocks will ensure that the resulting biases can be corrected for. Here it should be stressed, that the terms "blocks" and "batches" are used interchangeably in the literature. The former is commonly used to describe groups of samples at the stage of experimental design, while batches commonly mean factual groups as recorded in sample annotation and used at the correction stage.

2.3 Replication and Power Estimates

Another important consideration at the stage of experimental design is *replication*. Replicated measurements of biological subjects are essential for the statistical analysis [14], network inference [15] and other computational analysis approaches [16, 17]. Replication is required to mitigate the effects of variation caused by natural differences between samples (biological variation) and by technical or analytical variation.

High number of *biological replicates*, or study subjects, adds an undisputed advantage for biological signal detection. Increasing the number of biological replicates, or *sample size*, allows to quantify naturally occurring variance so that we can both generalize our conclusions and robustly determine the uncertainty of statistical estimates [18]. Adding biological replicates to the study can be extremely difficult, especially when it involves enrolling more patients. However, in cell line perturbation experiments, one prepares additional cell plates for the basal state, preferably grown several weeks apart or by different researchers. Variance recorded for the basal state can then be compared to the magnitude of change due to the perturbation. However, one should keep in mind that strictly speaking, a true biological replicate in a cell line based experiment would be a different cell line, that has been derived from an independent patient. Thus, the distinction between biological and technical replication is not always self-evident.

Technical replication (replicating the protein digestion and/or MS measurement) allows control of the variance due to nonbiological factors, primarily the quality of sample preparation or instrument stability. However, technical replication does not improve the sensitivity of statistical inference, and thus it is sometimes suggested that it can be omitted [19]. However, in proteomics technical variance is often a major noise factor. Therefore, if the cost of technical replication is low, one should consider it [8, 9, 19]. With respect to the required number of technical replicates, once should prefer running a small subset of samples in triplicates, over running each sample in duplicate, while approximately maintaining constant cost. Such replication scheme allows to estimate the variance which in turn is required by many data analysis approaches such as Bayesian network inference or logical modelling. Keeping the total number of samples low also helps to reduce the chance of bias due to stochastic events such as room temperature shift. Finally, for large-scale datasets, where instrument drift is expected, it is useful to measure a representative sample every 7–15 runs. Such a sample should be sufficiently similar to the rest of the samples so as not to affect peptide identification and run alignment. Hence, a good option is a pooled mix of several samples. In some cases, more than one such sample might be required to account for biologically relevant sample heterogeneity.

Replication is a major factor, determining the cost of the study. Biological replicates can increase costs dramatically, especially when large patient cohorts need to be profiled. To resolve the trade-off between the cost of the study and signal detection sensitivity, *statistical power estimates* are required. The power estimate models evaluate the number of samples required to achieve the desired sensitivity, based on biological and technical variation. The variation can be assessed in pilot studies or estimated from previous studies with a similar experimental setup. Power estimates are of critical importance for "omic" studies, where hundreds of genes and proteins are measured and thus chances for false positives are higher [20]. Multiple statistical models and computational pipelines have been proposed to facilitate the power estimates [8, 21, 22]. These models are especially sophisticated for complex samples, such as blood plasma [21]. Visually assisted approaches for sample size estimation have recently been developed for RNA-seq studies [23], and need to be adapted for proteomic data. Application of power estimate models allows for optimization of experimental costs and is essential for the reproducibility of the study.

Generally speaking, complex experiments, that is, with high sample heterogeneity, low signal-to-noise ratio, and multiple biological and technical covariates, increase requirements for the total number of samples [9, 20, 21]. Appropriate choice of number of technical and biological replicates, especially when supported by power estimates, one of the biggest challenges at experimental design stage.

2.4 Sample Annotation and Batch Effects

Having settled on an experimental design that takes into account batch effects, data acquisition can begin. During this phase of a project, it is of key importance that all technical covariates, such as reagents and personnel changes or machine interruptions are tracked and recorded in the sample annotation. This ensures that in the next step the effect of each factor can be assessed and adjusted for.

2.5 Experimental Design: Summary

In conclusion, experimental design is essentially the first step in data analysis, as it determines which approaches can be used for signal extraction from a given dataset. Special care should be taken for large-scale projects, in which multiple hospitals or research groups are involved, as well as inter-species studies, as they typically involve different research teams that use different sample preparation protocols. In these cases, it is a good idea to acquire an independent set of samples profiling the same tissue, disease or animal species from a different laboratory. These can then either be processed together with the majority of the data or used for independent validation of the analysis pipeline.

Experimental design should be carefully thought through since sample confounding or unbalanced design can undermine the power of predictive models [24], whereas careful sample randomization, blocking and tracking of all technical factors can ensure that the unavoidable biases can be corrected and the biological information preserved.

3 Correction of Batch Effects

3.1 Batch Effect Diagnostics

Batch effects are typically diagnosed visually. Popular batch effects *diagnostic methods* include boxplots, hierarchical clustering dendrograms and principal components biplots [12, 25] (*see* Fig. 3a). Note that these representations only provide a somewhat subjective inspection of the severity of batch effects. More precise quantification of batch effects can be achieved by approaches such as Principal Variance Component Analysis [26]. To facilitate the diagnostics of batch effects, several software tools have been proposed [27–29]. If present in the data, batch effects need to be corrected.

3.2 Normalization

Normalization is the first step of data correction. Normalization is required to make the samples more comparable by aligning their mean/median, variance or the whole distribution [12, 30]. Multiple normalization approaches have been proposed through the years, primarily for microarray studies [31–33], however, the same methods are widely applied in proteomics as well. Early microarray studies corrected the bias in the data by pairwise LOESS curve fitting [34], which is still widely used as a reference for method comparison (cited 6735 times in Web of Science as of October 2018). With growing sample numbers, these approaches have become less practical as they are computationally intensive and prone to overfitting [35]. Hence, approaches that align the sample distributions globally, for example, quantile normalization [33], have gained wide acceptance (5086 citations as of October 2018) and are among the commonly recommended methods also in proteomics [36]. In reality, new methods for "omic" data

A   B

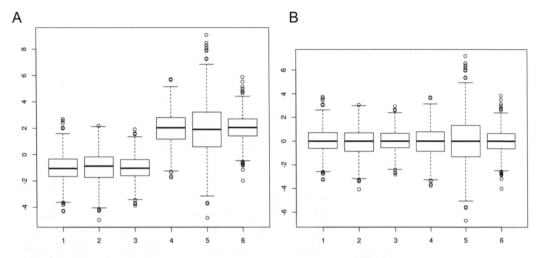

Fig. 3 Visualizing batch effects. Boxplots are commonly used to visually inspect data for the presence of batch effects. (**a**) In these six simulated datasets the presence of a batch effect is hinted at by the difference in the medians of the first and last three samples. (**b**) After median normalizing the samples from (**a**) no further apparent batch effect can be seen in the boxplot visualization

normalization are constantly proposed, despite two decades of research on this topic [37–41].

Note that global normalization methods, such as quantile normalization, rely on the assumption that the majority of measured entities (genes, proteins or peptides) don't change. This assumption does not hold when comparing samples with dramatic differences (e.g., certain tumor samples); thus, other methods have to be applied [42]. This is potentially a more acute problem in proteomics, as current methods allow to measure only a fraction of the proteome, thus proteins that could contribute to background distribution estimates are not measured. To facilitate the choice of the normalization procedure, statistical tests such as "quantro" have been developed [43].

3.3 Batch Effects Correction

Batch effects correction is the second step in data correction. This is required since normalization, which aligns major distribution patterns [12, 30], cannot account for biases differentially affecting distinct groups of genes or peptides [12, 44]. The presence of such effect is usually identified by the diagnostic methods described above, as a result samples tend to cluster by batch factor or the technical factors dominate the variance, as indicated by PVCA. Note that most plots, such as PCA or hierarchical clustering will look different before and after normalization and thus need to be plotted again before the batch effects correction. Batch effects correction aims to eliminate protein/peptide-level biases and create a "batch-free" data matrix.

Batch effects correction approaches can be subdivided by underlying mathematical apparatus into location-scale adjustment and matrix factorization based methods [25]. Location-scale adjustment methods rely on the observation that protein/peptide means (location) and variances (scale) are batch-dependent and thus their alignment will make the samples more comparable. Included in this methods group are batch mean centering [45], distance-weighted discrimination [46] and "ComBat" [47]. Alternatively, matrix factorization methods aim to identify leading components of variation and to associate them with known or unknown sources of technical variance [48, 49]. These methods are typically used to obtain a "batch free" data matrix.

However, the practice of "batch effects correction" has been criticized, and it has been suggested that methods incorporating batches in the predictive model should instead be preferred [13]. Such methods are typically implemented as variations of linear models and can be found in the R packages "limma" [50] and "MSstats" [51]. Unfortunately, these models cannot always be used in combination with other analysis methods, such as many machine learning or network inference. Hence, batch correction methods that produce a "batch-free matrix" and thus can be combined with multiple downstream analysis approaches are widely popular. Among these, "ComBat" [47] consistently performs well in comparative studies [30, 52], which, in combination with clear underlying assumptions, explains its popularity as a batch effect correction method (1726 citation in Web of Science, October 2018).

Note that both terms—"normalization" and "batch effects correction"—are used interchangeably to describe a specific step in data correction and for the whole process of adjusting for technical bias.

3.4 Assessment of Batch Effects Correction

It is often argued, whether batch effect correction wipes out the signal, including the biological variance in the data [42, 53]. It is known that sometimes normalization and batch effect correction introduce new, artifactual effects [53–56], that can affect downstream analysis [57]. Thus, evaluating the efficacy of a batch effect correction method is required. This evaluation is perhaps more challenging than the correction itself [25]. It is, however, an integral step in the batch effect correction workflow and should not be skipped under the assumption that once a batch correction method has been applied the data are "clean." The efficacy of normalization or batch effects correction is commonly achieved by visual comparison (*see* Fig. 3) of diagnostic plots before and after correction. It should be noted that most of these methods judge batch effect correction only by removal of the confounding. However, the real improvement in data quality is much harder to assess, and approaches incorporating positive controls need to be developed for batch effect analysis.

4   Conclusion

In general, publications addressing the batch effect problem in mass spectrometry research are not yet very common. Nevertheless, with the recent increase in sample throughput in MS-based proteomics, we expect the topic to gain more attention also in this field. In the meantime, refer to [58–60] for publications addressing batch effects in metabolomics, and to [61, 62] for publications demonstrating the benefit of batch correction in proteomics or comparing different normalization methods for proteomic data [63].

Fortunately, many aspects of batch effect correction in proteomics are similar to those from the genomic field. Nevertheless, some MS specific effects exist (e.g., MS signal drift) and a comprehensive comparative evaluation of the application of genomics batch effect correction methods in proteomics is still missing.

As discussed, batch effect correction is a multistep workflow that starts at the experimental design phase and continues through the data acquisition to batch effect diagnostics, correction and quality control of the correction, before the data can be used in the downstream analysis (*see* Fig. 4). At the correction step, one should note that normalization is a step preceding batch correction for most methods [47], so assessment of both components—normalization and batch adjustment procedure is required to choose the optimal method for the data. Furthermore, evaluation of batch

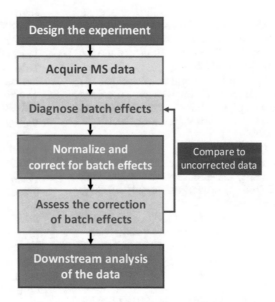

Fig. 4 Graphical overview of batch effects in large-scale experiment workflows. Various stages (highlighted in blue) of proteomic workflows require statistical expertise to prevent, diagnose, and correct batch effects

effects in nonnormalized data by sample clustering and principal components should be repeated after normalization as the clustering of the samples is likely to change dramatically. After the data are corrected, one needs to control for the quality of correction, however, only a few authors describe quality control metrics [25]. Even if mentioned [28] quality control is not highlighted as an essential, distinct step of the workflow, leading to evaluation of methods by "improved clustering of conditions" [28] that are not objective and thus cannot be recommended for broader applications.

In conclusion, whereas it is impossible to perform large-scale proteomics experiments that are batch effect free, with careful planning and processing it is possible to perform proteomics experiments in which batch effects are controlled. This is, in turn, a critical component in maximizing the biological signal one can extract from the experiment.

Acknowledgments

J.Č. was supported by funding from the European Union Horizon 2020 research and innovation program under grant agreement No 668858 and the Swiss State Secretariat for Education, Research and Innovation (SERI) under contract number 15.0324-2. P.P. was supported by SNF grant no. SNF IZLRZ3_163911.

References

1. Schubert OT, Röst HL, Collins BC et al (2017) Quantitative proteomics: challenges and opportunities in basic and applied research. Nat Protoc 12:1289–1294. https://doi.org/10.1038/nprot.2017.040

2. Mertins P, Mani DR, Ruggles KV et al (2016) Proteogenomics connects somatic mutations to signalling in breast cancer. Nature 534:55–62. https://doi.org/10.1038/nature18003

3. Zhang H, Liu T, Zhang Z et al (2016) Integrated proteogenomic characterization of human high-grade serous ovarian cancer. Cell 166:755–765. https://doi.org/10.1016/j.cell.2016.05.069

4. Collins BC, Hunter CL, Liu Y et al (2017) Multi-laboratory assessment of reproducibility, qualitative and quantitative performance of SWATH-mass spectrometry. Nat Commun 8:291. https://doi.org/10.1038/s41467-017-00249-5

5. Sajic T, Liu Y, Arvaniti E et al (2018) Similarities and differences of blood N-glycoproteins in five solid carcinomas at localized clinical stage analyzed by SWATH-MS. Cell Rep 23:2819–2831.e5. https://doi.org/10.1016/j.celrep.2018.04.114

6. Liu Y, Buil A, Collins BC et al (2015) Quantitative variability of 342 plasma proteins in a human twin population. Mol Syst Biol 11:786–786. https://doi.org/10.15252/msb.20145728

7. Ransohoff DF (2005) Bias as a threat to the validity of cancer molecular-marker research. Nat Rev Cancer 5:142–149. https://doi.org/10.1038/nrc1550

8. Oberg AL, Vitek O (2009) Statistical design of quantitative mass spectrometry-based proteomic experiments. J Proteome Res 8:2144–2156

9. Krzywinski M, Altman N (2014) Points of significance: analysis of variance and blocking. Nat Methods 11:699–700

10. Hu J, Coombes KR, Morris JS, Baggerly KA (2005) The importance of experimental design in proteomic mass spectrometry experiments: some cautionary tales. Brief Funct Genomic Proteomic 3:322–331

11. Gilad Y, Mizrahi-Man O (2015) A reanalysis of mouse ENCODE comparative gene expression data. F1000Res 4:121. https://doi.org/10.12688/f1000research.6536.1

12. Leek JT, Scharpf RB, Bravo HCHC et al (2010) Tackling the widespread and critical impact of batch effects in high-throughput data. Nat Rev Genet 11:733–739. https://doi.org/10.1038/nrg2825

13. Nygaard V, Rødland EA, Hovig E (2016) Methods that remove batch effects while retaining group differences may lead to exaggerated confidence in downstream analyses. Biostatistics 17:29–39. https://doi.org/10.1093/biostatistics/kxv027

14. Krzywinski M, Altman N (2013) Significance, P values and t-tests. Nat Methods 10:1041–1042. https://doi.org/10.1038/nmeth.2698

15. Ravikumar P, Wainwright MJ, Lafferty JD (2010) High-dimensional Ising model selection using ℓ1-regularized logistic regression. Ann Stat 38:1287–1319. https://doi.org/10.1214/09-AOS691

16. Martignetti L, Calzone L, Bonnet E et al (2016) ROMA: representation and quantification of module activity from target expression data. Front Genet 7:18. https://doi.org/10.3389/fgene.2016.00018

17. Kairov U, Cantini L, Greco A et al (2017) Determining the optimal number of independent components for reproducible transcriptomic data analysis. BMC Genomics 18:712. https://doi.org/10.1186/s12864-017-4112-9

18. Altman N, Krzywinski M (2015) Sources of variation. Nat Methods 12:5–6. https://doi.org/10.1038/nmeth.3224

19. Blainey P, Krzywinski M, Altman N (2014) Replication. Nat Methods 11:879–880. https://doi.org/10.1038/nmeth.3091

20. Krzywinski M, Altman N (2013) Points of significance: power and sample size. Nat Methods 10:1139–1140. https://doi.org/10.1038/nmeth.2738

21. Skates SJ, Gillette MA, LaBaer J et al (2013) Statistical design for biospecimen cohort size in proteomics-based biomarker discovery and verification studies. J Proteome Res 12:5383–5394. https://doi.org/10.1021/pr400132j

22. Cohen Freue GV, Meredith A, Smith D et al (2013) Computational biomarker pipeline from discovery to clinical implementation: plasma proteomic biomarkers for cardiac transplantation. PLoS Comput Biol 9:e1002963.

https://doi.org/10.1371/journal.pcbi.1002963

23. Shao Z (2018) ERSSA: empirical RNA-seq sample size analysis. R package version 1.0.0. https://github.com/zshao1/ERSSA

24. Parker HS, Leek JT (2012) The practical effect of batch on genomic prediction. Stat Appl Genet Mol Biol 11:Article 10

25. Lazar C, Meganck S, Taminau J et al (2013) Batch effect removal methods for microarray gene expression data integration: a survey. Brief Bioinform 14:469–490. https://doi.org/10.1093/bib/bbs037

26. Li J, Bushel PR, Chu T-M, Wolfinger RD (2012) Principal variance components analysis: estimating batch effects in microarray gene expression data. In: Batch effects and noise in microarray experiments. Wiley, Chichester, UK, pp 141–154

27. Manimaran S, Selby HM, Okrah K et al (2016) BatchQC: interactive software for evaluating sample and batch effects in genomic data. Bioinformatics 32:3836–3838. https://doi.org/10.1093/bioinformatics/btw538

28. Chawade A, Alexandersson E, Levander F (2014) Normalyzer: a tool for rapid evaluation of normalization methods for omics data sets. J Proteome Res 13:3114–3120. https://doi.org/10.1021/pr401264n

29. Chang C, Xu K, Guo C et al (2018) PANDA-view: an easy-to-use tool for statistical analysis and visualization of quantitative proteomics data. Bioinformatics. https://doi.org/10.1093/bioinformatics/bty408

30. Luo J, Schumacher M, Scherer A et al (2010) A comparison of batch effect removal methods for enhancement of prediction performance using MAQC-II microarray gene expression data. Pharmacogenomics J 10:278–291. https://doi.org/10.1038/tpj.2010.57

31. Quackenbush J (2002) Microarray data normalization and transformation. Nat Genet 32(Suppl):496–501. https://doi.org/10.1038/ng1032

32. Bilban M, Buehler LK, Head S et al (2002) Normalizing DNA microarray data. Curr Issues Mol Biol 4:57–64

33. Bolstad BM, Irizarry RA, Astrand M, Speed TP (2003) A comparison of normalization methods for high density oligonucleotide array data based on variance and bias. Bioinformatics 19:185–193

34. Irizarry RA, Hobbs B, Collin F et al (2003) Exploration, normalization, and summaries of high density oligonucleotide array probe level data. Biostatistics 4:249–264. https://doi.org/10.1093/biostatistics/4.2.249

35. Kreil DP, Russell RR (2005) There is no silver bullet--a guide to low-level data transforms and normalisation methods for microarray data. Brief Bioinform 6:86–97

36. Callister SJ, Barry RC, Adkins JN et al (2006) Normalization approaches for removing systematic biases associated with mass spectrometry and label-free proteomics. J Proteome Res 5:277–286. https://doi.org/10.1021/pr0503001

37. Ni TT, Lemon WJ, Shyr Y, Zhong TP (2008) Use of normalization methods for analysis of microarrays containing a high degree of gene effects. BMC Bioinformatics 9:505. https://doi.org/10.1186/1471-2105-9-505

38. Calza S, Valentini D, Pawitan Y (2008) Normalization of oligonucleotide arrays based on the least-variant set of genes. BMC Bioinformatics 9:140. https://doi.org/10.1186/1471-2105-9-140

39. Landfors M, Philip P, Rydén P, Stenberg P (2011) Normalization of high dimensional genomics data where the distribution of the altered variables is skewed. PLoS One 6:e27942. https://doi.org/10.1371/journal.pone.0027942

40. Ghandi M, Beer MA (2012) Group normalization for genomic data. PLoS One 7:e38695. https://doi.org/10.1371/journal.pone.0038695

41. Cheng L, Lo LY, Tang NLS et al (2016) Cross-Norm: a novel normalization strategy for microarray data in cancers. Sci Rep 6:18898. https://doi.org/10.1038/srep18898

42. Lovén J, Orlando DA, Sigova AA et al (2012) Revisiting global gene expression analysis. Cell 151:476–482. https://doi.org/10.1016/j.cell.2012.10.012

43. Hicks SC, Irizarry RA (2015) quantro: a data-driven approach to guide the choice of an appropriate normalization method. Genome Biol 16:117. https://doi.org/10.1186/s13059-015-0679-0

44. Phan JH, Quo CF, Cheng C, Wang MD (2012) Multiscale integration of -omic, imaging, and clinical data in biomedical informatics. IEEE Rev Biomed Eng 5:74–87. https://doi.org/10.1109/RBME.2012.2212427

45. Sims AH, Smethurst GJ, Hey Y et al (2008) The removal of multiplicative, systematic bias allows integration of breast cancer gene expression datasets – improving meta-analysis and prediction of prognosis. BMC Med Genet 1:42. https://doi.org/10.1186/1755-8794-1-42

46. Benito M, Parker J, Du Q et al (2004) Adjustment of systematic microarray data biases. Bioinformatics 20:105–114

47. Johnson WE, Li C, Rabinovic A (2007) Adjusting batch effects in microarray expression data using empirical Bayes methods. Biostatistics 8:118–127. https://doi.org/10.1093/biostatistics/kxj037

48. Alter O, Brown PO, Botstein D (2000) Singular value decomposition for genome-wide expression data processing and modeling. Proc Natl Acad Sci U S A 97:10101–10106. https://doi.org/10.1073/PNAS.97.18.10101

49. Leek JT, Storey JD (2007) Capturing heterogeneity in gene expression studies by surrogate variable analysis. PLoS Genet 3:1724–1735. https://doi.org/10.1371/journal.pgen.0030161

50. Smyth GK (2004) Linear models and empirical Bayes methods for assessing differential expression in microarray experiments. Stat Appl Genet Mol Biol 3:1–25. https://doi.org/10.2202/1544-6115.1027

51. Clough T, Thaminy S, Ragg S et al (2012) Statistical protein quantification and significance analysis in label-free LC-MS experiments with complex designs. BMC Bioinformatics 13 (Suppl 1):S6. https://doi.org/10.1186/1471-2105-13-S16-S6

52. Chen C, Grennan K, Badner J et al (2011) Removing batch effects in analysis of expression microarray data: an evaluation of six batch adjustment methods. PLoS One 6:e17238. https://doi.org/10.1371/journal.pone.0017238

53. Goh WWB, Wong L (2017) Advanced bioinformatics methods for practical applications in proteomics. Brief Bioinform. https://doi.org/10.1093/bib/bbx128

54. Hornung R, Boulesteix A-L, Causeur D (2016) Combining location-and-scale batch effect adjustment with data cleaning by latent factor adjustment. BMC Bioinformatics 17:27. https://doi.org/10.1186/s12859-015-0870-z

55. Welsh EA, Eschrich SA, Berglund AE, Fenstermacher DA (2013) Iterative rank-order normalization of gene expression microarray data. BMC Bioinformatics 14:153. https://doi.org/10.1186/1471-2105-14-153

56. Giorgi FM, Bolger AM, Lohse M, Usadel B (2010) Algorithm-driven artifacts in median polish summarization of microarray data. BMC Bioinformatics 11:553. https://doi.org/10.1186/1471-2105-11-553

57. Lim WK, Wang K, Lefebvre C, Califano A (2007) Comparative analysis of microarray normalization procedures: effects on reverse engineering gene networks. Bioinformatics 23:i282–i288. https://doi.org/10.1093/bioinformatics/btm201

58. Wang SY, Kuo CH, Tseng YJ (2013) Batch normalizer: a fast total abundance regression calibration method to simultaneously adjust batch and injection order effects in liquid chromatography/time-of-flight mass spectrometry-based metabolomics data and comparison with current calibration met. Anal Chem 85:1037–1046. https://doi.org/10.1021/ac302877x

59. Kuligowski J, Sánchez-Illana Á, Sanjuán-Herráez D et al (2015) Intra-batch effect correction in liquid chromatography-mass spectrometry using quality control samples and support vector regression (QC-SVRC). Analyst 140:7810–7817. https://doi.org/10.1039/C5AN01638J

60. Kuligowski J, Pérez-Guaita D, Lliso I et al (2014) Detection of batch effects in liquid chromatography-mass spectrometry metabolomic data using guided principal component analysis. Talanta 130:442–448. https://doi.org/10.1016/j.talanta.2014.07.031

61. Tracy MB, Cooke WE, Gatlin CL et al (2011) Improved signal processing and normalization for biomarker protein detection in broad-mass-range TOF mass spectra from clinical samples. Proteomics Clin Appl 5:440–447. https://doi.org/10.1002/prca.201000095

62. Gregori J, Villarreal L, Méndez O et al (2012) Batch effects correction improves the sensitivity of significance tests in spectral counting-based comparative discovery proteomics. J Proteome 75:3938–3951. https://doi.org/10.1016/j.jprot.2012.05.005

63. Välikangas T, Suomi T, Elo LL (2018) A systematic evaluation of normalization methods in quantitative label-free proteomics. Brief Bioinform 19:1–11. https://doi.org/10.1093/bib/bbw095

Chapter 17

Using the Object-Oriented PowerShell for Simple Proteomics Data Analysis

Yassene Mohammed and Magnus Palmblad

Abstract

Scripting languages such as Python and Bash are appreciated for solving simple, everyday tasks in bioinformatics. A more recent, object-oriented command shell and scripting language, PowerShell, has many attractive features: an object-oriented interactive command line, fluent navigation and manipulation of XML files, ability to explore and consume Web services from the command line, consistent syntax and grammar, rich regular expressions, and advanced output formatting. The key difference between classical command shells and scripting languages, such as bash, and object-oriented ones, such as PowerShell, is that in the latter the result of a command is a structured object with inherited properties and methods rather than a simple stream of characters. Conveniently, PowerShell is included in all new releases of Microsoft Windows and is available for Linux and macOS, making any data processing script portable. In this chapter we demonstrate how PowerShell in particular allows easy interaction with mass spectrometry data in XML formats, connection to Web services for tools such as BLAST, and presentation of results as formatted text or graphics. These features make PowerShell much more than "yet another scripting language."

Key words PowerShell, XML parsing, Web services, Object-oriented scripting

1 Introduction

In bioinformatics we are often faced with the task of efficiently managing and processing many and large datasets using existing software. This can be achieved through scripting, using common shells like bash in Linux/macOS, or more sophisticated languages such as Perl or Python, with their respective extensions BioPerl [1] and BioPython [2]. Specialized programming environments like MATLAB or R (with BioConductor [3]) are also frequently used in bioinformatics. The data may be from sequencing projects, microarray or RNA-Seq gene expression measurements in transcriptomics, mass spectrometry in proteomics, or NMR in metabolomics. Structured data formats based on XML or JSON are becoming increasingly common in all of these fields as well as in the life sciences in general. Examples include SeqXML for sequence

Rune Matthiesen (ed.), *Mass Spectrometry Data Analysis in Proteomics*, Methods in Molecular Biology, vol. 2051,
https://doi.org/10.1007/978-1-4939-9744-2_17, © Springer Science+Business Media, LLC, part of Springer Nature 2020

and orthology data [4], NeXML [5] and PhyloXML [6] for phylogenetic relationships; MAGE-ML [7] for microarray data; SRA XML [8, 9] for next-generation sequencing data; and mzXML [10], mzML [11, 12], pepXML [13], or mzIdentML [14] in proteomics. These are all data formats designed for saving, depositing, and sharing data between researchers, and would often be the input data in a bioinformatics analysis.

Windows PowerShell is a command line shell and scripting language built using Microsoft's .NET Framework and shipped originally with all Microsoft Windows operating systems since 2007 [15]. Currently it is available freely under MIT license for all major operating systems including Windows, Linux, and macOS [16]. While originally developed to provide IT professionals with a powerful tool for system administration and designed without any consideration of uses in fields such as bioinformatics, PowerShell, now in its 6.1 release, exhibits a number of features that deserve attention and that can be employed directly in advanced bioinformatics. Practically convenient, PowerShell is already installed on all Windows laptops and desktop computers and can be installed on Linux or macOS machines. This means members of a research organization and classroom computers are likely to be able to directly run, share and improve PowerShell scripts. PowerShell has a rich list of commands of verb-noun style cmdlets or "command-lets" that are often self-explanatory. It has inherited some of the capabilities of the .NET Framework, including object orientation and access to large repository of libraries. The output of a cmdlet has *properties*, *methods* and *events*, which in turn can be accessed from the command line or in a script. For example, a string variable in PowerShell would have a property `Length` that is the length of the string and a method `Contains` `()` that checks whether a string `$x` contains a specified substring `$y` with `$x.Contains($y)`. The properties and methods can be called and navigated in PowerShell from within the command line with the auto completion functionality using the tab button, which means they do not need to be memorized. In addition to common variable types such as `byte`, `char`, `string`, `int` and `long`, `single` and `double` precision floating point numbers and arrays of all of these. Structured data like XML and JSON are also a built-in data types in PowerShell. This makes it easy—even for the novice user—to navigate, display, and manipulate any dataset in XML or JSON. Complex self-defined types and structures are also possible, as well as system autogenerated types. This is very useful for accessing and interacting with Web services as well as in more advanced data processing. Web service proxies can be initiated and interacted with from the command line without the need to write or use a dedicated Web service client. Like most UNIX shells, PowerShell can combine several cmdlets in pipelines using the pipe operator | (vertical bar). However, a crucial difference compared to UNIX pipelines is that data is transferred between cmdlets as objects or an array of objects rather than as a stream of characters. This gives the commands in the pipeline access to the various objects specific properties and operations. In summary, the object orientation and feature richness make PowerShell a versatile tool for many data analysis and automation tasks, especially for XML and JSON data and when using Web services.

Power and flexibility of a scripting or programming language often come at the expense of simplicity. In our experience, teaching PowerShell alongside bash and common UNIX tools to students with little or no prior experience with UNIX, scripting, or programming, the majority of students find PowerShell easier to learn than the bash shell and other UNIX tools. Previous comparisons and listings of bioinformatics tools have not included PowerShell [17, 18]. Here we illustrate the use of PowerShell for teaching and simple research purposes. We demonstrate how PowerShell makes it easy to work with structured data, interact with Web services and produce elegantly formatted output directly on the command-line. Although focused on these features distinguishing PowerShell from other shells, it should be stressed that PowerShell is a general shell and scripting language that can automate any data processing or system management tasks.

2 Simple Uses of PowerShell

PowerShell is probably best suited for interactive and simple automated tasks, such as fetching and managing data, file manipulation, and other day-to-day practices. Although one can access and use . NET libraries easily from within PowerShell, it is not a general tool for advanced mathematical or statistical analysis, or a language for developing complex programs. The PowerShell Integrated Scripting Environment—ISE—is a lightweight integrated development tool that supports writing and debugging PowerShell scripts. Table 1 is a subjective summary of the strengths and weaknesses of using PowerShell in bioinformatics. We will show a selection of simple uses of PowerShell from our own teaching and practical work. These examples can provide building blocks for or interfaces to more advanced programs. All scripts are available online at https://cpm.lumc.nl/yassene/powershell/.

2.1 Object Properties and Methods

PowerShell cmdlets result in objects that can be displayed in the output or manipulated further. For instance, to query how many characters and lines there are in a FASTA file, the contents of the file can be piped to the `Measure-Object` cmdlet: `get-Content file.fasta | Measure-Object -line -char`. To search for lines that contain a specific substring, one can use `Where-Object` cmdlet: `get-Content file.fasta | Where-Object {$_.Contains("agtcgt")}`, where $_ refers to the current object in the

Table 1
Advantages and disadvantages of Windows PowerShell in bioinformatics

Pros	Cons
• Preinstalled on Windows and available for Linux and macOS • Open source • Object-oriented, not text-centered • Fluent handling and manipulation of XML and JSON files • Consistent grammar and style • Rich regular expressions • Advanced output formatting (tables, colors) • PowerShell Integrated Scripting Environment (ISE) • Fast and easy learning procedure for beginners • Aliases for common UNIX shell commands	• Still new to the bioinformatics community • Significantly different from most UNIX shells • Designed for system administration rather than bioinformatics • Limited userbase—small knowledgebase, nevertheless it is increasing steadily

pipeline, that is, a string object containing the current line from the FASTA file. Although grammatically different due to the object-oriented nature of PowerShell, the end result is equivalent to using the familiar `grep` command in UNIX shells. We can store the contents in a variable by: `$x=Get-Content file.fasta`, and then extract the information of interest. For instance, the file length in lines is a property of `$x`: `$x.Length` (`$x` in this case is an array of strings). We can check whether the fifth line of a FASTA file contains a specific substring: `$x[5].contains("agtcgt")` or extract all strings in the FASTA file containing that substring by using the `Select-String` cmdlet: `$x | Select-String "agtcgt"`. Regular expressions can be used to search for nucleotide or amino acid sequence motifs. For example, to find all N-linked glycosylation motifs—a tripeptide consisting of asparagine followed by any amino acid except proline, followed by serine or threonine—we can pipe the content of an amino acid FASTA file to `Select-String` with the motif in PowerShell's regular expression syntax: `$x | Select-String -pattern N[^P][ST]`. Similarly, most SUMO-modified proteins [19] contain the tetrapeptide consensus motif Ψ-K-x-E, where Ψ is an aliphatic residue [20], K the lysine conjugated to SUMO, and x any amino acid. As the aliphatic amino acids are isoleucine (I), leucine (L), methionine (M), and valine (V), the regular expression corresponding to the common SUMOylation motif is `[ILMV]K.E` and we can display all lines containing the motif in a protein sequence by selecting the corresponding strings: `Get-Content file.fasta | Select-String -all -Pattern [ILMV]K.E`. As regular expression matching operates on strings and each line in the FASTA file is its

Fig. 1 Output of PowerShell script counting and highlighting sequence motifs. In this example, the PowerShell Script 1 was used to find and count SUMOylation motifs in a FASTA sequence using regular expressions, here for Ikaros DNA-binding protein [Swiss-Prot:Q13422]. Using the output formatting options in the Write-Host cmdlet, the motifs can be highlighted directly in the PowerShell window

own string, special care must be taken to match motifs spanning line breaks. This is easiest done by merging all lines for one protein sequence into a single, long string and then searching this string for the motif. When combined with the `Write-Host` cmdlet, the matching motifs can be highlighted in a sequence (Fig. 1). Script 1 contains the corresponding PowerShell script.

Each PowerShell object has different associated methods, just like in other object-oriented programming languages. Useful and self-explanatory methods for string handling include `CompareTo ()`, `Contains()`, `IndexOf()`, `IndexOfAny()`, `Split()`, `Substring()`, `ToUpper()`, and `ToLower()`. A list of all properties and methods of an object can be retrieved by `Get-Member`: `$x | Get-Member`. One can use common object-oriented practice when using these methods and properties. For instance, if we want to find all lines which contain two patterns `$pat1` and `$pat2` we can use the and logical `--and` operator: `get-content file.fasta | Where-Object {$_.Contains($pat1) -and $_.Contains ($pat2)}`. It is also possible to define corresponding objects to any .NET class, enabling a new way of interactive data processing on the command line. In the following we will show how to make use of these features in some chosen example.

2.2 Importing and Navigating XML Files

While XML parsers such as DOM and SAX are available as libraries or plug-ins for most common programming languages, PowerShell already comes with XML support built-in. This allows us to navigate an XML object's structure interactively or through a script, extracting, manipulating, and exporting data. Although convenient and powerful, a possible limitation is that the entire XML file will be loaded into memory by default. Nevertheless, individual XML files

Script 1. PowerShell script for counting and highlighting SUMOylation motifs in an input FASTA file

```
#### Counting and highlighting motifs in a FASTA file ####

#1: Define the regular expression of the motif we are searching for
$pattern=[regex] "[ILMV]K.E" # SUMOylation regex

#2: Get content of the input file
$fasta=Get-Content $args[0]

#3: Set the output to be 60 characters wide to match FASTA file format
$ui = (Get-Host).UI.RawUI;
$ws = $ui.WindowSize; $oldWidth=$ws.Width; $ws.Width = 60; $ui.WindowSize = $ws
$bs = $ui.BufferSize; $oldBSWidth=$bs.Width ; $bs.Width = 60; $ui.BufferSize =
$bs

#4: Define function to output patterns highlighted as red text on green
background
function WriteOutput {
    param($patternMatches)
    if([int]$patternMatches.Matches.Count -gt 0) {
        $index=0
        foreach($pm in $patternMatches.Matches) {
            Write-Host    $patternMatches.line.SubString($index,    $pm.Index    -
$index) -nonewline
            Write-Host    $pm.Value    -nonewline    -ForegroundColor    "Red"    -
BackgroundColor "Green"
            $index = $pm.Index + $pm.Length
        }
        Write-Host                        $patternMatches.line.SubString($index,
$patternMatches.line.length-$index)
    }
}

#5: Count the motifs and highlight them in the output using WriteOutput
$proteinName=$fasta[0].split("|")[1]
for ($i = 1; $i -le $fasta.Length-1; $i++)
{
    if ( !$fasta[$i].contains(">") )
    {        $sequence = $sequence+$fasta[$i]       }
    else
    {
        $patternMatches = $sequence| Select-String -all -Pattern $pattern
        ""; $proteinName + " contains " + `
            [int]$patternMatches.matches.count + " SUMOylation motif(s):"

        $proteinName = $fasta[$i].split("|")[1]
        if([int]$patternMatches.Matches.Count -gt 0)
        { WriteOutput($patternMatches)    }
        clear-variable sequence
    }
}
$patternMatches = $sequence| Select-String -all -Pattern $pattern
""; $proteinName + " contains " + [int]$patternMatches.matches.count + "
SUMOylation motif(s):"
WriteOutput($patternMatches)

#6: Restore window and buffer sizes
$bs.Width = $oldBSWidth; $ui.BufferSize = $bs
$ws.Width = $oldWidth; $ui.WindowSize = $ws
```

can be easily handled nowadays on a regular desktop computer. The XML browsing in PowerShell can also be helpful in prototyping XML processing logic on smaller files for later implementation in a specialized software. In the following, we show how to use Power-Shell to work with mass spectrometry data from proteomics experiments in the open and common mzXML format [10].

PowerShell provides an XML type and we can import an mzXML file directly into an XML object by `$xml=[xml](Get--Content mzXMLfile.mzXML)`, where the square brackets perform a typecasting. We can now navigate to any point in the mzXML file, for instance to display the software used in creating the file: `$xml.mzXML.msRun.dataProcessing.software`. The children of a particular node in the XML tree can be retrieved by listing the members of type "property" of this node: `Get-Member -Member-Type "property"`. A common task when dealing with XML files is extracting a defined part of the data for further processing. For liquid chromatography-mass spectrometry data in mzXML (or mzML), we may be interested in quickly generating a base peak chromatogram to visually inspect the quality of the data and the chromatographic separation. To do this, we `Select-Object basePeakIntensity, retentionTime` of the corresponding scan nodes and then export the results to a target file (*see* Script 2). To visualize the base peak chromatogram directly from PowerShell we can use the Microsoft Chart Control for .NET Framework, MSChart, freely available from Microsoft (Fig. 2a). Such basic visualization is achieved by extracting the data we want to plot from the input file, loading the .NET library for plotting, initializing a Chart object and a ChartArea object, and placing everything into a Form object. Each of these objects can be modified to control the appearance of the chart. A detailed tutorial on MSChart can be found online [21]. Other mass spectrometry XML formats are handled analogously. For instance, the 98th tandem mass spectrum from a Bruker Daltonics DataAnalysis Compound XML file can be retrieved by `$xml=[xml](Get-Content compounds.xml)` and `$spectrum = $xml.root.analysis.DataAnalysis.peptide_database_querydata.compounds.cmpd[97].ms_spectrum[1].ms_peaks.pk`. This spectrum can be plotted similarly to the base peak chromatogram (Fig. 2b) or identified using a search engine such as SpectraST [22].

2.3 Accessing SOAP Web Services from PowerShell

Organizations such as The National Center for Biotechnology Information and the European Bioinformatics Institute provide access to diverse data resources and analysis tools as remote Web services [23]. Most of these services are provided with software clients written in programming or scripting languages like Perl, Java, Python, or C/C#/C++. Some of the Web services are also provided with (example) workflows for Taverna [24]. The user can download a software client for a specific service and use it from the

Script 2. PowerShell script for extracting and plotting a base peak chromatogram from an mzXML file using the MSChart library

```
#### Chart base peak intensity vs retention time  ####
#Microsoft Chart Control for Microsoft .NET Framework 3.5 is needed
#to run this example. To download this go to:
#www.microsoft.com/download/en/details.aspx?displaylang=en&id=11001

#1: Load mzXML file and extract data
$mzXML =[xml](Get-Content $args[0])
$peaks   =   ($mzXML.mzXML.msRun.scan   |   Select-Object   basePeakIntensity,
retentionTime)
$x = @(foreach($peak in $peaks){$peak.retentionTime})
$x = [int[]]($x| %{$_.Substring(2,$_.Length-3)})
$y = @(foreach($peak in $peaks){$peak.basePeakIntensity})

#2: Load the required .NET library
[void][Reflection.Assembly]::LoadWithPartialName("System.Windows.Forms")
[void][Reflection.Assembly]::LoadWithPartialName("System.Windows.Forms.DataVisua
lization")

#3: Define and initialize a Chart object
$chart = New-object System.Windows.Forms.DataVisualization.Charting.Chart
$Chart.Width = 600
$Chart.Height = 340
$Chart.Left = 10
$Chart.Top = 10

#4: Define and initialize a ChartArea object
$ChartArea                              =                              New-Object
System.Windows.Forms.DataVisualization.Charting.ChartArea
$Chart.ChartAreas.Add($ChartArea)
[void]$Chart.Series.Add("Data")
$Chart.Series["Data"].Points.DataBindXY($x,$y)
$Chart.Anchor = [System.Windows.Forms.AnchorStyles]::Bottom -bor
                [System.Windows.Forms.AnchorStyles]::Right -bor
                [System.Windows.Forms.AnchorStyles]::Top -bor
                [System.Windows.Forms.AnchorStyles]::Left

[void]$Chart.Titles.Add("Base Peak Chromatogram of " + $args[0])
$ChartArea.AxisX.Title = "retention time (s)"
$ChartArea.AxisY.Title = "base peak intensity"
$ChartArea.AxisY.LabelStyle.Enabled = 0
$ChartArea.AxisX.Minimum = 0
$ChartArea.AxisX.Interval = 600
$chartArea.AxisX.MajorGrid.LineDashStyle = "NotSet"
$chartArea.AxisY.MajorGrid.LineDashStyle = "NotSet"
$chartArea.AxisY.MajorTickMark.Enabled = $false

#5: Define and initialize a Form object
$Form = New-Object Windows.Forms.Form
$Form.Text = "Base Peak Chromatogram"
$Form.Width = 640
$Form.Height = 400
$Form.controls.add($Chart)
$Form.Add_Shown({$Form.Activate()})
$Form.ShowDialog()
```

command line or download the source code and use it to implement own software and scientific logic. To use the software clients, the user has to learn the basic syntax of the Web service and how to call it from the software client, how to submit data, and how to retrieve the results. We have used the familiar BLAST [25]

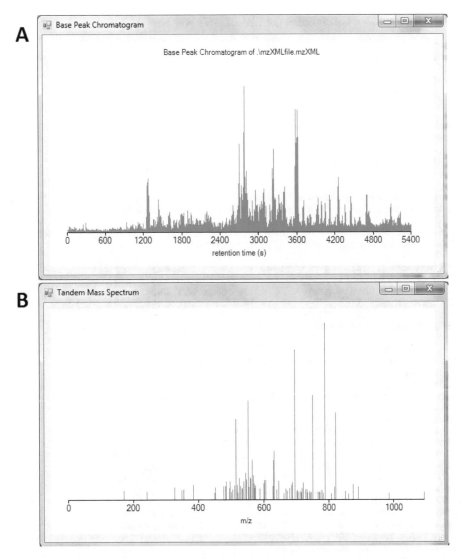

Fig. 2 (A) Script 2 extracts the base peak chromatogram, the intensity of the most intense peak in the mass spectrum as function of retention time, from the mzXML file and use MSChart from the .NET Framework to generate the graph. **(B)** Compound tandem mass spectrum from an E. coli whole cell lysate LC-MS/MS dataset, or an average of two tandem mass spectra acquired in a short time window, created by DataAnalysis 4.0 (Bruker Daltonics) and exported as a Compound XML file. The precursor and tandem mass spectrum match the triply protonated peptide VAALEGDVLGSYQHGAR from the elongation factor Ts in the NIST E. coli spectral library using SpectraST with a dot product of 0.741.

alignment tool to illustrate the use of Web services in PowerShell scripts—for instance the EBI Web service for BLAST [26]. This is a less trivial exercise than the previous examples, so we will describe it in more detail. It should be emphasized that Web services are directly accessed through PowerShell, without any additional Web service client software. Script 3 contains the complete PowerShell script for accessing and using the BLAST Web service.

Script 3. PowerShell script to consume the EBI BLAST Web service

```
#### Using EBI BLAST Web service ####
#This script is meant to be run stepwise on the PowerShell command line

#1: Create the Web service proxy
$url = "http://www.ebi.ac.uk/Tools/services/soap/ncbiblast?wsdl"
$blastWS = New-WebServiceProxy -uri $url

#2: List all methods and properties
$blastWS | Get-Member

#3: Display how to run BLAST
$blastWS.run

#4a: Create the input parameters object
#the auto generated line is different for each web service object proxy.
$param                    =                    new-object
Microsoft.PowerShell.Commands.NewWebserviceProxy.AutogeneratedTypes.WebServicePr
oxy1s_services_soap_ncbiblast_wsdl.InputParameters

#4b: Convert run() call to a string, split at spaces,
#and take the fifth entry, which is the automatically generated
#type in 4a. This should also work with other EBI Web services
$param1= New-Object (([string]($blastWS.run.OverloadDefinitions)).split())[5]

#5: List the input parameters
$param

#6: Populate the input parameters with the required data
$param.program = "blastp"
$param.stype = "protein"
$param.sequence = ">sp|P0A6Y8|DNAK_ECOLI Chaperone protein DnaK OS=Escherichia
coli (strain K12) GN=dnaK PE=1 SV=2
MGKIIGIDLGTTNSCVAIMDGTTPRVLENAEGDRTTPSIIAYTQDGETLVGQPAKRQAVT
NPQNTLFAIKRLIGRRFQDEEVQRDVSIMPFKIIAADNGDAWVEVKGQKMAPPQISAEVL
KKMKKTAEDYLGEPVTEAVITVPAYFNDAQRQATKDAGRIAGLEVKRIINEPTAAALAYG
LDKGTGNRTIAVYDLGGGTFDISIIEIDEVDGEKTFEVLATNGDTHLGGEDFDSRLINYL
VEEFKKDQGIDLRNDPLAMQRLKEAAEKAKIELSSAQQTDVNLPYITADATGPKHMNIKV
TRAKLESLVEDLVNRSIEPLKVALQDAGLSVSDIDDVILVGGQTRMPMVQKKVAEFFGKE
PRKDVNPDEAVAIGAAVQGGVLTGDVKDVLLLDVTPLSLGIETMGGVMTTLIAKNTTIPT
KHSQVFSTAEDNQSAVTIHVLQGERKRAADNKSLGQFNLDGINPAPRGMPQIEVTFDIDA
DGILHVSAKDKNSGKEQKITIKASSGLNEDEIQKMVRDAEANAEADRKFEELVQTRNQGD
HLLHSTRKQVEEAGDKLPADDKTAIESALTALETALKGEDKAAIEAKMQELAQVSQKLME
IAQQQHAQQQTAGADASANNAKDDDVVDAEFEEVKDKK"
$param.database = "uniprotkb_human"

#7: Run BLAST
$blastJobId = $blastWS.run("y.mohammed@lumc.nl","DNAK_ECOLI", $param)

#8: Check job status
$blastWS.getStatus($blastJobId)

#9: Display how to retrieve the results
$blastWS.getResult

#10: Display the different possible results types
$blastWS.getResultTypes($blastJobId)

#11: Retrieve the command line output. $x is only for the command syntax
$output=$blastWS.getResult($blastJobId,"out",$x)

#12: Change the retrieved output to readable text
$results = New-Object String ( , $output ); $results > resultsDNAK.txt
```

PowerShell provides a dedicated cmdlet to create Web service proxy objects that allow using and managing Web services: `New-WebServiceProxy`. Once connected to the BLAST Web service we can find out exactly what methods are available by listing the members of the Web service proxy with `Get-Member`. We will use the two basic methods here: `run()` and `getResults()`. These two methods are specific for the EBI Web service. For the NCBI Web service, `SubmitSearch()` and `GetSearchResults()` should be used. Invoking a method without brackets displays information on how to use it, for example if our BLAST proxy object is named `$blastWS`, `$blastWS.run` will show information on how to use the `run()` method. This method needs three inputs: two strings, `email` and `title`, and `parameters` of a type defined by the Web service proxy [26]. In order to create and populate an instance of this type, we use the `New-Object` cmdlet: `$param = new-object Microsoft.PowerShell.Commands.NewWeb-serviceProxy.AutogeneratedTypes.WebServiceProx-y1s_services_soap_ncbiblast_wsdl.InputParameters`. This step is normally a main obstacle in designing clients to consume Web services. The structure and the current value of any variable can be viewed by calling it from the PowerShell command line. After assigning values to the variables in `$param` (*see* Script 3), we can `run()` BLAST by calling the Web service object we created with the three needed inputs, for example `$blastWS.run ("user@lumc.nl", "DNAK", $param)`. This will return a string containing the BLAST job identifier. We can check the status of our submitted job by passing the job identifier to the `getStatus()` method and when the job is finished, the results can be retrieved by `getResults()`. The BLAST Web service offers 16 different results to choose from. All result fields can be listed by `getRe-sultTypes($blastJobId)`. The command line output of BLAST can be retrieved by `getResults()` with the appropriate job identifier and result type: `$blastWS.getResult($blas-tJobId, "out", $x)`. The results of the BLAST Web service are base64 encoded, which are converted to a byte array (`byte[]`) in PowerShell. In order to convert the results to a human-readable version, we can cast those to character array (`char[]`) and then to a `String`.

This demonstrates direct and interactive access to a BLAST Web service via the PowerShell command line, for use as is or as part of a larger script or bioinformatics workflow. Other Web services can be integrated in a similar way. One can also consider connecting two or more Web services using PowerShell by passing the output of one Web service to the input of another, or pipelining Web services, local scripts, and locally installed third-party software.

2.4   Querying Data Repositories Using PowerShell

Beside the possibilities of navigating and consuming SOAP Web services that were demonstrated in Subheading 2.3, PowerShell allows easy interaction and data retrieval using REST Web services. REST services are used by many bioinformatics services and data repositories, for example those from the European Institute for Bioinformatics—EBI. Usually EBI REST services return results to queries in JSON format, which PowerShell supports out of the box like XML. Here we use the interface of the EBI PRoteomics IDEntifications archive—PRIDE to obtain information on various projects of interest. Script 4 contains an example code for accessing and using the PRIDE REST interface.

In this example we are interested in finding which projects submitted to PRIDE have identifications of a specific protein of interest. In the example we consider cellular tumor antigen p53 with the UniProtKB accession number P04637 and we are interested in cancer. The interaction with PRIDE is performed using the REST interface which is documented on the EBI website (www.ebi.ac.uk/pride/help/archive/access/webservice). In the first step we retrieve the number of all PRIDE submissions using `Invoke-RestMethod` with the project count function `www.ebi.ac.uk:443/pride/ws/archive/project/count`. It is worth mentioning that all communications are performed on a secure channel (i.e., https) and PowerShell takes care of this in background without additional steps from the user to establish the channel (as is needed in various programming languages like Java). In the second step we use another PRIDE REST function to obtain additional information on all projects that are related to cancer in human. The function returns various metadata, including the title of the project, a short description, the publication date, the species, the tissues used, the instrument used, and the unique PRIDE project accession number. All these information entries are reported as properties for each project and can easily be filtered using the cmdlet `Select-Object --Property`. In order to retrieve information on the protein identifications in a specific project, PRIDE interface requires the unique project accession number, which we can obtain by filtering the projects metadata for the property "`accession`". At the time of writing there were 523 projects in PRIDE that were annotated with cancer and human.

Using the accession number of our protein of interest (i.e., P04637) and the list of project accession numbers, in step 3 of Script 4 we retrieve the number of TP53 identifications in each of the 523 projects, from which 14 were reported to have one or more identification. Using the indices of these projects we can display all the corresponding metadata on each project by simple call to the matrix object of all projects `$results2.list[$index3]`. At this stage, one can be interested in obtaining a list of all reported protein identifications in the projects where one or more TP53 identification was listed, or even in downloading all raw and submission data regarding for further analysis locally on own

Script 4. PowerShell script to interact with the PRIDE repository REST Web services and download mass spectrometry data to local computer

```
#### Using PRIDE REST Web service ####
#This script is meant to be run stepwise on the PowerShell command line (this
script requires powershell 3 or above)

cd c:\tmp
#1: Obtain the total number of projects in PRIDE using the REST interface
$url = "https://www.ebi.ac.uk:443/pride/ws/archive/project/count"
$nrOfProjects=Invoke-RestMethod -Uri ($url); $nrOfProjects
#2: Obtain information on all projects that are related to cancer in human.
$organism=9606 ; $disease="cancer"; $nrOfResults=$nrOfProjects
$url2 = "https://www.ebi.ac.uk:443/pride/ws/archive/project/list?"
$param = "show="+$nrOfResults+"&speciesFilter="+$organism+ "&diseaseFilter="+
$disease
$results2=Invoke-RestMethod -Uri ($url2+$param)
$project_accessions=$results2.list | Select-Object -Property accession
$project_accessions.Count

#3: Ask PRIDE which project has identifications of the protein of interest.
$prot_accession="P04637" #tp53
$results3 = New-Object System.Collections.Generic.List[System.Object]
foreach ($proj_acc in  $project_accessions){
    Write-Output($proj_acc)
    $counts=Invoke-RestMethod -Uri
("https://www.ebi.ac.uk:443/pride/ws/archive/protein/count/project/" +
$proj_acc.accession +"/protein/" +$prot_accession)
    $results3.Add($counts)
    Start-Sleep -S 0.55 # a delay because PRIDE has a limit of queries per
minute
}
$index3=(0..($results3.Count-1)) | where {$results3[$_] -gt 0}
$index3.Count; $results2.list[$index3]

#4: Obtain all protein identification in the projects of interest
$proj_accessions_with_prot_of_interest=$results2.list[$index3] | Select-Object -
Property accession
$url4 = "https://www.ebi.ac.uk:443/pride/ws/archive/protein/list/project/"
$nrOfResults4=1000 # PRIDE has a limit of 10000, we here retrieve the first 1000
for demonstration
$results4 = New-Object System.Collections.Generic.List[System.Object]
foreach ($proj_acc in  $proj_accessions_with_prot_of_interest){
    $prot_id=Invoke-RestMethod -Uri
($url4+$proj_acc.accession+"?show="+$nrOfResults4)
    $prot_id_accnr=$prot_id.list | Select-Object -Property accession
    $results4.Add($prot_id_accnr)
}
$results4[0] #example of results4

#5: Download raw data (here the first submission in our list, i.e. PXD009738)
$url5 = "https://www.ebi.ac.uk:443/pride/ws/archive/file/list/project/"
$results5=Invoke-RestMethod -Uri
($url5+$proj_accessions_with_prot_of_interest[0].accession)
$download_links=$results5.list| Select-Object -Property downloadLink
$download_links_mztab=$download_links -match "mztab"
$mztab_filenames=New-Object System.Collections.Generic.List[System.Object]
foreach ($download_link in  $download_links_mztab)
{
    $filename=$download_link.downloadLink.Split("/")[-1]
    Invoke-WebRequest -Uri $download_link.downloadLink -OutFile $filename
    $mztab_filenames.add($filename)
}

#6: Processing downloaded data to obtain information on PSMs
Function ungzip{ #this function is a modified version from an online example.
    Param($input_file,$output_file)
```

```
    $in_filestream = New-Object System.IO.FileStream $input_file,
([IO.FileMode]::Open), ([IO.FileAccess]::Read), ([IO.FileShare]::Read)
    $out_filestream = New-Object System.IO.FileStream $output_file,
([IO.FileMode]::Create), ([IO.FileAccess]::Write), ([IO.FileShare]::None)
    $gzipStream = New-Object System.IO.Compression.GzipStream $in_filestream ,
([IO.Compression.CompressionMode]::Decompress)
    $buf = New-Object byte[](1024)

    while($true){
        $read = $gzipStream.Read($buf, 0, 1024)
        if ($read -le 0){break}
        $out_filestream.Write($buf, 0, $read)
        }
    $gzipStream.Close();$out_filestream.Close();$in_filestream.Close()
}

$gz_in_file=(ls -path $mztab_filenames[0]).FullName
$gz_out_file=$gz_in_file.Replace(".gz","")
ungzip $gz_in_file $gz_out_file
$mztab = Import-Csv -Delimiter "`t" -Path $gz_out_file
$PSMs=$mztab | Select-String -Pattern "PSM"
```

computer. In steps 4 and 5 in Script 4 both these scenarios are performed, and we used again the cmdlet: `Invoke-RestMethod` with the corresponding PRIDE REST functions and parameters, as well as simple manipulation of the JSON response to obtain the final results we want.

In step 6, we attempt to show how advanced data processing and pipeline building in PowerShell is possible. Here we use .NET functions to unzip the PRIDE mztab.gz files, importing the content into PowerShell as tab-delimited-file, and then select and display the rows containing the Peptide-Spectrum-Matches—PSM.

2.5 Writing Mass Spectrometry mzIdentML Files

In the second example we demonstrated how PowerShell supports advanced handling of XML data, allowing interactive navigation of the XML structure as well as extracting, manipulating, and exporting data using few commands. In addition to these reading capability, and through its native support to access .NET libraries, PowerShell offers easy but advance XML writing possibilities. In this example we show how complex XML documents, like mzIdentML, can be written from scratch using peptide identification list in a spread sheet. As the mzIdentML format is fairly rich to allow comprehensive documentation of peptide and protein identifications from mass spectrometry data, we have limited our example here to the basics of writing the essential sections of an mzIdentML document. This should demonstrate the essential PowerShell methods for this task, while keeping the possibly to extend the code to a full parser open. In Script 5 we have included an example code for writing an mzIdentML file of a list of peptide identifications. The peptide identification table in this example is a CSV file containing 13 columns: the peptide sequence, the location of modification (if present), the monoisotopic mass delta of the modification (if present), protein accession, protein description, spectrum id, rank of identification, charge state, experimental *m/z*, pass threshold (a binary entry for if the identification passed a specific threshold), the database use in the search, the database location, the database name, the spectra data location, and the spectra data file name. An example of the CSV file as well as the code for this example can be downloaded from this chapter web page.

This example code starts by defining few functions that are helpful in writing various, repetitive sections of an mzIdentML file. These include writing entries in the controlled vocabulary list—**cvList** section, controlled vocabulary parameter—cvParam parts of the various entries, as well as dbsequence, peptide, and peptideEvidence entries in the **SequenceCollection** section, and finally SpectrumIdentificationResult and its nested Spectrum-IdentificationItem entries in the **AnalysisData** section. Each of these function and the code in this example depend largely on four PowerShell (general .NET) methods, namely XmlWriter.WriteStartElement, XmlWriter.WriteAttributeString, XmlWriter.WriteValue, and XmlWriter.WriteEndElement, which have representative names of their functions.

Writing XML file starts by choosing the appropriate settings, especially the XML encoding and assign these to a System.Xml.XmlWriterSettings object, followed by initializing the XmlWriter by creating a System.XML.XmlWriter object. Generally, all various XML elements can be added using the rich built-in list of methods of this object, like WriteStartElement and WriteValue.

After initializing the XML document, we write first the **header** of the mzIdentML, followed by the **cvList** section. These are essential, but rather standard sections in the structure of most of mass spectrometry XML formats. The following **SequenceCollection** section contains information on (1) the list of identified proteins in the DBSequence entries, (2) the list of identifies peptides in the Peptide entries, and (3) the relation between the protein and peptide identification in the PeptideEvidence entries. In our example we consider that each row in the (CSV) identification table represents one protein, and we use three For-loops to iterate over all rows to add the DBSequence, Peptide, and PeptideEvidence entries accordingly.

In the follow-up step, we write the two sections on the **AnalysisCollection** and **AnalysisProtocolCollection**. The first contains information on the analyses performed to obtain the results and can contain SpectrumIdentification (spectra to peptides), which is used our case, or ProteinDetection (peptides to proteins), which is not included in our example but can be added using similar code section. The **AnalysisProtocolCollection** lists the parameters and settings of the software tools used to obtain current results. In these two sections, and for the purpose of the simple demonstration we intended, we used place holders and kept entries to their minimum.

The final essential part of an mzIdentML file is the **DataCollection** section, with its `Inputs` and `AnalysisData` parts. The `Inputs` in our case contains references to the `SpectraData`, which is a requirement in the mzIdentML specifications and refers to the mass spectrometry data. It also references the `SearchDatabase`, which is not a requirement, but important to have, and refers to the peptide database used in the search.

The `AnalysisData` lists the assignments generated by the analyses and contain the set of all search results represented in `SpectrumIdentificationList`, which is a requirement and part of our document, and `ProteinDetectionList` which is not a requirement and left out in our example. The code writes all the `SpectrumIdentificationResult`(s) in the `SpectrumIdentificationList` part by looping over the rows of our table and using the function of the same names introduced above.

Using these steps, a minimal mzIdentML document can be built with the entries required by the format specifications. The final step is to `WriteEndDocument` corresponding to the `mzIdentML` start element, followed by closing the document which will write it to the location specified in the beginning of the example code.

3   Concluding Remarks

Through few examples we have illustrated how PowerShell can accomplish tasks of low to medium complexity in a logical and consistent style and with little effort. The availability as open-source implementation, uniform cmdlet syntax, object orientation, and built-in support of XML, JSON, and Web service make PowerShell an accessible and powerful tool for many basic data manipulation tasks in bioinformatics. Moreover, due to the possibility to use all . NET functions and libraries simply from the commend line, PowerShell allows many opportunities of writing simple yet very useful program to read, reformat, write complex XML mass spectrometry data like mzIdentML. Due to its compatibility with virtually all common platforms including Windows, Linux, and macOS, it is a practical shell with easy scripting language for teaching and instruction. We do not expect that PowerShell will replace any other scripting language, but it can fill a gap and be used as an interactive tool. PowerShell allows new ways of processing XML and JSON data and interacting with external Web services without advanced experience on parsing semistructured data or prior knowledge of the exact implementation of the Web services. PowerShell is a convenient tool for prototyping and interactively developing bioinformatics analysis pipelines, or fetching data from bioinformatics repositories through common Web services APIs. All examples mentioned in this chapter are available for download from https://cpm.lumc.nl/yassene/powershell/.

References

1. Stajich JE (2007) An introduction to BioPerl. Methods Mol Biol 406:535–548

2. Cock PJ et al (2009) Biopython: freely available Python tools for computational molecular biology and bioinformatics. Bioinformatics 25 (11):1422–1423

3. Reimers M, Carey VJ (2006) Bioconductor: an open source framework for bioinformatics and computational biology. Methods Enzymol 411:119–134

4. Schmitt T et al (2011) Letter to the editor: SeqXML and OrthoXML: standards for sequence and orthology information. Brief Bioinform 12(5):485–488

5. NeXML - phylogenetic data as xml (2012). http://www.nexml.org

6. Han MV, Zmasek CM (2009) phyloXML: XML for evolutionary biology and comparative genomics. BMC Bioinformatics 10:356

7. Spellman PT et al (2002) Design and implementation of microarray gene expression markup language (MAGE-ML). Genome Biol 3(9):RESEARCH0046

8. Picardi E et al (2011) ExpEdit: a webserver to explore human RNA editing in RNA-Seq experiments. Bioinformatics 27(9):1311–1312

9. SRA Format (2012). http://www.ebi.ac.uk/ena/about/sra_format

10. Pedrioli PG et al (2004) A common open representation of mass spectrometry data and its application to proteomics research. Nat Biotechnol 22(11):1459–1466

11. Deutsch E (2008) mzML: a single, unifying data format for mass spectrometer output. Proteomics 8(14):2776–2777

12. Martens L et al (2011) mzML—a community standard for mass spectrometry data. Mol Cell Proteomics 10(1):R110.000133

13. Keller A et al (2005) A uniform proteomics MS/MS analysis platform utilizing open XML file formats. Mol Syst Biol 1:2005.0017

14. Eisenacher M (2011) mzIdentML: an open community-built standard format for the results of proteomics spectrum identification algorithms. Methods Mol Biol 696:161–177

15. Microsoft Powershell Documentation (2018). https://docs.microsoft.com/en-us/powershell/

16. GitHub Powershell Project (2018). https://github.com/PowerShell/PowerShell

17. Fourment M, Gillings MR (2008) A comparison of common programming languages used in bioinformatics. BMC Bioinformatics 9:82

18. Dudley JT, Butte AJ (2009) A quick guide for developing effective bioinformatics programming skills. PLoS Comput Biol 5(12): e1000589

19. Hay RT (2005) SUMO: a history of modification. Mol Cell 18(1):1–12

20. Aasland R et al (2002) Normalization of nomenclature for peptide motifs as ligands of modular protein domains. FEBS Lett 513 (1):141–144

21. MacDonald R (2011) Charting with PowerShell. http://blogs.technet.com/b/richard_macdonald/archive/2009/04/28/3231887.aspx

22. Lam H et al (2007) Development and validation of a spectral library searching method for peptide identification from MS/MS. Proteomics 7(5):655–667

23. Camacho C, Madden T (2011) SOAP-based BLAST Web Service. http://www.ncbi.nlm.nih.gov/books/NBK55699/

24. Oinn T et al (2004) Taverna: a tool for the composition and enactment of bioinformatics workflows. Bioinformatics 20(17):3045–3054

25. Altschul SF et al (1990) Basic local alignment search tool. J Mol Biol 215(3):403–410

26. NCBI BLAST (SOAP) (2011). http://www.ebi.ac.uk/Tools/webservices/services/sss/ncbi_blast_soap

Chapter 18

Considerations in the Analysis of Hydrogen Exchange Mass Spectrometry Data

Michael J. Eggertson, Keith Fadgen, John R. Engen, and Thomas E. Wales

Abstract

A major component of a hydrogen exchange mass spectrometry experiment is the analysis of protein and peptide mass spectra to yield information about deuterium incorporation. The processing of data that are produced includes the identification of each peptic peptide to create a master table/array of peptide identity that typically includes sequence, retention time and retention time range, mass range, and undeuterated mass. The amount of deuterium incorporated into each of the peptides in this array must then be determined. Various software platforms have been developed in order to perform this specific type of data analysis. We describe the fundamental parameters to be considered at each step along the way and how data processing, either by an individual or by software, must approach the analysis.

Key words Deuterium, Software, Algorithm, Protein dynamics, Isotope

1 Introduction

Hydrogen exchange mass spectrometry (HDX-MS) is a biophysical technique with roots in analytical chemistry [1, 2]. HDX-MS utilizes the naturally occurring exchange of the backbone amide hydrogens for hydrogens or deuterium in the solvent water [3–8]. In an HDX-MS experiment, a protein is exposed to "heavy water" (D_2O) for seconds to hours or days and within that time the protein incorporates deuterium as a function of the protein's physicochemical properties [9, 10]. HDX-MS is a medium resolution structural technique, although the exchangeable hydrogen positions that are measured (the backbone amide hydrogens) are found at each amino acid throughout the primary structure, except for proline which has no backbone amide hydrogen [5, 11]. Some of the most valuable pieces of information extracted from an HDX-MS experiment can come from observing a protein in two or more different conformational states [12, 13]. For example, if HDX of a wild-type and mutant form of a protein are compared and a difference in deuterium uptake behavior is observed, one can

Rune Matthiesen (ed.), *Mass Spectrometry Data Analysis in Proteomics*, Methods in Molecular Biology, vol. 2051,
https://doi.org/10.1007/978-1-4939-9744-2_18, © Springer Science+Business Media, LLC, part of Springer Nature 2020

conclude that there must be some change in structure or change in backbone dynamics as a result of the mutation.

With the rapid expansion and utilization of hydrogen exchange mass spectrometry as a tool to study protein conformation and dynamics, there is an increased demand for both robust data acquisition and data analysis tools. Unification of data analysis tools and processing methodology will increase the general reliability of HDX-MS overall and allow for more direct and valid comparisons of HDX-MS data across laboratories that are investigating the same or similar proteins or protein systems.

A comprehensive list of what was considered "state of the art" for HDX-MS data analysis tools, with brief descriptions, was published in 2012 [14]. In the 7 years since that publication, many of the software listed have received incremental updates to improve usability and features while others have become less visible in published literature. The present chapter does not serve as a manual for any of the listed software, nor does it favor one over the other; rather, the presentation here of the analysis of HDX-MS data stands as a foundation for understanding which parameters are considered during data processing and how most software approaches the analysis. All software algorithms need to be sensitive to restrictive parameters defined in a list of peptide identities, including: monoisotopic and average mass, m/z range, charge state(s) of the peptide ions of interest, retention time and retention time windows, peptide isotope distributions for peptide isotope intensity above the background noise (for both the undeuterated and deuterated peptides) and peptide ion isotope overlap.

The most complex software is designed to extract raw m/z data from a typical LC-MS run, analyze the HDX-MS data by determining deuterium incorporation information either through a center of mass approach or through theoretical modeling of predictive isotope patterns (via statistical analyses), and finally output both two- and three-dimensional results for visualization. There are a myriad of noncommercial software solutions as well as several commercial platforms for the analysis of HDX-MS data. The software include those with basic input—for example, HXExpress, which requires only raw m/z versus intensity data to be pasted into an excel macro in order to produce deuterium incorporation and peak width plots [15, 16]—to both web-based (HDX Finder [17]) and stand-alone tools (Hydra/MassSpecStudio [18], HeXicon [19, 20], and AUTOHD [21]) that search raw LC-MS data before completing the data analysis. These software tools can all be used in combination with other software (e.g., MSTools) [22], as required in order to complete data visualization in both two and three dimensions. Other freely available software—HDX finder [17], and HDX Analyzer [23]—start with the raw LC-MS hydrogen exchange data and finish with complete data visualization. There are now four commercial software solutions: HDExaminer [24], DynamX [25],

HDX workbench [26], and HDX Workflow in Byos™ (Protein Metrics). HDExaminer, HDX workbench, and HDX Workflow solutions work with multiple mass spectrometer output data formats while DynamX is a Waters-specific product [27].

Despite the growing number of software solutions that have become available over the past decade, there is no single fully automated solution for the analysis of HDX-MS data. Each involves some degree of user input for complete analysis and visualization. For example, all software packages require some means of supplying peptide identity (m/z, m/z range, z, retention time range, etc.) when performing pepsin-digestions HDX experiments. We believe that user interaction with the data, in some fashion, is desirable both from the standpoint of forcing the user to actually look at the MS data and from the perspective of quality control. Because a good understanding of the HDX-MS data processing workflow is required to generate experimental data that will yield robust processing results, we will next present the workflow for the manual analysis of continuous labeling HDX-MS data, with a focus on processing local/peptide level data.

2 Materials

The materials needed to process HDX-MS data include typical desktop/laptop PCs, data storage capabilities, and of course, the actual HDX-MS data. These data may often come from the most common exchange experiment known as continuous labeling (for a recent review please *see* ref. 8). Figure 1 illustrates an overview of the hydrogen exchange experiment workflow indicating various options at each step (isotopic labeling, LC/MS analysis, and data analysis).

It is important to understand the origin of HDX-MS data, so we have chosen to briefly describe a typical experiment starting from a protein at equilibrium and under its native buffer conditions. Acquisition of continuous labeling local level HDX-MS data is composed of three logical steps: isotopic labeling of exchangeable hydrogens and reaction quenching; enzymatic digestion of the quenched samples (for localizing deuterium incorporation); reversed-phase peptide separation and mass spectrometric measurement of the mass of each peptide ion.

Isotopic labeling of the exchangeable hydrogens is the first step in the experiment. Traditionally, labeling is initiated either via a dilution or fast buffer switching (minimizing dilution) into a buffer prepared with D_2O [28]. Dilution into a buffer containing D_2O is the most common method and dilutions range from 1:1 $H_2O{:}D_2O$ (designed to reduce spectral complexity) to as high as 1:25 or greater if possible. The larger excess of deuterium drives the forward labeling reaction. The reaction is allowed to proceed for

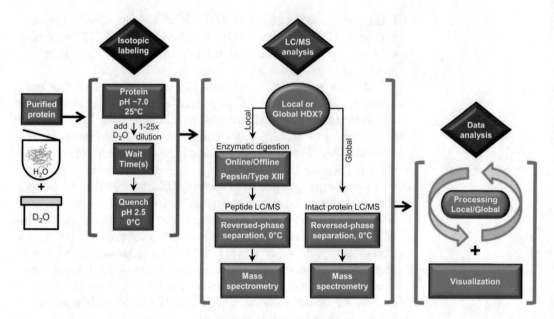

Fig. 1 Overview of typical hydrogen exchange experiment workflow indicating various options at each step: isotopic labeling, LC/MS analysis, and data analysis

specific amounts of time (from milliseconds to days), and is then quenched through a reduction of the solution pH to 2.5 and temperature of 0 °C [11]. These specific quench conditions are achieved either using a strong buffer system (10× the strength of the labeling and equilibration buffers) or a small volume of a concentrated acid. Quenching conditions can also include chaotropic agents (guanidinium chloride, urea) and/or reducing agents (TCEP, online electrochemical reduction, etc.) in an attempt to facilitate a more productive proteolytic digestion, see below [29–33].

The next step in the experiment is the enzymatic digestion of the protein in quench buffer. Remembering that specific conditions (low pH and temperature) need to be maintained in order to preserve the deuterium labeling, the digestion also needs to be performed under these quench conditions. Additionally, the digestion step needs to be performed as fast as possible in order to maintain as much of the deuterium label as possible. There are two widely used proteolytic enzymes that meet these criteria: pepsin [11], and aspergillopepsin (protease type XIII from *Aspergillus saitoi*) [34]. The digestion step can be performed in one of two ways: in solution using either free or immobilized pepsin or performed online using a digestion column that is packed with pepsin that has been immobilized onto a bead/particle [30, 35].

Reversed-phase chromatography of the peptides produced during digestion (known hereafter as "local analysis"), or if enzymatic digestion was not performed a simple LC desalting of the intact protein (hereafter referred to as "global analysis") is the next step in the experiment. Like digestion, the chromatography step also needs to be performed quickly under quench conditions in order to minimize the loss of deuterium label. The length of the LC separation can be extended if temperatures below 0 °C are employed: similar levels of deuterium retention can be achieved with a 10 min separation at 0 °C and a 60 min separation at -20 °C [36]. Peptide separation is either achieved using traditional RP-HPLC [37] or more recently RP-UPLC technologies [27, 38]. Separation of the peptides significantly reduces the complexity of the resulting mass spectra, especially for larger proteins and protein complexes.

Figure 2 illustrates elements of raw HDX-MS data that any software tool or human equivalent would encounter after the data are acquired and processing is about to begin. There are data like this for the undeuterated control and typically between 4 and 12 hydrogen–deuterium exchange samples (a greater number of samples are required when the desired processing outputs include exchange rates, *see* Subheading 3.5.2). There are total ion chromatograms (TIC), each with corresponding m/z spectra acquired with scan rates that are user defined and generally dictated by the speed of the chromatographic separation. As an example, consider a single protein, local HDX-MS experiment with a 10-min chromatography cycle time, a mass spectrometry scan rate of five scans per second, and a total of six LC-MS runs. In this example, there would be 18,000 mass spectra with m/z values ranging from 50 to 2000 m/z. If there are 130 unique peptide ions to follow in the data analysis, this number of samples and peptides would result in 2.34×10^6 total peptide ions that will have to be queried in the mass-to-charge range (50–2000 m/z). A human sitting at a computer and manually processing these data may encounter fewer spectra to investigate as they might sum scans together and immediately reject the analysis of peaks that are obvious artifacts, noise or adduct peaks. However, an algorithm will have to deal with all the data to sort out what is good and what is not. Criteria for rejecting spectra are a definitive aspect of an algorithm's behavior and can include evaluation for an unexpected isotope distribution, evidence of overlap with another isotope distribution, *see* **Note 1**, among other properties. Additionally, as the protein system that is being investigated becomes increasingly more complex (in terms of unique sequence) and perhaps includes multiple proteins binding on to one another, the number of peptide ions that have to be dealt with becomes more than a single human user can handle within a reasonable amount of time. The magnitude of the data that possibly could be collected has been prohibitive in recent years, limiting the size of a protein/protein system a person was comfortable processing in a reasonable amount of time. Streamlining the processing, increasing the efficiency of data interrogation, and automation of

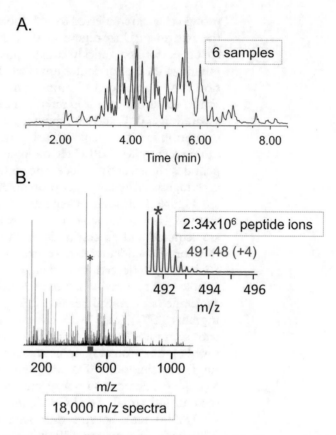

Fig. 2 Visual inspection of HDX LC-MS data. (**a**) Base peak intensity chromatogram (BPI) for the online digestion of a model 97 kDa protein separated using UPLC with a water–acetonitrile gradient in 10 min. (**b**) Summed data for the *m/z* range of 150–1050 *m/z* for the red highlighted portion of the BPI in panel **a**. The starred inset is taken from a zoom in on the 491–496 *m/z* range and displays a single natural isotopic distribution for a (4+) ion in this modeled undeuterated control sample. Dashed rectangles indicate the amount of data that are produced in a typical HDX-MS experiment

many of the steps has been key to the development of HDX-MS data analysis tools.

3 Methods

The scope of data analysis in HDX-MS experiments is illustrated in Fig. 3. This flow chart illustrates the basic workflow of data analysis for both global (or intact protein level, Fig. 3a) and local (or peptide level, Fig. 3b) analysis. Global analysis and local analysis are different in complexity yet maintain a similarity in the core processes of displaying data, deuterium pattern analysis, and the calculation of average deuterium incorporation (Fig. 3 dashed

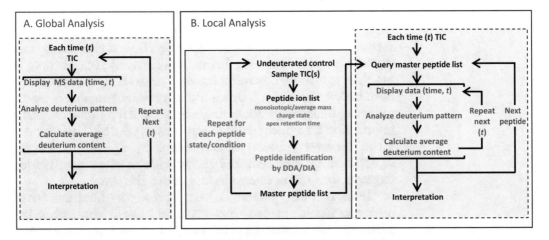

Fig. 3 Typical data analysis workflows for both (**a**) global or intact protein and (**b**) local or peptide level hydrogen–deuterium exchange data analysis. The repetitive nature of the analysis is emphasized at each step. Dashed lined boxes indicate the common steps between the two workflows

boxes). While a global HDX-MS experiment is easier to perform, the information that can be gleaned is limited. Global analysis measurements can report on the solution-phase dynamics of the protein backbone solely at the whole protein level. As the exact protein sequence is known, the theoretical amount of deuterium incorporation can be compared to the measured deuterium incorporation level to reveal overall characteristics of the protein [2]. Specifically, the number of residues in a protein which are part of stable structures can be inferred from the difference between the theoretical maximum deuterium incorporation for the protein and the maximum observed incorporation at the longest labeling time points (from hours to days). Similarly, a measurement of the residues which are likely in unstructured or very dynamic parts of the protein can be estimated by measuring the uptake at very short exposure times (10 s or less) [39, 40]. The shape of the uptake curve can also yield information on the timescales of various conformational dynamics in the protein. With the limited amount of biophysical information that can be obtained from global analysis, the frequency of global analysis experiments is low. For this reason, we will devote the remainder of the chapter to local analysis, and note that many of the principles presented for local analysis also apply to global analysis.

The significant preference for local analysis over global analysis is due to the amount of biophysical information that can be obtained. Local analysis can resolve deuterium incorporation information to specific locations within a protein [11, 41, 42]. Deuterium incorporation can be localized to peptides that range in size from 5 to 20 amino acids. Measurement of peptides with overlapping sequences can often be used to further resolve uptake to individual backbone amide hydrogens along the primary structure of the analyte protein. Some researchers are investigating the use of peptide fragments produced by the nonergodic fragmentation methods electron transfer dissociation (ETD) and electron capture dissociation (ECD) to even further localize uptake information. It has been shown that by using carefully tuned ETD conditions for both bottom-up and top-down MS, deuterium scrambling can be minimized, allowing at times single amino acid resolution of uptake information if fragment data are considered [43–48]. These experiments however are not yet commonplace, especially when LC separations are required, and the analysis of deuterium labeled fragment ions will not be addressed in this chapter.

In the following paragraphs we will present the fundamentals of how one would manually process local analysis data, what key parameters are considered at each step in the process, and describe what special considerations that need to be addressed by an algorithm trying to mimic in an automated fashion what a person would do when presented with data.

3.1 Displaying HDX-MS Data

The first step in the analysis is the visualization of the raw LC-MS data (total ion chromatograms and the corresponding/linked mass spectra). This inspection serves as the initial quality control to validate that it makes sense to proceed and to complete data analysis. For example, if the separation is of poor quality or the signal-to-noise ratio is poor, it may not make sense to move on to the subsequent steps. A typical progression of HDX-MS data inspection is illustrated in Fig. 2 where a model reversed-phase UPLC chromatogram at 0 °C (Fig. 2a) and the complicated m/z spectrum (Fig. 2b) that can be observed from a single peak of the chromatogram (shaded region of Fig. 2a) for an online pepsin digestion (in this case of a 97 kDa protein) are shown. Focusing on the shaded region of the chromatogram, it is obvious that there is a significant amount of peptide signal in the mass spectrometer from any one chromatographic peak. Each one of the ions corresponds to a peptide from the pepsin digestion. Despite the full m/z spectrum being complicated and crowded, zooming into smaller m/z ranges will often show well resolved peptides with no interfering ions nearby. For example, the inset in Fig. 2b shows a range of 491–496 m/z containing a single +4 peptide ion. The inspection of the data continues in this manner across the LC and m/z scales until it is clear that there are no issues with the peptide data such as low ion intensity, dynamic range, poor chromatographic resolution of ions (*see* **Note 1**), and lack of deuterium label (see below and Fig. 4), etc. The higher the fidelity of the reversed-phase separation of peptides, the easier it is to pinpoint and identify which deuterium labeled peptides goes with which unlabeled version of the peptide.

It is at this point in the inspection of the data that the evidence for the successful labeling of the protein backbone with deuterium is determined. This inspection of the deuterated protein samples is done to be certain deuterium was actually incorporated, the isotope

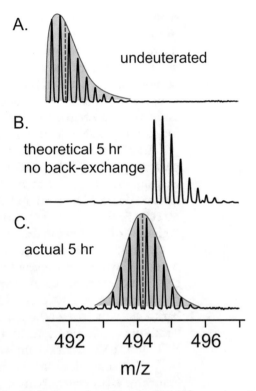

Fig. 4 Isotopic distributions of peptides in HDX-MS experiments. The natural abundance isotope distribution for peptides in the undeuterated control is shown in (**a**). The isotopic distributions of the same peptide after 5 h of exposure to deuterium are shown for the theoretical case of no back-exchange (**b**) and the actual distribution for the peptide after experiencing back-exchange (**c**)

distributions are as expected (see below) and that the signal-to-noise is acceptable for processing. Consider the example shown in Fig. 4 for a +4 ion (the same as shown in Figs. 2b and 6a) of a peptic peptide. The natural isotopic abundance, dictated by amino acid composition, for the undeuterated version of this peptide appears as shown in panel a of Fig. 4. Theoretically, after 5 h of hydrogen exchange in D_2O, with the assumption of uniform labeling and no back exchange of the deuterium label to hydrogen, the resultant isotope distribution would look as shown in Fig. 4b. As the incorporated deuterium (2.014101 amu) weighs more than hydrogen (1.007825 amu), for each deuterium atom incorporated into the peptide, the natural abundance isotope distribution will shift to a higher mass dictated by the mass difference between deuterium and hydrogen isotopes, equal to 1.006 Da/z. However, deuterated peptides undergo back exchange in the fully protonated chromatographic step (for a complete discussion of back-exchange please refer to [11, 49–51]), and this deuterium loss is random. As a result, the isotopic distribution of a deuterated peptide in HDX-MS is not the "normal" isotope distribution one expects, but rather it is a more symmetrical "Gaussian-like" distribution in shape (Fig. 4c). As a protein spends more time exposed to deuterium, this isotope distribution moves to higher m/z values but maintains this distribution through its increase in m/z. Movement will continue until the peptide reaches a maximum possible incorporation that is dictated both by experimental conditions (i.e., pH, temperature, labeling time, and percent excess of deuterium in the sample) and protein conformation. *See* **Note 2** for the calculation of maximum possible deuterium incorporation.

Another important consideration during data inspection is the peak shape. It will be readily apparent if there is evidence that exchange has occurred through either or both of the two kinetic limits of HDX (for a complete discussion of these two limits, please *see* refs. 5, 9, 10, 52, 53). Briefly, most proteins follow the EX2 regime for backbone hydrogen exchange. In EX2, proteins make many visits to the exchange competent state before deuteration occurs. The result is a near-binomial distribution in the mass spectrum that slowly increases in m/z, as described above. The much more rare type of kinetics is EX1. In EX1, a region(s) of the protein backbone visits the exchange competent state and exchanges all backbone amide positions in a coordinated fashion. Importantly, EX1 and EX2 kinetics are visually differently in mass spectra [54]. For simplification, we will only discuss EX2 data in the following sections (*see* **Note 3** for a discussion of obvious as well as less-obvious EX1 data as well as how to best plot these deuterium incorporation data; *see* **Note 4** for a discussion on false EX1 as a result of run-to-run peptide retention and how best to avoid erroneous EX1).

3.2 Peptide Identification

One main feature of an HDX-MS experiment on the local level is the creation of peptides. The identity of each peptide (i.e., the amino acids contained within) must be determined. The mass spectrometer, however, just measures mass. There must be a correlation step in which specific peptide ions—each with a measured monoisotopic mass, charge state, and chromatographic retention time—are attributed to amino acids of the protein. For example, an ion with a monoisotopic mass of 563.98, a charge state of 3+ and chromatographic retention time of 2.79 min may be eventually assigned to amino acids 11–23 of the protein being studied. The proper identification of peptides creates a master peptide list of peptides for which valid data interpretation can be obtained after the deuterium incorporation of each ion has been measured.

The correct identification of peptides is important if the final interpretation of the HDX-MS data is to be considered reliable. Peptides are identified using an undeuterated control sample that is prepared alongside the deuterated samples. Despite maintaining a preference for cleavages between hydrophobic residues, pepsin is a nonspecific acid protease and the cleavage pattern, although reproducible, is not predictable [55, 56]. The seemingly simplest, and considered by some wholly sufficient, way to identify a peptic peptide is use the exact mass of each peptide in a search of the protein sequence for all possible amino acid combinations that could produce the observed mass. For a single protein HDX-MS experiment, searching the known protein sequence may yield only one or two theoretical possibilities solely based on exact mass alone. From this, if you are lucky, one of the other hits for the exact mass will be beyond an error tolerance of the mass measurement. Figure 5a gives an example where a hypothetical search using the observed monoisotopic mass of 779.713 (+3) was performed against the entire amino acid sequence of a small isolated 8 kDa protein (Hck SH3) domain using an initial mass tolerance of 1.0 Da. The results of this search yield two possible peptide matches or hits for this ion. Both hits have very small deviations (−0.01 and 0.05 Da) from the measured monoisotopic mass. The two

A.

Monoisotopic mass: 779.713 (+3)
Mass tolerance: 1.00 Da
Number of Hits: 2

Hit #	Residues	Mass (Da)	Sequence	Deviation
1	37-55	2336.11	PWWRARDKNGQEGYIPSNY	-0.01
2	44-61	2336.17	YFILEESNLPWWRARDKN	0.05

B.

Monoisotopic mass: 779.713 (+3)
Mass tolerance: 1.00 Da
Number of Hits: 7

Hit #	Residues	Mass (Da)	Sequence	Deviation
1	29-46	2335.19	LFLLTVHKLSYYEYDFER	-0.93
2	28-45	2335.19	RLFLLTVHKLSYYEYDFE	-0.93
3	187-207	2335.27	PLPPTPEEDQILKKPLPPEPT	-0.85
4	128-146	2336.21	LKNVIRYNSDLVQKYHPCF	0.09
5	250-268	2336.11	PWWRARDKNGQEGYIPSNY	-0.01
6	241-258	2336.17	YFILEESNLPWWRARDKN	0.05
7	536-556	2336.12	KVSDFGLSRYVLDDEYTSSVG	0.00

Fig. 5 Identification of potential peptic peptides from exact mass matching. (**a**) A search of the amino acid sequence of an 8 kDa protein domain for a peptic peptide that matches a monoisotopic mass of 779.713 for the 3+ charge state with a 1.0 Da tolerance yields a total of two possible peptides. (**b**) A search of the same peptic peptide in the full length 77 kDa protein (including the 8 kDa protein domain) using the same parameters in (**a**) yields a total of seven possible peptides

possibilities are within the error of the mass measurement. In order to determine which of the two remaining possibilities goes with the mass 779.713, tandem MS (MS/MS) experiments are necessary. It must be noted that for a small protein the number of potential hits is small, but this becomes less and less likely as the amount of unique protein sequence increases when analyzing larger proteins or considering multiprotein systems. Take as an example Fig. 5b. In this case, the observed monoisotopic mass from above is searched against the full length 77 kDa protein (full-length Hck) in which the 8 kDa SH3 domain is contained, using identical parameters. Here the potential peptide matches jump to seven; four of the hits have very small deviations (including the original two from above) from the measured monoisotopic mass of the peptide and three have significant deviations (-0.93, -0.93, and -0.85 Da). These three potential hits with the highest deviation from the true mass are discarded as the accuracy of mass determination in a well calibrated modern mass spectrometer is much better than these errors. As a consequence, the task of peptide identification becomes more difficult and relies more heavily on MS/MS analysis. We advocate the identification of all peptic peptides using MS/MS, and the application of strict criteria before assignment is made. There are few things worse in HDX-MS than making an interpretation of deuterium incorporation data to later find there was a peptide that had been misassigned.

Identification of peptides by MS/MS can be performed using two different MS/MS methods, data dependent acquisition (DDA) or data independent acquisition (DIA or MS^E [57] and SWATH-MS [58] as they are commercially known). Note that the identification of peptic peptides is a separate experiment from that of the determination of deuterium content for HDX samples (Fig. 3b workflow in the solid box). There is no way to avoid this separate experiment. Using DDA, the first LC-MS run of the undeuterated control sample is used to survey both the quality of the digestion and the ions that are produced from the pepsin digestion. Additional undeuterated controls are then needed to perform the MS/MS analyses, either including ions to be interrogated by MS/MS or creating exclusion lists of ions not to look for in subsequent runs. Using DIA, all survey and MS/MS data are acquired simultaneously. Regardless of which method is used to identify the peptic peptides, it is only necessary that the fragment ions that are produced allow for an accurate identification of the peptide between the hypothetical choices; total sequencing of each peptide is not necessary.

The final result of the peptide identification step is the creation of a master peptide list that is used within the software as a map of where to look within the deuterium exchange samples to locate the peptides for which positive identification is known (i.e., the correlation between ion and amino acid sequence has been made). This list will include peptide sequence information including the number of amino acids, the beginning and ending sequence number, the number of proline residues, the monoisotopic and average masses of each ion, the charge state(s), an m/z window within which to look for deuterated peptides (determined by knowing the maximum amount of deuterium that can be incorporated, which is known once the sequence of amino acids is known), the retention time (at the apex of each chromatographic peak) and the retention time window.

3.3 Determining Deuterium Incorporation

Deuterium incorporation is measured relative to an unlabeled sample acquired along with the labeled samples. In the case of EX2 kinetics, measuring the deuterium uptake is a simple matter of tracking the mass of the peptide as a function of increasing exposure time. For the individual processing of HDX-MS data by hand, there is a deep reliance on visual cues (consistent charge state, alignment of the isotopes on the m/z scale between undeuterated and labeled samples, the increase in mass of the isotope distribution, etc.) when determining the amount of deuterium in any one peptide. A data processing algorithm needs more guidance and limitations.

Before an algorithm can determine the amount of deuterium that has been incorporated into a peptide, it is necessary to define the maximum mass increase possible and therefore the window of m/z within which a partially deuterated ion may be found (the window stretches from the undeuterated peptide m/z to the maximally deuterated peptide m/z). We will call this the deuteration window. If there is good chromatographic alignment (i.e., reproducible elution time) of peptides between undeuterated and deuterated samples, it becomes much easier to be confident that a peptide appearing in a deuteration window is in fact the peptide that has become deuterated rather than another species/peptide that simply has an m/z that happens to fall within the defined deuteration window. A person can quickly look and find the deuterated ions of interest often without even knowing the upper boundary of the deuteration window. Software, on the other hand, needs to be given an upper limit to the deuteration window so it can perform peak detection only within the deuteration window. The peak detection to find the deuterated isotope distribution within the window must then be done in light of a few other variables. Consider the isotope patterns in Fig. 4a, c. The natural isotopic abundance of a peptide is dominated largely by peaks spaced according to the varying number of ^{13}C isotopes in the population. The m/z spacing of the peaks is approximately a function of the mass difference between the ^{13}C and ^{12}C isotopes, equal to 1.003 Da/z. For each deuterium atom incorporated into the peptide, the entire natural abundance isotope distribution will shift to a higher mass. The amount of shift is dictated by the number of deuterium incorporated, realizing that the difference between ^{2}H

and ^1H isotopes is equal to 1.006 Da/z. An example is shown in Fig. 4b, where in the deuterated form, the isotope distribution remains the same as that found in the undeuterated peptide, only shifted to higher mass as the result of exchanging several hydrogens for deuterium atoms. Software must be able to recognize when a deuterium isotopic distribution begins and when it ends, in order to proceed to the next step. Peak detection, including isotope peak detection within the cluster, must consider both intensity and spacing as just defined.

In practice, the observed deuterated peptide spectrum becomes more complex as different numbers of both ^{13}C and ^2H isotopes define the population of ions contributing to the peptide spectrum. The spectrum is most easily defined as a series of natural abundance distributions offset by multiples of the $^2H/^1H$ mass delta, the intensity of each offset being defined by the fraction of ions with the corresponding number of ^2H atoms which have been incorporated. It is imperative that any algorithm consider these parameters. The different mass shifts arising from $^{13}C/^{12}C$ (1.003 Da/z) versus $^2H/^1H$ (1.006 Da/z) mass deltas will for the most part be unresolvable on time-of-flight or ion trap-based mass spectrometers; however, they must still be accounted for when performing isotopic peak detection with low mass tolerances. For example, consider a deuterated isotope distribution in which the first peak in the distribution will be the monoisotopic peak (containing all isotopes with the lowest mass number, i.e., ^1H, ^{12}C, ^{14}N, ^{16}O, and ^{32}S) with a well-characterized mass of $[M]$ Da/z. The third peak in the distribution, located approximately at $[M + 2]$ Da/z will be observed as a superposition of three ion populations each with two heavy isotopes: $^{13}C_2{}^2H_0$ at $[M + 2.006]$ Da/z, $^{13}C_1{}^2H_1$ at $[M + 2.009]$ Da/z, and $^{13}C_0{}^2H_2$ at $[M + 2.012]$ Da/z. The measured position and width of the peak will therefore lie somewhere in the range 2.006–2.012. This 6 mDa/z range might seem insignificant at first but amounts to 6 ppm for a 1000 Da peptide. The uncertainty in expected mass will get larger when considering peptides that can exchange more deuterium atoms and should be considered if searching spectra for accurate masses. Fortunately, the error expansion is bounded: instrument sensitivity will impose a practical limit to the maximum observable ^{13}C isotopes and the primary structure of a peptide dictates the maximum number of ^2H atoms which can exchange.

After the low and high ends of the m/z range for an isotopic distribution of a deuterated peptide ion have been determined, the centroid C or first moment of the cluster can be determined for the undeuterated control (UND), and each deuterium labeling time point (t):

$$C = \frac{\sum (m/z)_i \times I_i}{\sum I_i} \qquad (1)$$

where I is the spectral intensity at each m/z value. The summation is carried out over a range encompassing the entire isotopic distribution, which is defined by some intensity threshold on either side of the distribution. For a person doing this determination, one simply uses software (such as MagTran [59]) or the mass spectral processing software of the instrument vendor to find the first moment by drawing the boundaries of the m/z range for centroiding. Software designed for HDX-MS analysis makes a similar calculation but instead uses information it has obtained from peak detection as the low and high m/z boundaries. A centroid calculation is indicated in Fig. 4a, c with the shaded area underneath the defined curve describing the area that is being considered and with the calculated first moment indicated with the dashed line.

The first moment on an m/z scale can be converted to the mass only scale by removing the charge component:

$$M = z \times (C) - M_C \times (z - 1) \tag{2}$$

where we use M_C as the mass of the charge carrier (in most experiments this is a proton and equal to 1.007276 Da). The charge state of the ion is represented as z. Using mass only values instead of m/z allows comparison of data regardless of peptide charge state and makes it easier to average or combine data from several charge states of the same peptide (see below).

Figure 4c shows the spectra for a peptide that has been exposed to deuterium for 5 h. The relative deuterium level, defined as D_t, is calculated by subtracting the charge state corrected first moment (or the centroid value converted to a mass only scale) for the undeuterated peptide M_{UND} (Fig. 4a) from that of the deuterium labeled peptide M_t.

$$D_t = M_t - M_{UND} \tag{3}$$

The relative deuterium level is the fundamental measurement obtained in an HDX-MS experiment.

When presenting relative deuterium level results, it is desirable to include estimates of errors, usually in the form of standard deviations, or related properties such as variance or confidence intervals. These error estimates can arise when the D_t is determined from multiple charge states of the same peptide or by averaging multiple determinations. For a review of sources of errors in an HDX-MS experiment see ref. 60. For a collection of centroid mass measurements $(M_{t, i})$, each measurement can use a weighting factor (w_i) corresponding to the total intensity of all isotopes used to calculate each centroid. Lower intensity measurements often have higher errors, weighting the mean calculation can prevent these signals from having disproportionately large effects on proceeding calculations. An aggregate measurement of the centroid mass ($\bar{M}_t$), can then be calculated with a weighted mean:

$$\bar{M}_t = \frac{\sum w_i M_{t,i}}{\sum w_i} \qquad (4)$$

The corresponding weighted variance σ_t^2 is calculated as:

$$\sigma_t^2 = \frac{\sum w_i \left(M_{t,i} - \bar{M}_t\right)^2}{\sum w_i} \qquad (5)$$

Note an algorithm can just as easily apply equal weighting to each measurement, in which case Eqs. 4 and 5 reduce to their traditional unweighted forms. When calculating the relative deuterium level in Eq. 3, if the deuterated and undeuterated masses have associated variances of σ_t^2 and σ_{UND}^2 respectively, then the variance of the relative deuterium level $\sigma_{D,t}^2$ becomes:

$$\sigma_{D,t}^2 = \sigma_t^2 + \sigma_{UN}^2 \qquad (6)$$

For the individual processing HDX-MS data by hand, it is a task of visually identifying the isotope distributions for the undeuterated control sample as well as for all exchange time points. Then, the centroid values of all these isotope distributions are determined using an excel macro (such as HXExpress [15, 16]), the vendor specific mass spectrometer software, or a stand-alone program such as MagTran [59]. The measured values are then inserted into Eq. 3 to determine the relative deuterium level D_t.

Keeping the above in mind, the number of these calculations that must be performed in a typical HDX-MS experiment can become very large, and therefore become a major challenge to perform by hand. Each measurement of deuterium incorporation will be a function of three experimental parameters: the peptide itself, physical state of the sample (wild-type versus mutant, native versus denatured, free in solution versus ligand bound, etc.), as well as the deuterium exposure time. Consider a peptide which exists primarily in two charge states. To obtain reasonable estimates of variability, each HDX-MS experiment (from labeling to data analysis) is performed in triplicate [5]. Therefore, three deuterium incorporation determinations are performed for each charge state, or $2 \times 3 = 6$ total deuterium incorporation determinations. A simple HDX-MS experiment might be interested in measuring 50 peptides from pepsin digestion, under two separate equilibrium conditions, at four different deuterium labeling times (not to mention multiple undeuterated control samples). At this point $6 \times 50 \times 2 \times (4 + 1) = 3000$ individual deuterium incorporation determinations must then be performed. Software can make these determinations fairly rapidly, all the while taking advantage of the peptide identification information provided by the master peptide list. As more complex protein systems are analyzed, say, with perhaps

300 peptic peptides, 4 replicate determinations, 6 exchange points and for a wild-type and 3 mutant versions of the protein, the number of analyses becomes astronomically larger. When examining more complex proteins and protein systems it is common to have experiments involving tens of thousands of deuterium incorporation measurements. The availability of software that can perform automated calculations is very helpful when faced with such a large number of measurements. We also think that once software has made such determinations, it should display the final calculations on top of the raw data for quick visual inspection of the fidelity of the calculations. Whereas a human is much slower at making all the manipulations required to do the actual deuterium incorporation determination, a human can very rapidly provide quality control for thousands of software-determined deuterium incorporation determinations per hour, ascertaining if the software has performed well or not.

3.4 Plotting HDX-MS Data

The relative deuterium levels that have been determined for the peptides of interest at each labeling time are then plotted against exposure time in deuterium to create a deuterium incorporation plot, or uptake curve. The data are best plotted on a semilogarithmic plot as it allows for a clear view of the information from the earliest time points (seconds) to the longest exchange points (many hours to days). Figure 6 shows spectra and example uptake curves for a protein in two different states, wild-type as well as a mutant version (the mutant could be the result of a single-point amino acid mutation or an alteration to the protein native-state equilibrium as a result of denaturants or a ligand binding event). Panel a of Fig. 6 shows model data for this 4+ peptide in the undeuterated control as well as after 5 h of exposure to deuterium. The dashed lines show the shift in the centroid between the two isotope distributions. The wild-type protein had a much larger upward shift in centroid than the mutant, which is reflected in the uptake curve itself (Fig. 6b). For the data in this figure, we are not showing error bars but there is always an error associated (on the Y-axis) with each data point. Typical errors under well controlled experimental conditions such as precise sample preparation, stable mass spectrometer calibration, and short time in between replicate sample set collection, are in the range of ± 0.2 Da [61] (see also ref. 14 where a thorough review of errors in measurement was described). Looking at the data in both panels of Fig. 6, both the raw mass spectra and the uptake graphs, it is clear that something of interest occurs in this region of the protein as a result of mutation. There is a consistent retardation in deuterium incorporation in this part of the protein from the earliest time points. The absolute magnitude of this change is not always as important as the identification that this is a region of the protein that is sensitive to the mutation, as evidenced by the change in deuterium uptake.

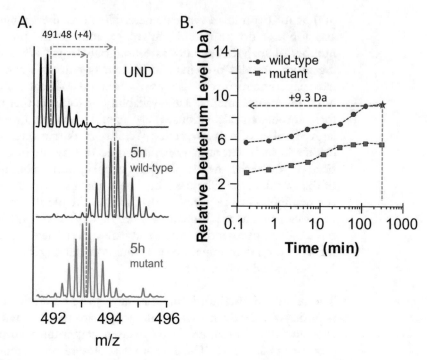

Fig. 6 From raw *m/z* data to a deuterium incorporation plot. (**a**) Mass spectra for a peptide 491.48 *m/z* (4+) for the undeuterated control as well as a 5 h deuterium exposure time point. Data are shown for both the peptide that comes from a wild-type protein as well as a mutant form of the same protein. This peptide does not contain the site of amino acid mutation. (**b**) The relative change in mass for the same 491.48 *m/z* peptide, both in the wild-type (circles) and mutant (squares) states. Deuterium incorporation data are plotted together on the same axes highlighting the differences at each time point

3.5 Post-data Analysis: Interpretation and Visualization

At the end of determining the deuterium incorporation into each peptide on the master peptide list, the result is a tabulation of deuterium level at each exchange point, along with the corresponding plots. This is, however, not "the end of the road" for these data. Most commonly there are various postanalysis steps, including several kinds of data manipulations, calculations of kinetic parameters, additional ways in which to visualize the deuterium incorporation versus time data, and/or biophysical interpretation of the HDX-MS data. Performing any or all of these final steps in the HDX-MS experiment depends upon the type of information that is desired. We have chosen to discuss four of the most common post-data analysis processing steps.

3.5.1 Back Exchange Correction

All the data processing steps that have been discussed up to this point have determined the relative deuterium level of a protein in deuterium relative to the same protein without deuterium. There has not been any correction for back exchange that occurs during the LC-MS analysis step. The data from many HDX-MS experiments can be corrected for back-exchange (to account for the loss of deuterium that occurs postquench), although at times the usefulness of the correction in many types of experiments has been questioned (*see* refs. 5, 6, 49). The correction is based on the first descriptions of HDX-MS [11] and has been explained in detail in [11, 51]. Briefly, a sample of the protein of interest is maximally deuterated (maximally deuterated control) by unfolding the protein (through a combination of pH, temperature, and/or denaturant) to expose all backbone amide hydrogens to D_2O. This sample is then considered to be deuterated at each available backbone amide position (or as dictated by the percent deuterium in the buffer after dilution) until injection into the LC-MS system for analysis. The most common correction [11] for back exchange is performed using the equation:

$$D = \frac{M_t - M_{\text{UND}}}{M_{100\%} - M_{\text{UND}}} \times N \qquad (7)$$

The deuterium level (D) is calculated at each time point (t) using experimentally determined centroid masses measured for the undeuterated control M_{UND}, the maximally deuterated control $M_{100\%}$, and by knowing the maximum number of deuterium atoms which can be incorporated N. The deuterium incorporation plots in Fig. 7 illustrate the application of this correction to produce the absolute deuterium level (corrected data) from the relative deuterium level data (observed data). There is very little added advantage to making this correction, except in cases where the absolute number of deuterium incorporated must be determined, for example if kinetic exchange rates are being sought or intrinsically disordered proteins are being studied, etc. For comparison of experiments between two or more state of the protein, this absolute number of deuterium information is often not required. There are also some caveats to the back-exchange correction. Occasionally, a peptide does not follow the theoretical considerations of Eq. 7. This correction only works for 92% of peptides, where the error is less than 10%; for the remaining 8% of peptides, the error of the correction is significantly higher [11]. Additionally, a maximally deuterated control must be prepared, and this is often not possible due to aggregation, resistance to unfolding for complete deuteration, and several other factors. In fact, one is never certain that a totally deuterated species has been prepared without analysis—but the analysis itself causes loss of deuterium meaning that the mass of the totally deuterated species can never be truly known. Back exchange corrections can also be performed on a run-to-run basis using mixtures of short peptides with fast HDX rates as internal standards [62]. However, this method of correction has yet to see widespread adoption.

Again, when using an HDX-MS experiment to determine if there is a region or regions of the protein that has undergone a change in deuterium incorporation upon mutation or an alteration

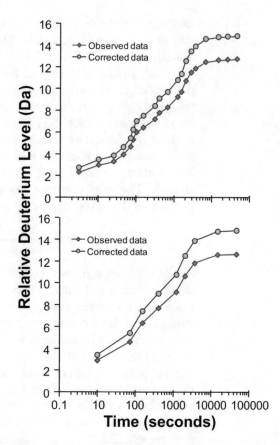

Fig. 7 Correcting HDX-MS data for back-exchange and fitting the data to obtain kinetic parameters. The data are plotted as both relative deuterium level (observed data, diamonds) as well as number of deuterium (corrected data, circles) after the data were corrected for back-exchange according to Eq. 7. Identical model data are shown in both the top and bottom panels but the graph at the top shows 20 exchange time points between 3 s and 12.5 h while the bottom graph is the same as the top but with 11 time points removed

to the native-state equilibrium, it is not necessary to convert the *relative deuterium* information to an absolute number of deuterium incorporated. The identified difference will persist regardless of whether or not the data have been corrected for back-exchange. It should be noted that amino acid mutations that involve proline residues have the added effect of either the addition or the removal of exchangeable backbone amide hydrogens. In these cases, care needs to be taken when considering differences for these peptides.

3.5.2 Extracting Kinetic Information

Fitting the back exchange corrected HDX-MS data in order to extract exchange rates can also be performed. The measured exchange is the sum of the exchange of each backbone amide hydrogen in a peptide, defined by a multiterm exponential equation that sums each of the exponential terms describing each of the backbone amide hydrogens [11]. As some backbone amide hydrogens may have similar exchange rates within a peptic peptide, the summation usually reduces to a series of terms describing several populations: slow, medium, and fast exchangers using Eq. 8 where D is the deuterium level of the peptide with N amide linkages, t is the deuterium exposure time and k_i the pseudo-first-order rate constant for exchange at each backbone amide linkage [3, 11].

$$D = N - \sum \exp(-k_i t) \qquad (8)$$

Multiple examples of determining the rate constants of exchange and how many amide hydrogens are in each category have been described (e.g., [63–65]). Despite fitting the deuterium incorporation data to the sum of exponentials, no individual amide specific information can be gleaned from this analysis at the peptide level. Importantly, the number of deuterium exchange time points influences the success of these kinds of analysis. Fitting the data to three exponentials will be difficult without being able to establish regions for slow, medium, and fast backbone amides on the corrected data. Figure 7 illustrates this point. The 20 exchange time points in the top panel easily show three transitions. Deleting 11 of these data points across the experiment time course yields the deuterium incorporation plot at the bottom of Fig. 7. Significantly, with fewer exchange time points, it is more difficult to see the transitions and the resultant fit of these data to Eq. 8 can be harder to perform correctly.

3.5.3 HDX-MS Data Visualization

The visual representation of the deuterium incorporation information can sometimes be the most difficult part of an HDX-MS experiment, as there are multiple ways in which the data can be represented and multiple values to doing it in different ways. Inexperience or naivety in this aspect of data analysis can lead to many erroneous conclusions. All of the current software tools take data analysis through to include a visualization step. The simplest two-dimensional visualization comes from deuterium incorporation plots (Figs. 6b and 7). These plots are informative as they can provide information about dynamics and solvent accessibility. All of the HDX-MS data that are contained in the deuterium incorporation plots can be translated into other two- and three-dimensional figures in order to describe the HDX-MS data in context of the whole protein or protein system. A partial list of all types of figures that can be created include: heat maps where deuterium levels are plotted onto peptide maps; 3D models where deuterium levels are plotted onto the NMR or X-ray crystal structure cartoon model of the protein if it is available; difference plots where when acquiring data for the protein of interest in two or more different states, the difference between the two states is shown (e.g., comparability plots [61], or various kinds of difference plots [29, 66, 67]). Visualizing where in the protein deuterium has gone, or not gone, or where it has gone differently in the case of comparisons, is vital to interpretation of the data.

3.5.4 HDX-MS Data Interpretation

Data interpretation is the part of the HDX-MS experiment that experiences the most variability; however, this topic is beyond the scope of this chapter. As a start, a description of the possible deuterium incorporation plots that can be encountered when performing an HDX-MS experiment has been described (Fig. 8 in Morgan and Engen [6]). The plots that are shown in this figure are examples of most if not all of types of curves that can be found with a deuterium incorporation plot and importantly what various shapes in deuterium incorporation plots mean. The discussion surrounding this figure encompasses a single protein experiment as well as for one that investigates the result of a protein being in an "excited" or "perturbed" state through modification of the native equilibrium.

4   Notes

1. Overlapping isotope distributions are a frequent problem in HDX-MS analyses of large proteins and protein complexes. Where there are a lot of peaks that must fit within limited peak capacity, overlap is inevitable. Often peptides will appear in multiple charge states, so if a protein's isotope distribution is obscured by overlap in one charge state, analysis can often be done by examining alternative charge states. However sometimes no other charge states are available, or there is just so much protein loaded on the system that all available charges are also obscured and one is unable to obtain a satisfactory HX measurement. In these situations, the user can choose to extend the chromatographic separation in an effort to resolve the peptides; however, this will be at the expense of increased back exchange. Another option with high resolution mass spectrometers is to use deconvolution methods to extract overlapping signals from one another [68, 69].

   Some researchers are turning to ion mobility to help resolve overlapping isotope distributions [70] wherein an ion mobility separation occurs postionization but before ion detection. If there is a sufficient difference between the peptides such that they are resolved via mobility, their overlap in spectral space becomes resolvable. For instruments capable of performing ion mobility separations, using this feature can help obtain measurements for protein systems that might otherwise prove difficult on traditional instrumentation.

Performing postionization fragmentation during local analysis experiments can also be beneficial in order to locate individual deuterium atoms within a peptic peptide. If peptide fragments are generated via collision induced dissociation, the fragments will not yield any further localization information due to scrambling of deuterium across the ion during gas-phase collisions [71–73]. However this can be beneficial for using the fragments to serve as surrogate measurements when the parent ions would otherwise not be measurable [74].

2. The maximum number of exchangeable backbone amide hydrogens in a peptide or protein (N) is calculated using Eq. 9 [11]

$$N = L_{peptide} - n_{Pro} - 1 \tag{9}$$

where $L_{peptide}$ is the length of the peptide and n_{pro} is the number of proline residues. The subtraction of 1 in the above equation arises from the fast exchange of the primary amine at the N-terminus of a protein or peptide and a loss of the deuterium label in the reversed-phase separation step. However, depending upon the amino acids bordering the penultimate amino acid, the rate of exchange for this amide hydrogen also can be rapid [5, 75, 76]. Subsequently, there will also be a loss of the deuterium label at this position in the LC-MS step. In these instances, the subtraction should be 2 instead of 1 in the above equation. Additionally, the maximum possible deuterium label will also be effected by the percent deuterium in the labeling reaction which is dictated by both the percent deuterium in the D_2O solvent supplied (depending upon vendor deuterium content can range from 70% to 99%) as well as the labeling dilution discussed above. Figure 8 illustrates the above corrections to the theoretical maximum label (Eq. 9) as well as indicating that there will also be loss of deuterium label due to back exchange in the LC-MS step. This then yields an experimental maximum possible label for each peptide. This experimental maximum should be considered when interpreting HDX-MS data.

3. For the majority of this chapter, we have only considered how to process HDX-MS data in which the isotope distributions indicated EX2 kinetics. EX1 kinetics generally are rare for proteins maintained under physiological conditions. However, this kinetic limit is a possibility and processing of EX1 data is not as straightforward as processing EX2 data [52]. Figure 9 illustrates four possible ways in which EX1 could manifest itself in the raw m/z or transformed mass spectra. EX1 can appear as either two discreet peaks or distributions of isotopes (Fig. 9a) or a combination of the two in varying degrees of overlap (Fig. 9b–d). Unfortunately, there is no consensus for the

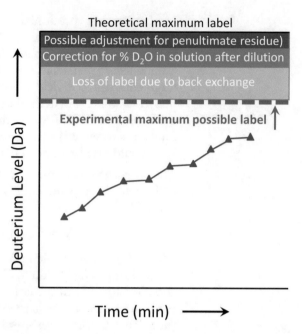

Fig. 8 The experimental maximum label. Three potential corrections to the theoretical maximum label (calculated using Eq. 9) are illustrated using a hypothetical deuterium incorporation plot

manner in which EX1 HDX data should be presented on the deuterium incorporation plot. Traditionally, if the centers of the two peaks are resolved (Fig. 9a, b) the deuterium incorporation data can be plotted for each peak (Fig. 9e). Here, the relative deuterium levels of each separate peak are plotted on the same axes, and there will be two data points for each time point where centroid values for the two populations are measurable and, in each case, the relative areas of each peak will be indicated by the size of the data point (*see* also ref. 16).

It is also possible that the isotope peaks for the two populations are too close for two resolved peak centers; there is simply evidence of isotopic peak widening (Fig. 9c) or peak tailing (Fig. 9d). In this case, significant evidence that there is EX1 kinetics comes from the creation of a peak width plot [52]. This plot is created by measuring the absolute width (usually at 20% height) of the isotopic distribution versus deuterium exposure time. If there is a difference in width across the time course (>4 Da for intact protein and >2 Da for peptides, *see* ref. 52) this is evidence for EX1 kinetics. For these data the centroid value over the entire widened isotope distribution is plotted versus deuterium exposure time (Fig. 9f).

4. Run-to-run carryover sometimes cannot be avoided. "Sticky" peptides that are retained on reversed-phase media can persist into the next exchange sample that is being analyzed.

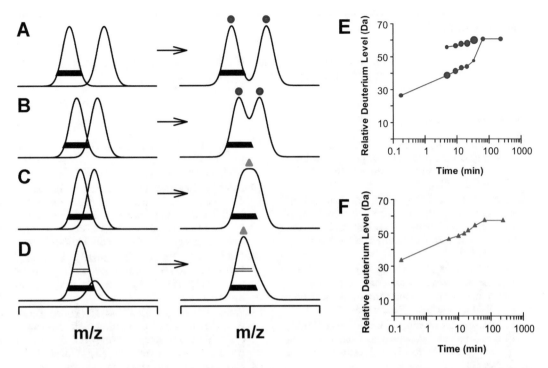

Fig. 9 Manifestation of EX1 kinetics in *m/z* spectra. The merging of two binomial isotopic distributions of equal width but varying centers (in *m/z* units) are shown in (**a–d**). The separated distributions are shown on the left and the merged shown on the right. The centroid of the lower-mass distribution (no EX1 unfolding) is indicated by the filled diamond and that of the higher-mass distribution (EX1 unfolding and deuteration) is indicated with the filled circles. As a visual aid, the solid bar is set at 20% maximum peak intensity and the open bar at 50% maximum peak intensity. (**e**) Model data plotted assuming the isotope patterns shown in (**a**) and (**b**). The data color and shape of the data points coordinate with those that are indicated on the isotope patterns in both (**a**) and (**b**). The relative sizes of the data points between 5 and 20 min coordinates with relative proportions of calculated peak areas for each of the two isotope patterns in this time region in a manner similar to ref. 16. (**f**) Model data plotted assuming (**c**). Parts of this figure have been reproduced from [52], with permission from Springer (reference number 500661560)

Carryover can cause two complications with data analysis: the amount of deuterium incorporated into the persistent peptide cannot be accurately determined, and/or there is false EX1. Run to run carryover within the LC system has been shown to mimic EX1 kinetic data [77]. In HDX-MS data analysis it is best to minimize the amount of carryover either through loading a smaller amount of protein into the system, or efficient system washing in between analyte sample runs. There are published methods that have been designed to reduce the amount of peptide carryover [77, 78]. If there are indications that the analyte protein might be exhibiting EX1 kinetics, it is important that blank injections be performed in between samples to confirm the presence of EX1 and not simply the presence of peptide carry-over.

Acknowledgments

The authors' work with HDX-MS is partially supported by the NIH (R01-AI043957 and R35-CA197583).

References

1. Chowdhury SK, Katta V, Chait BT (1990) Probing conformational-changes in proteins by mass-spectrometry. J Am Chem Soc 112 (24):9012–9013

2. Katta V, Chait BT (1991) Conformational-changes in proteins probed by hydrogen-exchange electrospray-ionization mass-spectrometry. Rapid Commun Mass Spectrom 5(4):214–217

3. Smith DL, Deng YZ, Zhang ZQ (1997) Probing the non-covalent structure of proteins by amide hydrogen exchange and mass spectrometry. J Mass Spectrom 32(2):135–146

4. Hoofnagle AN, Resing KA, Ahn NG (2003) Protein analysis by hydrogen exchange mass spectrometry. Annu Rev Biophys Biomol Struct 32:1–25

5. Wales TE, Engen JR (2006) Hydrogen exchange mass spectrometry for the analysis of protein dynamics. Mass Spectrom Rev 25 (1):158–170

6. Morgan CR, Engen JR (2009) Investigating solution-phase protein structure and dynamics by hydrogen exchange mass spectrometry. Curr Protoc Protein Sci Chapter 17, Unit 17 16 11–17

7. Konermann L, Pan JX, Liu YH (2011) Hydrogen exchange mass spectrometry for studying protein structure and dynamics. Chem Soc Rev 40(3):1224–1234

8. Masson GR, Jenkins ML, Burke JE (2017) An overview of hydrogen deuterium exchange mass spectrometry (HDX-MS) in drug discovery. Expert Opin Drug Discov 12 (10):981–994

9. Hvidt A, Nielsen SO (1966) Hydrogen exchange in proteins. Adv Protein Chem 21:287–386

10. Englander SW, Downer NW, Teitelbaum H (1972) Hydrogen exchange. Annu Rev Biochem 41:903–924

11. Zhang ZQ, Smith DL (1993) Determination of amide hydrogen-exchange by mass-spectrometry - a new tool for protein-structure elucidation. Protein Sci 2(4):522–531

12. Engen JR (2003) Analysis of protein complexes with hydrogen exchange and mass spectrometry. Analyst 128(6):623–628

13. Brier S, Lemaire D, DeBonis S, Forest E, Kozielski F (2006) Molecular dissection of the inhibitor binding pocket of mitotic kinesin Eg5 reveals mutants that confer resistance to anti-mitotic agents. J Mol Biol 360(2):360–376

14. Iacob RE, Engen JR (2012) Hydrogen exchange mass spectrometry: are we out of the quicksand? J Am Soc Mass Spectrom 23:1003

15. Weis DD, Engen JR, Kass IJ (2006) Semi-automated data processing of hydrogen exchange mass spectra using HX-Express. J Am Soc Mass Spectrom 17(12):1700–1703

16. Guttman M, Weis DD, Engen JR, Lee KK (2013) Analysis of overlapped and noisy hydrogen/deuterium exchange mass spectra. J Am Soc Mass Spectrom 24(12):1906–1912

17. Miller DE, Prasannan CB, Villar MT, Fenton AW, Artigues A (2012) HDXFinder: automated analysis and data reporting of deuterium/hydrogen exchange mass spectrometry. J Am Soc Mass Spectrom 23(2):425–429

18. Slysz GW, Baker CA, Bozsa BM, Dang A, Percy AJ, Bennett M et al (2009) Hydra: software for tailored processing of H/D exchange data from MS or tandem MS analyses. BMC Bioinformatics 10:1–14

19. Lou X, Kirchner M, Renard BY, Kothe U, Boppel S, Graf C et al (2010) Deuteration distribution estimation with improved sequence coverage for HX/MS experiments. Bioinformatics 26(12):1535–1541

20. Kreshuk A, Stankiewicz M, Lou XH, Kirchner M, Hamprecht FA, Mayer MP (2011) Automated detection and analysis of bimodal isotope peak distributions in H/D exchange mass spectrometry using HeXicon. Int J Mass Spectrom 302(1–3):125–131

21. Palmblad M, Buijs J, Hakansson P (2001) Automatic analysis of hydrogen/deuterium exchange mass spectra of peptides and proteins using calculations of isotopic distributions. J Am Soc Mass Spectrom 12(11):1153–1162

22. Kavan D, Man P (2011) MSTools-Web based application for visualization and presentation of HXMS data. Int J Mass Spectrom 302 (1–3):53–58

23. Liu SM, Liu LT, Uzuner U, Zhou X, Gu MX, Shi WB et al (2011) HDX-Analyzer: a novel package for statistical analysis of protein structure dynamics. BMC Bioinformatics 12(Suppl 1):S43

24. Hamuro Y, Coales SJ, Southern MR, Nemeth-Cawley JF, Stranz DD, Griffin PR (2003) Rapid analysis of protein structure and dynamics by hydrogen/deuterium exchange mass spectrometry. J Biomol Tech 14(3):171–182

25. Wei H, Ahn J, Yu YQ, Tymiak A, Engen JR, Chen G (2012) Using hydrogen/deuterium exchange mass spectrometry to study conformational changes in granulocyte colony stimulating factor upon PEGylation. J Am Soc Mass Spectrom 23(3):498–504

26. Pascal BD, Willis S, Lauer JL, Landgraf RR, West GM, Marciano D et al (2012) HDX workbench: software for the analysis of H/D exchange MS data. J Am Soc Mass Spectrom 23(9):1512–1521

27. Wales TE, Fadgen KE, Gerhardt GC, Engen JR (2008) High-speed and high-resolution UPLC separation at zero degrees Celsius. Anal Chem 80(17):6815–6820

28. Engen JR, Smith DL (2000) Investigating the higher order structure of proteins. Hydrogen exchange, proteolytic fragmentation, and mass spectrometry. Methods Mol Biol 146:95–112

29. Tiyanont K, Wales TE, Aste-Amezaga M, Aster JC, Engen JR, Blacklow SC (2011) Evidence for increased exposure of the Notch1 metalloprotease cleavage site upon conversion to an activated conformation. Structure 19 (4):546–554

30. Ehring H (1999) Hydrogen exchange electrospray ionization mass spectrometry studies of structural features of proteins and protein/protein interactions. Anal Biochem 267 (2):252–259

31. Woods Jr VL (2001) Methods for the high-resolution identification of solvent-accessible amide hydrogens in polypeptides or proteins and for characterization of the fine structure of protein binding sites. US Patent 6,291,189

32. Mysling S, Salbo R, Ploug M, Jorgensen TJ (2014) Electrochemical reduction of disulfide-containing proteins for hydrogen/deuterium exchange monitored by mass spectrometry. Anal Chem 86(1):340–345

33. Trabjerg E, Jakobsen RU, Mysling S, Christensen S, Jorgensen TJ, Rand KD (2015) Conformational analysis of large and highly disulfide-stabilized proteins by integrating online electrochemical reduction into an optimized H/D exchange mass spectrometry workflow. Anal Chem 87(17):8880–8888

34. Cravello L, Lascoux D, Forest E (2003) Use of different proteases working in acidic conditions to improve sequence coverage and resolution in hydrogen/deuterium exchange of large proteins. Rapid Commun Mass Spectrom 17 (21):2387–2393

35. Wang L, Pan H, Smith DL (2002) Hydrogen exchange-mass spectrometry: optimization of digestion conditions. Mol Cell Proteomics 1 (2):132–138

36. Wales TE, Fadgen KE, Eggertson MJ, Engen JR (2017) Subzero Celsius separations in three-zone temperature controlled hydrogen deuterium exchange mass spectrometry. J Chromatogr A 1523:275–282

37. Johnson RS, Walsh KA (1994) Mass-spectrometric measurement of protein amide hydrogen-exchange rates of apo-myoglobin and holo-myoglobin. Protein Sci 3 (12):2411–2418

38. Wu Y, Engen JR, Hobbins WB (2006) Ultra performance liquid chromatography (UPLC) further improves hydrogen/deuterium exchange mass spectrometry. J Am Soc Mass Spectrom 17(2):163–167

39. Mandell JG, Falick AM, Komives EA (1998) Identification of protein-protein interfaces by decreased amide proton solvent accessibility. Proc Natl Acad Sci U S A 95 (25):14705–14710

40. Dharmasiri K, Smith DL (1996) Mass spectrometric determination of isotopic exchange rates of amide hydrogens located on the surfaces of proteins. Anal Chem 68 (14):2340–2344

41. Rosa JJ, Richards FM (1979) An experimental procedure for increasing the structural resolution of chemical hydrogen-exchange measurements on proteins: application to ribonuclease S peptide. J Mol Biol 133(3):399–416

42. Englander JJ, Rogero JR, Englander SW (1985) Protein hydrogen exchange studied by the fragment separation method. Anal Biochem 147(1):234–244

43. Zubarev RA, Kelleher NL, McLafferty FW (1998) Electron capture dissociation of multiply charged protein cations. A nonergodic process. J Am Chem Soc 120(13):3265–3266

44. Rand KD, Adams CM, Zubarev RA, Jorgensen TJD (2008) Electron capture dissociation proceeds with a low degree of intramolecular migration of peptide amide hydrogens. J Am Chem Soc 130(4):1341–1349

434   Michael J. Eggertson et al.

45. Zehl M, Rand KD, Jensen ON, Jorgensen TJD (2008) Electron transfer dissociation facilitates the measurement of deuterium incorporation into selectively labeled peptides with single residue resolution. J Am Chem Soc 130 (51):17453–17459

46. Pan J, Zhang S, Borchers CH (2016) Comparative higher-order structure analysis of antibody biosimilars using combined bottom-up and top-down hydrogen-deuterium exchange mass spectrometry. Biochim Biophys Acta 1864(12):1801–1808

47. Pan J, Zhang S, Parker CE, Borchers CH (2014) Subzero temperature chromatography and top-down mass spectrometry for protein higher-order structure characterization: method validation and application to therapeutic antibodies. J Am Chem Soc 136 (37):13065–13071

48. Jensen PF, Larraillet V, Schlothauer T, Kettenberger H, Hilger M, Rand KD (2015) Investigating the interaction between the neonatal Fc receptor and monoclonal antibody variants by hydrogen/deuterium exchange mass spectrometry. Mol Cell Proteomics 14 (1):148–161

49. Marcsisin SR, Engen JR (2010) Hydrogen exchange mass spectrometry: what is it and what can it tell us? Anal Bioanal Chem 397 (3):967–972

50. Kipping M, Schierhorn A (2003) Improving hydrogen/deuterium exchange mass spectrometry by reduction of the back-exchange effect. J Mass Spectrom 38(3):271–U278

51. Hoofnagle AN, Resing KA, Ahn NG (2004) Practical methods for deuterium exchange/mass spectrometry. Methods Mol Biol 250:283–298

52. Weis DD, Wales TE, Engen JR, Hotchko M, Ten Eyck LF (2006) Identification and characterization of EX1 kinetics in H/D exchange mass spectrometry by peak width analysis. J Am Soc Mass Spectrom 17(11):1498–1509

53. Englander SW, Kallenbach NR (1983) Hydrogen exchange and structural dynamics of proteins and nucleic acids. Q Rev Biophys 16 (4):521–655

54. Miranker A, Robinson CV, Radford SE, Aplin RT, Dobson CM (1993) Detection of transient protein-folding populations by mass-spectrometry. Science 262(5135):896–900

55. Fruton JS, Bergmann M (1938) The specificity of pepsin action. Science 87(2268):557

56. Ahn J, Cao MJ, Yu YQ, Engen JR (2013) Accessing the reproducibility and specificity of pepsin and other aspartic proteases. Biochim Biophys Acta 1834(6):1222–1229

57. Geromanos SJ, Vissers JP, Silva JC, Dorschel CA, Li GZ, Gorenstein MV et al (2009) The detection, correlation, and comparison of peptide precursor and product ions from data independent LC-MS with data dependant LC-MS/MS. Proteomics 9(6):1683–1695

58. Gillet LC, Navarro P, Tate S, Rost H, Selevsek N, Reiter L et al (2012) Targeted data extraction of the MS/MS spectra generated by data-independent acquisition: a new concept for consistent and accurate proteome analysis. Mol Cell Proteomics 11(6): O111.016717

59. Zhang Z, Marshall AG (1998) A universal algorithm for fast and automated charge state deconvolution of electrospray mass-to-charge ratio spectra. J Am Soc Mass Spectrom 9 (3):225–233

60. Engen JR, Wales TE (2015) Analytical aspects of hydrogen exchange mass spectrometry. Annu Rev Anal Chem (Palo Alto, Calif) 8:127–148

61. Houde D, Berkowitz SA, Engen JR (2011) The utility of hydrogen/deuterium exchange mass spectrometry in biopharmaceutical comparability studies. J Pharm Sci 100 (6):2071–2086

62. Zhang Z, Zhang A, Xiao G (2012) Improved protein hydrogen/deuterium exchange mass spectrometry platform with fully automated data processing. Anal Chem 84(11):4942–4949

63. Tsutsui Y, Liu L, Gershenson A, Wintrode PL (2006) The conformational dynamics of a metastable serpin studied by hydrogen exchange and mass spectrometry. Biochemist 45(21):6561–6569

64. Zheng X, Wintrode PL, Chance MR (2008) Complementary structural mass spectrometry techniques reveal local dynamics in functionally important regions of a metastable serpin. Structure 16(1):38–51

65. Busenlehner LS, Armstrong RN (2005) Insights into enzyme structure and dynamics elucidated by amide H/D exchange mass spectrometry. Arch Biochem Biophys 433 (1):34–46

66. Tsutsumi S, Mollapour M, Prodromou C, Lee CT, Panaretou B, Yoshida S et al (2012) Charged linker sequence modulates eukaryotic heat shock protein 90 (Hsp90) chaperone activity. Proc Natl Acad Sci U S A 109 (8):2937–2942

67. Street TO, Lavery LA, Verba KA, Lee CT, Mayer MP, Agard DA (2012) Cross-monomer substrate contacts reposition the Hsp90 N-terminal domain and prime the chaperone activity. J Mol Biol 415(1):3–15

68. Kazazic S, Zhang HM, Schaub TM, Emmett MR, Hendrickson CL, Blakney GT et al (2010) Automated data reduction for hydrogen/deuterium exchange experiments, enabled by high-resolution Fourier transform ion cyclotron resonance mass spectrometry. J Am Soc Mass Spectrom 21(4):550–558

69. Zhang Z, Li W, Logan TM, Li M, Marshall AG (1997) Human recombinant [C22A] FK506-binding protein amide hydrogen exchange rates from mass spectrometry match and extend those from NMR. Protein Sci 6 (10):2203–2217

70. Iacob RE, Murphy JP III, Engen JR (2008) Ion mobility adds an additional dimension to mass spectrometric analysis of solution-phase hydrogen/deuterium exchange. Rapid Commun Mass Spectrom 22(18):2898–2904

71. Demmers JA, Rijkers DT, Haverkamp J, Killian JA, Heck AJ (2002) Factors affecting gas-phase deuterium scrambling in peptide ions and their implications for protein structure determination. J Am Chem Soc 124(37):11191–11198

72. Jorgensen TJ, Gardsvoll H, Ploug M, Roepstorff P (2005) Intramolecular migration of amide hydrogens in protonated peptides upon collisional activation. J Am Chem Soc 127 (8):2785–2793

73. Jorgensen TJ, Bache N, Roepstorff P, Gardsvoll H, Ploug M (2005) Collisional activation by MALDI tandem time-of-flight mass spectrometry induces intramolecular migration of amide hydrogens in protonated peptides. Mol Cell Proteomics 4(12):1910–1919

74. Percy AJ, Slysz GW, Schriemer DC (2009) Surrogate H/D detection strategy for protein conformational analysis using MS/MS data. Anal Chem 81(19):7900–7907

75. Bai Y, Milne JS, Mayne L, Englander SW (1993) Primary structure effects on peptide group hydrogen exchange. Proteins 17 (1):75–86

76. Connelly GP, Bai Y, Jeng MF, Englander SW (1993) Isotope effects in peptide group hydrogen exchange. Proteins 17(1):87–92

77. Fang J, Rand KD, Beuning PJ, Engen JR (2011) False EX1 signatures caused by sample carryover during HX MS analyses. Int J Mass Spectrom 302(1–3):19–25

78. Majumdar R, Manikwar P, Hickey JM, Arora J, Middaugh CR, Volkin DB et al (2012) Minimizing carry-over in an online pepsin digestion system used for the H/D exchange mass spectrometric analysis of an IgG1 monoclonal antibody. J Am Soc Mass Spectrom 23 (12):2140–2148

INDEX

Printed in the United States
By Bookmasters